全国高等院校应用型创新规划教材·计算机系列

Java EE 应用开发案例教程

卢守东　编著

清华大学出版社

北　京

内容简介

本书以技术需求为导向，以技术应用为核心，以开发模式为主线，以应用开发为重点，以能力提升为目标，全面介绍了基于 Java EE 的企业级应用开发的关键技术、流行框架、主要模式、实施要点与方法步骤。全书共 10 章，包括 Java EE 概述、JSP 基础、JDBC 技术、JavaBean 技术、Servlet 技术、Struts 2 框架、Hibernate 框架、Spring 框架、Ajax 应用与 Web 应用案例等内容，并附有相应的思考题与实验指导。

本书内容全面、实例翔实、案例丰富，注重应用开发能力的培养，可作为各高校本科或高职高专计算机、电子商务、信息管理与信息系统及相关专业软件开发技术课程的教材或教学参考书，也可作为 Java EE 应用开发与维护人员的技术参考书以及初学者的自学教程。

本书所有示例的代码均已通过调试，并能成功运行，其开发环境为 Windows7、JDK 1.7.0_51、Tomcat 7.0.50、MyEclipse 10.7.1 与 SQL Server 2008。

图书在版编目(CIP)数据

Java EE 应用开发案例教程/卢守东编著. —北京：清华大学出版社，2017(2025.1 重印)
(全国高等院校应用型创新规划教材·计算机系列)
ISBN 978-7-302-46310-8

Ⅰ. ①J…　Ⅱ. ①卢…　Ⅲ. ①JAVA 语言—程序设计—高等学校—教材　Ⅳ. ①TP312.8

中国版本图书馆 CIP 数据核字(2017)第 021365 号

责任编辑： 孟　攀
封面设计： 杨玉兰
责任校对： 吴春华
责任印制： 杨　艳
出版发行： 清华大学出版社
　　网　　址： https://www.tup.com.cn, https://www.wqxuetang.com
　　地　　址： 北京清华大学学研大厦 A 座　　**邮　　编：** 100084
　　社 总 机： 010-83470000　　**邮　　购：** 010-62786544
　　投稿与读者服务： 010-62776969, c-service@tup.tsinghua.edu.cn
　　质量反馈： 010-62772015, zhiliang@tup.tsinghua.edu.cn
　　课件下载： https://www.tup.com.cn, 010-62791865
印 装 者： 三河市龙大印装有限公司
经　　销： 全国新华书店
开　　本： 185mm×260mm　　**印　张：** 27.75　　**字　数：** 675 千字
版　　次： 2017 年 3 月第 1 版　　**印　次：** 2025 年 1 月第 8 次印刷
定　　价： 79.00 元

产品编号：073043-02

前　言

党的二十大报告明确指出，“教育、科技、人才是全面建设社会主义现代化国家的基础性、战略性支撑。”同时强调“必须坚持科技是第一生产力、人才是第一资源、创新是第一动力，深入实施科教兴国战略、人才强国战略、创新驱动发展战略，开辟发展新领域新赛道，不断塑造发展新动能新优势”，要求“坚持教育优先发展、科技自立自强、人才引领驱动，加快建设教育强国、科技强国、人才强国，坚持为党育人、为国育才，全面提高人才自主培养质量，着力造就拔尖创新人才，聚天下英才而用之”。这充分体现了党中央对教育、科技与人才工作的高度重视与深远谋划，同时也为高校的发展与教学的改革确立了行动纲领，提供了根本遵循。众所周知，高校是教育、科技与人才的结合点，其首要职能就是人才培养。党的二十大已吹响了全面建成社会主义现代化强国的号角，如何为国家的发展、民族的振兴与社会的进步培养各行各业所需要的各类高素质人才，全国各高校均责无旁贷，且任重道远。

Java EE 是当前构建企业级 Java Web 应用的主流开发技术与平台之一，其应用已相当广泛。为满足社会不断发展的实际需求，并提高学生或学员的专业技能与就业能力，多数高校的计算机、电子商务、信息管理与信息系统等相关专业以及各地的相关培训机构均开设了 Java EE 程序设计或 Java EE 开发技术等课程。

本书以技术需求为导向，以技术应用为核心，以开发模式为主线，以应用开发为重点，以能力提升为目标，遵循案例教学、任务驱动与项目驱动的思想，结合教学规律与开发规范，按照由浅入深、循序渐进的原则，精心设计，合理安排，全面介绍了基于 Java EE 的企业级应用开发的关键技术、流行框架、主要模式、实施要点与方法步骤。全书实例翔实、案例丰富、编排合理、循序渐进、结构清晰，共分为 10 章，包括 Java EE 概述、JSP 基础、JDBC 技术、JavaBean 技术、Servlet 技术、Struts 2 框架、Hibernate 框架、Spring 框架、Ajax 应用与 Web 应用案例。各章均有“本章要点”“学习目标”与“本章小结”，既便于抓住重点、明确目标，也利于温故而知新。书中的诸多内容均设有相应的“说明”“注意”“提示”等知识点，以便于读者的理解与提高，并为其带来“原来如此”“豁然开朗”的美妙感觉。此外，各章均安排有相应的思考题，以便于读者的及时回顾与检测。书末还附有全面的实验指导，以便于读者的上机实践。

本书内容全面，结构清晰，语言流畅，通俗易懂，准确严谨，颇具特色，集系统性、条理性、针对性于一身，融资料性、实用性、技巧性于一体，注重应用开发能力的培养，可充分满足课程教学的实际需要，适合各个层面、各种水平的读者，既可作为各高校本科或高职高专计算机、电子商务、信息管理与信息系统及相关专业软件开发技术课程的教材或教学参考书，也可作为 Java EE 应用开发与维护人员的技术参考书以及初学者的自学教程。

本书所有示例的代码均已通过调试，并能成功运行，其开发环境为 Windows 7、JDK 1.7.0_51、Tomcat 7.0.50、MyEclipse 10.7.1 与 SQL Server 2008。

本书的写作与出版，得到了作者所在单位及清华大学出版社的大力支持与帮助，在此表示衷心感谢。在紧张的写作过程中，自始至终也得到了家人、同事的理解与支持，在此也一并深表谢意。

由于作者经验不足、水平有限，且时间较为仓促，书中不妥之处在所难免，恳请广大读者多加指正、不吝赐教。

编　者

目　录

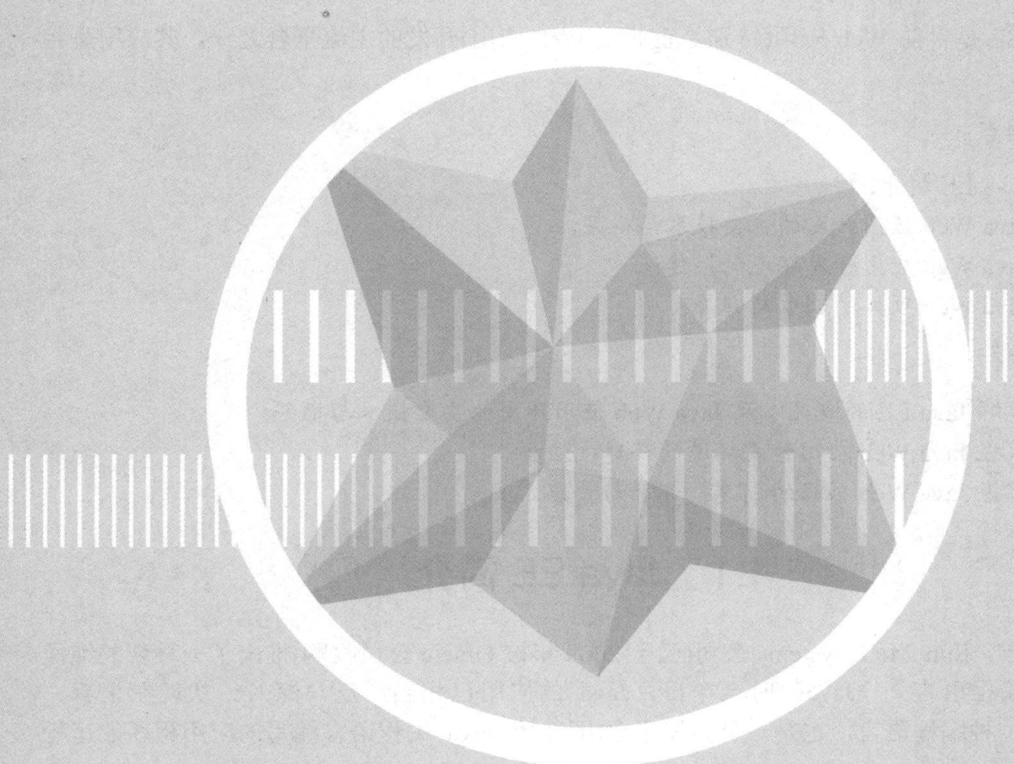

第 1 章

Java EE 概述

Java EE 是目前 Web 应用(特别是企业级 Web 应用)开发的主流平台之一，其应用是相当普遍的。

本章要点：

- Java EE 简介；
- Java Web 应用开发的主要技术与框架；
- Java Web 应用开发环境的搭建；
- Java Web 项目的创建与部署。

学习目标：

- 了解 Java EE 的概况以及 Java Web 应用开发的主要技术与框架；
- 掌握 Java Web 应用开发环境的搭建方法；
- 掌握 Java Web 项目的创建、部署与管理方法。

1.1 Java EE 简介

1996 年，Sun Microsystems 公司(已于 2009 年被 Oracle 公司收购)推出了一种新的纯面向对象的编程语言——Java。时至今日，Java 的应用已相当广泛与深入。从现在来看，Java 既是一种编程语言，也是一种开发平台(即使用 Java 编程语言编写的应用程序的运行环境)。

根据应用领域的不同，Java 平台分为 3 种，即 Java SE、Java EE 与 Java ME。其中，Java SE 为 Java 平台标准版(Java Platform Standard Edition)，提供了 Java 编程语言的核心功能，主要用于开发一般的台式机应用程序(或桌面应用程序)；Java EE 为 Java 平台企业版(Java Platform Enterprise Edition)，是基于 Java SE 的一种企业级应用标准平台，主要用于快速设计、开发、部署和管理企业级的大型软件系统；Java ME 为 Java 平台微型版(Java Platform Micro Edition)，包含了 Java SE 应用程序接口(Application Programming Interface，API)的一个子集，以及一个可运行于移动手机等微型设备之上的小型 Java 虚拟机(Java Virtual Machine，JVM)，主要用于开发掌上电脑、智能手机等移动设备使用的嵌入式系统。在此，企业级应用特指具备多层结构、可扩展的、高可靠性的、大规模的网络应用软件，而 Java ME 应用软件则通常作为 Java EE 应用软件的客户端。

由于 Java 的所有平台都包含有一个 Java 虚拟机和一套应用程序接口，因此任何基于这些平台开发的应用都具备 Java 编程语言的优点，包括跨平台、安全性、可扩展性等。

说明： Sun Microsystems 公司在 1998 年推出 JDK 1.2 时，使用了新名称 Java 2 Platform(即 Java 2 平台)，修改后的 JDK 称为 Java 2 Platform Software Developing Kit(即 J2SDK)，并分为标准版(Standard Edition，J2SE)、企业版(Enterprise Edition，J2EE)与微型版(Micro Edition，J2ME)。其中，J2SE 包含于 J2EE 中，而 J2ME 则包含了 J2SE 的核心，并添加了一些专有类。2005 年 6 月召开 Java One 大会时，Sun Microsystems 公司发布 Java SE 6。此时，Java 的各种版本已经更名，取消了其中的数字 2，分别为 Java SE、Java EE 与 Java ME。

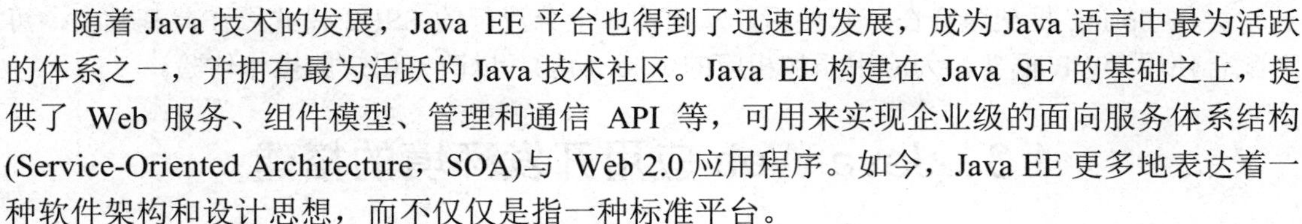

随着 Java 技术的发展，Java EE 平台也得到了迅速的发展，成为 Java 语言中最为活跃的体系之一，并拥有最为活跃的 Java 技术社区。Java EE 构建在 Java SE 的基础之上，提供了 Web 服务、组件模型、管理和通信 API 等，可用来实现企业级的面向服务体系结构(Service-Oriented Architecture，SOA)与 Web 2.0 应用程序。如今，Java EE 更多地表达着一种软件架构和设计思想，而不仅仅是指一种标准平台。

1.2　Java Web 应用开发的主要技术与框架

C/S(Client/Server，客户机/服务器)与 B/S(Browser/Server，浏览器/服务器)是目前应用程序的两种主要架构或模式。与基于 C/S 架构的 Windows 应用程序不同，Web 应用程序是基于 B/S 架构的。Web 应用程序部署在 Web 服务器上，因此易于升级和维护。同时，Web 应用程序的访问是通过浏览器进行的，因此只需在客户机上安装一个标准的浏览器即可，而无须安装专门的客户端程序。此外，由于 Web 应用程序的数据分析与处理工作主要是在服务器中完成的，因此对客户机的配置要求不高，特别适合“瘦客户端”的运行环境。随着 Internet 的快速发展，Web 应用程序的使用将更加广泛与深入。

要开发 Java Web 应用，需要掌握一系列相关的技术。必要时，可应用相应的框架，以提高 Java Web 应用的开发效率，并确保其可扩展性与可维护性。

Java Web 应用开发的主要技术包括 HTML/XHTML、XML、JavaScript、Java、JDBC、JSP、JavaBean、Servlet、Ajax 等。其中，JDBC(Java Data Base Connectivity，Java 数据库连接)是一种用于执行 SQL 语句的 Java API(Application Programming Interface，应用编程接口或应用程序编程接口)，由一组用 Java 语言编写的类和接口组成，可让开发人员以纯 Java 方式连接数据库，并完成相应的操作。JSP(Java Server Pages)是由 Sun Microsystems 公司倡导、许多公司参与建立的一种动态网页技术标准(类似于 ASP 与 PHP 技术)，可在 HTML 文档(*.htm、*.html)中嵌入 Java 脚本小程序(Scriptlet)和 JSP 标记(tag)等元素，从而形成 JSP 文件(*.jsp)。JavaBean 是 Java 中的一种可重用组件技术(类似于微软的 COM 技术)，其本质是一种通过封装属性和方法而具有某种功能的 Java 类。Servlet 是一种用 Java 编写的与平台无关的服务器端组件，其实例化对象运行在服务器端，可用于处理来自客户端的请求，并生成相应的动态网页。Ajax(Asynchronous JavaScript and XML，异步 JavaScript 和 XML)是一种创建交互式网页应用的开发技术，其核心理念为使用 XMLHttpRequest 对象发送异步请求，可实现 Web 页面的动态更新。

Java Web 应用开发的流行框架主要有 Struts 2、Hibernate、Spring、DWR 等。其中，Struts 2 是一个基于 MVC 设计模式的 Web 应用框架，本质上相当于一个 Servlet，可在 MVC 设计模式中作为控制器(Controller)建立模型(Mode)与视图(View)之间的数据交互。Hibernate 是一个开源的 ORM(Object-Relation Mapping，对象关系映射)框架，通过对 JDBC 进行轻量级封装(未完全封装)，让 Java 编程者可以使用面向对象的编程思想来操纵数据库，是一种十分成功的 ORM 解决方案。Spring 是一个由 Rod Johnson 创建的开源框架，具有简单、可测试与松耦合的特征，可有效简化企业级应用的开发。DWR 是一种用于创建 Ajax Web 站点的开源框架，可让开发人员无须具备专业的 JavaScript 知识就可以轻松实现 Ajax 功能，并有效减少代码的编写量。目前，在实际的应用开发中，Struts 2、Hibernate

通常会与 Spring 框架整合在一起使用，从而构成十分流行的 SSH2 模式。在此基础上，可再配合使用 DWR 框架，为应用添加相应的 Ajax 功能，以改善用户的操作体验。

1.3 Java Web 应用开发环境的搭建

要进行 Java Web 应用的开发，首先要搭建好相应的开发环境。为此，需逐一完成 JDK 开发包、Web 服务器、IDE 开发工具以及数据库管理系统的安装与配置。在此，JDK 开发包使用 jdk1.7.0_51，Web 服务器使用 Tomcat 7.0.50，IDE 开发工具使用 MyEclipse 10.7.1，数据库管理系统使用 SQL Server 2008。下面，简要介绍一下在 Windows 7 中构建开发环境的基本步骤与关键配置。

1.3.1 JDK 的安装与配置

Java Web 应用程序的开发需要 JDK(Java Development Kit，Java 开发工具包)的支持。JDK 是整个 Java 的核心，内含 JRE(Java Runtime Environment，Java 运行环境)、Java 工具、Java 基础类库以及相关范例与文档，是一种用于构建在 Java 平台上发布的应用程序、Applet 与组件的开发环境。如今，Oracle 公司已取代 Sun 公司负责 JDK 的维护与升级工作，并定期在其官网上发布最新的版本。

1. 下载

各种版本的 JDK 均可从其官网免费下载，地址为

```
http://www.oracle.com/technetwork/java/javase/downloads/index.htm
```

或者

```
http://java.sun.com/javase/downloads/index.jsp
```

下载 JDK 7 Update 51 版，其安装文件为 jdk-7u51-windows-i586.exe。

2. 安装

JDK 的安装过程非常简单，只需双击安装文件 jdk-7u51-windows-i586.exe，启动 JDK 的安装向导(见图 1.1)，并根据向导完成后续的有关操作即可。在此，分别将 JDK 与 JRE 的安装目录设置为 C:\Program Files\Java\jdk1.7.0_51 与 C:\Program Files\Java\jre7，如图 1.2、图 1.3 所示。

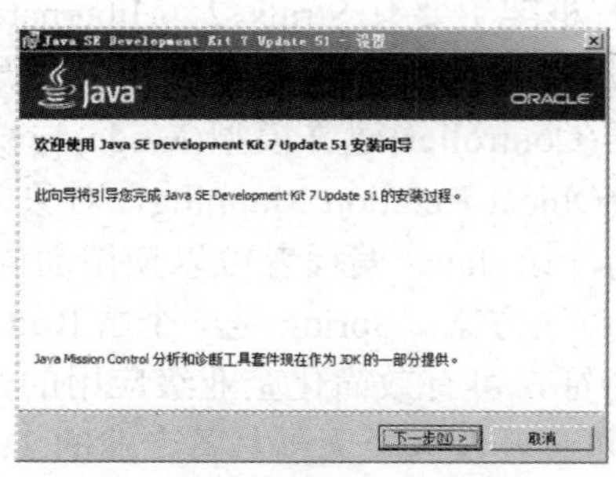

图 1.1 JDK 的安装向导

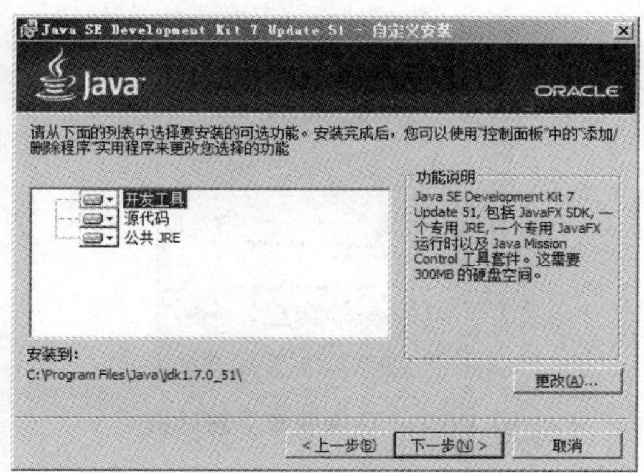

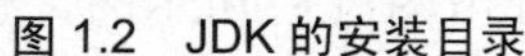
图 1.2 JDK 的安装目录

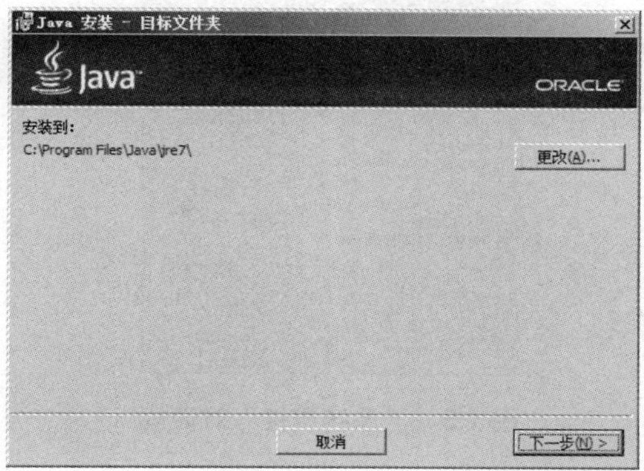

图 1.3 JRE 的安装目录

3．配置

JDK 安装完毕后，还需进行相应的配置，主要是创建或设置有关的系统变量，以告知 Windows 系统 JDK 与 JRE 的安装位置。

以系统变量 JAVA_HOME 为例，其创建的基本步骤如下。

(1) 在桌面右击“计算机”图标，选择其快捷菜单中的“属性”菜单项，打开图 1.4 所示的“控制面板\系统和安全\系统”窗口。

图 1.4 “控制面板\系统和安全\系统”窗口

(2) 单击“高级系统设置”链接，弹出图 1.5 所示的“系统属性”对话框。

(3) 单击“环境变量”按钮，弹出图 1.6 所示的“环境变量”对话框。

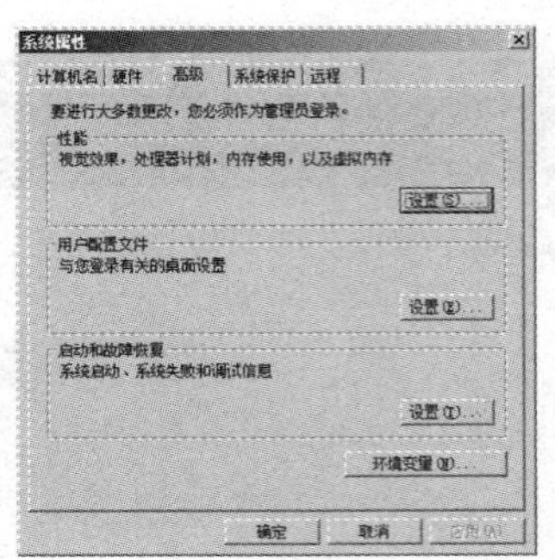

图 1.5 “系统属性”对话框

图 1.6 “环境变量”对话框

(4) 单击“系统变量”列表框下的“新建”按钮，弹出图 1.7 所示的“新建系统变量”对话框，并在其中输入变量名 JAVA_HOME、变量值“C:\Program Files\Java\jdk1.7.0_51”，最后单击“确定”按钮。

(5) 单击“确定”按钮，关闭“环境变量”对话框。

(6) 单击“确定”按钮，关闭“系统属性”对话框。

图 1.7 “新建系统变量”对话框

类似地，新建环境变量 Path，将其值设置为“.;%JAVA_HOME%\bin”(若该变量已经存在，则在其值前添加“.;%JAVA_HOME%\bin;”)；新建环境变量 CLASSPATH，将其值设置为“.;%JAVA_HOME%\lib”(若该变量已经存在，则在其值前添加“.;%JAVA_HOME%\lib;”)；新建环境变量 JRE_HOME，将其值设置为“C:\Program Files\Java\jre7”(若该变量已经存在，则在其值前添加“C:\Program Files\Java\jre7”)。

4．测试

配置完毕后，为检验 JDK 是否正常，可按以下步骤进行操作。

(1) 单击“开始”按钮，选择“开始”菜单中的“运行”菜单项，弹出“运行”对话框(见图 1.8)，然后输入 cmd 命令，单击“确定”按钮，打开命令提示符窗口(见图 1.9)。

(2) 输入并执行 java –version 命令，若能显示当前 JDK 的版本(在此为 java version “1.7.0_51”)，则表示一切正常(见图 1.9)。

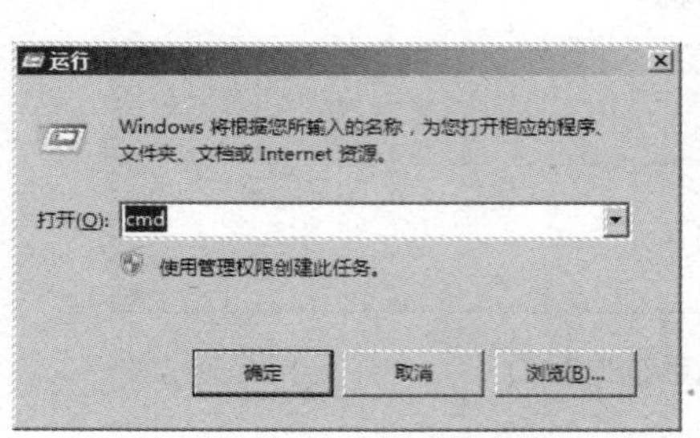

图 1.8 “运行”对话框

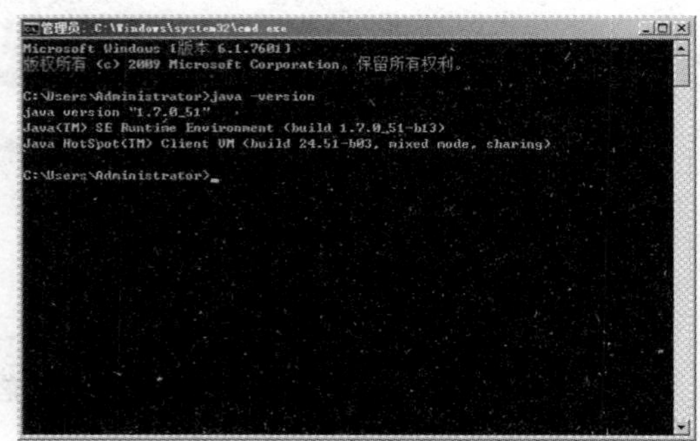

图 1.9 命令提示符窗口

1.3.2　Tomcat 的安装与配置

Tomcat 是 Apache 软件基金会(Apache Software Foundation)Jakarta 项目中的一个核心子项目，最初由 Apache、Sun 与其他一些公司及个人共同开发而成。由于有了 Sun 的参与和支持，最新的 Servlet 与 JSP 规范总是能够在 Tomcat 中得到体现。正因为 Tomcat 技术先进、性能稳定，而且是免费开源的，因此深受广大 Java 爱好者的喜爱并得到了部分软件开发商的认可，最终成为目前最为流行的 Web 应用服务器之一。作为一种轻量级应用服务器，Tomcat 在中小型系统与并发访问用户不是很多的场合下被普遍使用，是开发与调试 JSP 或 Java EE 程序的首选。

1．下载

各种版本的 Tomcat 均可从其官网免费下载，地址为 http://tomcat.apache.org/。

下载安装版的 Tomcat-7.0.50，其安装文件为 apache-tomcat-7.0.50.exe。

2．安装

Tomcat 的安装过程非常简单，只需双击安装文件 apache-tomcat-7.0.50.exe，启动 Tomcat 的安装向导(见图 1.10)，并根据向导完成后续的有关操作即可。在此，选择 Normal 安装类型(见图 1.11)，采用默认的端口设置，创建管理员登录账号 admin，并将其密码设置为 12345(见图 1.12)，同时选定 JRE 的安装路径为 C:\Program Files\Java\jre7(见图 1.13)，指定 Tomcat 的安装目录为 C:\Program Files\Apache Software Foundation\Tomcat 7.0 (见图 1.14)。

图 1.10　Tomcat 的安装向导

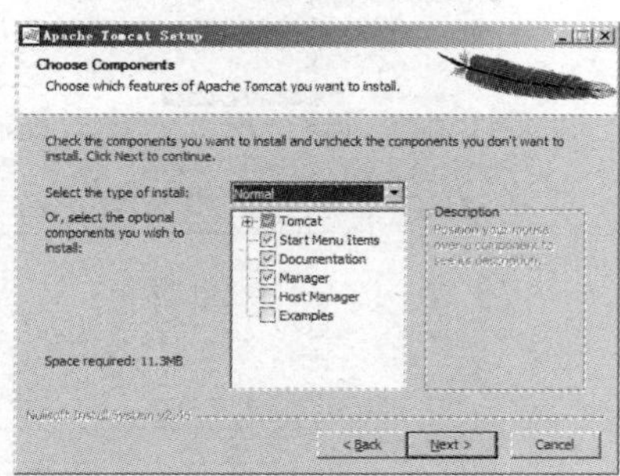

图 1.11　Tomcat 的安装类型

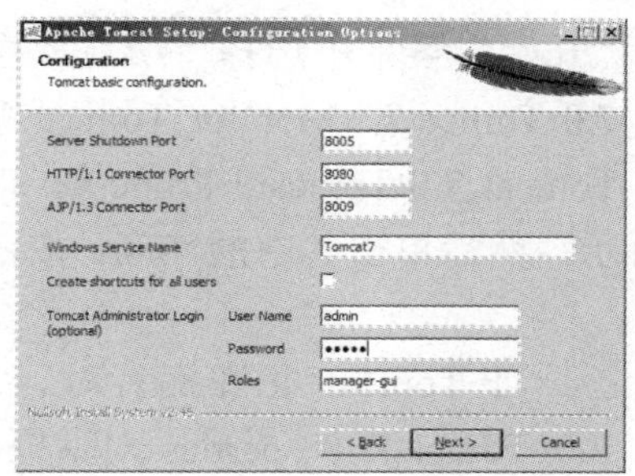

图 1.12　Tomcat 的基本配置

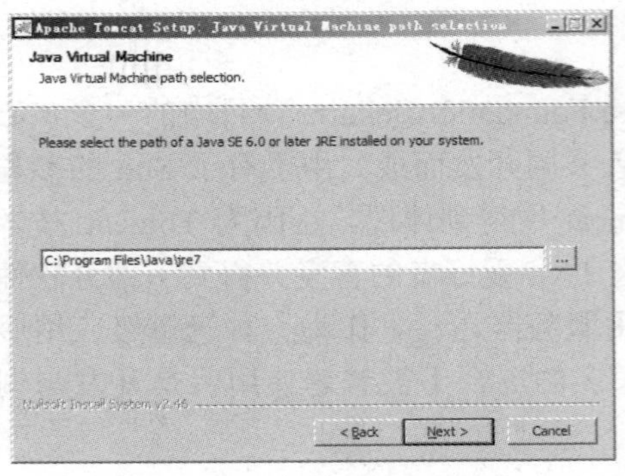

图 1.13　Tomcat 的 JRE 路径

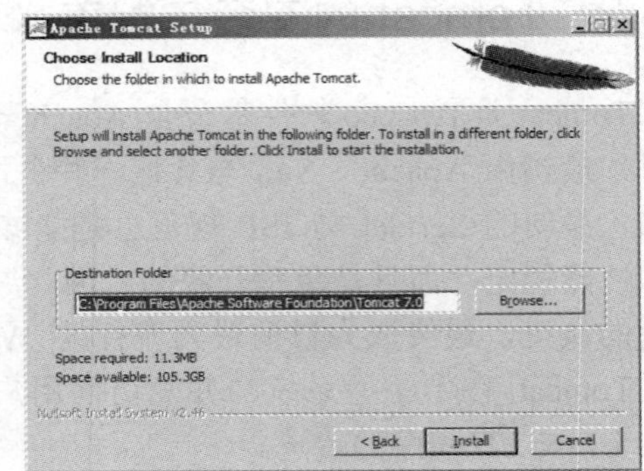

图 1.14　Tomcat 的安装路径

3. 配置

Tomcat 安装完毕后，必要时可对其进行相应的配置。

1)　HTTP 端口号的修改

Tomcat 默认的 HTTP 端口号为 8080，若要使用其他端口号，可直接打开 Tomcat 的配置文件 C:\Program Files\Apache Software Foundation\Tomcat 7.0\conf\server.xml，找到如下代码，并将其中的 8080 改为所需要的端口号(如 8090 等)即可。

```
<Connector port="8080" protocol="HTTP/1.1"
           connectionTimeout="20000"
           redirectPort="8443" />
```

2)　管理员登录账号的设置

Tomcat 管理员登录账号的信息保存在 C:\Program Files\Apache Software Foundation\Tomcat 7.0\conf\tomcat-users.xml 文件中。通过修改该文件，即可实现管理员登录账号的添加、修改或删除。例如，以下代码表明当前 Tomcat 的管理员登录账号为 admin，密码为 12345。

```
<tomcat-users>
<role rolename="manager-gui"/>
<user username="admin" password="12345" roles="manager-gui"/>
</tomcat-users>
```

4. 启动与关闭

在“开始”菜单中选择“所有程序”→Apache Tomcat 7.0 Tomcat7→Monitor Tomcat 菜单项，打开 Apache Tomcat 7.0 Tomcat7 Properties 对话框，如图 1.15 所示，然后单击其中的 Start 按钮，即可启动 Tomcat。

反之，当 Tomcat 正在运行时，单击 Apache Tomcat 7.0 Tomcat7 Properties 对话框中的 Stop 按钮，可随即关闭 Tomcat。

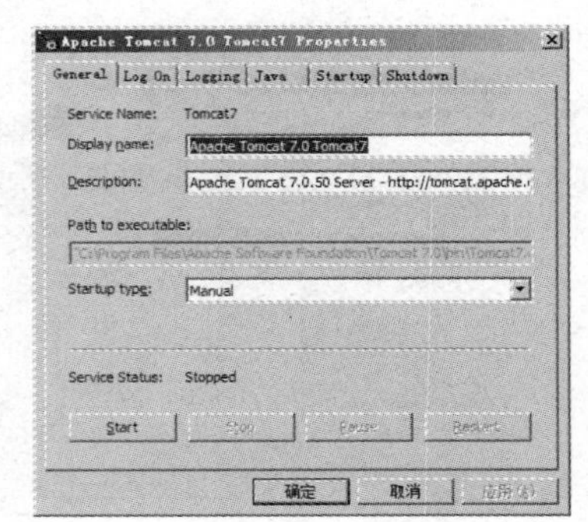

图 1.15　Apache Tomcat 7.0 Tomcat 7 Properties 对话框

5．测试

为测试 Tomcat 是否正常，应先启动 Tomcat，然后打开浏览器，在地址栏输入“http://localhost:8080/”并按 Enter 键。若显示图 1.16 所示的 Apache Tomcat/7.0.50 页面，则表明 Tomcat 一切正常。单击其中的 Manage App 按钮，弹出图 1.17 所示的“Windows 安全”对话框，并在其中输入用户名 admin 与密码 12345，再单击“确定”按钮，即可打开图 1.18 所示的 Tomcat Web Application Manager 页面。在该页面中，可对当前发布的有关应用进行相应的管理，包括启动、停止等。

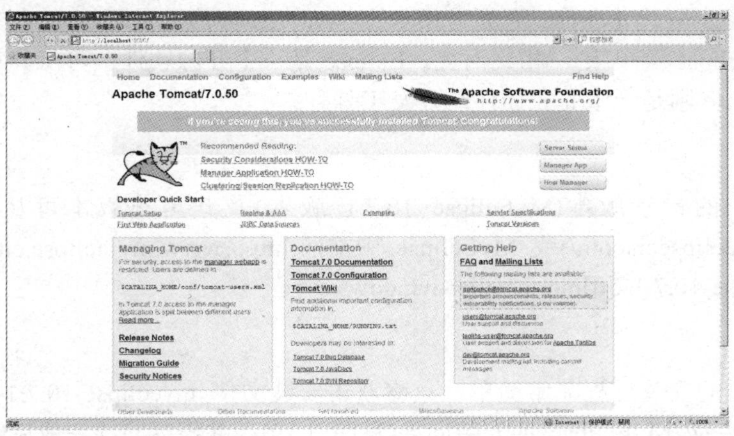

图 1.16　Apache Tomcat/7.0.50 页面

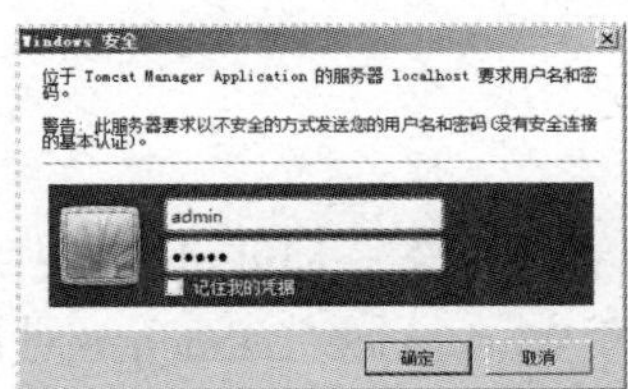

图 1.17　“Windows 安全”对话框

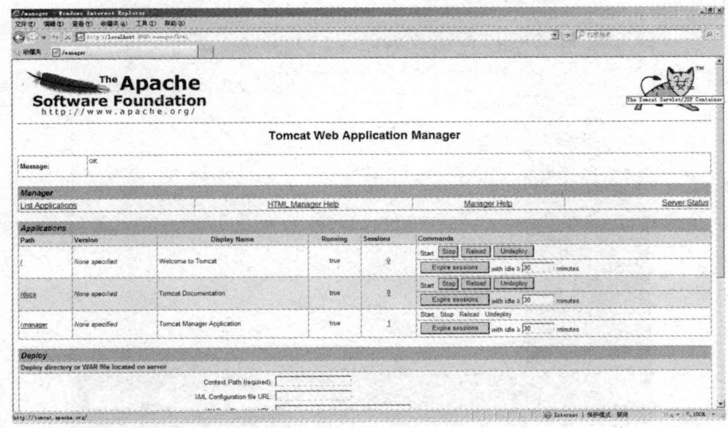

图 1.18　Tomcat Web Application Manager 页面

1.3.3　MyEclipse 的安装与配置

MyEclipse 是 MyEclipse Enterprise Workbench(MyEclipse 企业级工作平台)的简称，是一种功能极其丰富的 Java EE 集成开发环境，包括完备的编码、调试、测试与发布功能，并完整支持 HTML、CSS、JavaScript、Ajax、Servlet、JSP、JSF、EJB、JDBC、SQL、Struts、Hibernate、Spring 等各种 Java EE 相关技术的标准与框架。借助于 MyEclipse，可在数据库与 Java EE 的开发、发布以及应用程序服务器的整合方面极大地提高工作效率。实际上，MyEclipse(6.0 版之前)原本只是作为 Eclipse 的一个插件存在的，后来随着其功能的日益强大，逐步取代 Eclipse 而成为独立的 Java EE 集成开发环境(但在其主菜单中至今仍保留着 MyEclipse 菜单项)。相比而言，Eclipse 是一个开源软件，可从其官网免费下载安装，而 MyEclipse 则是一个商业插件或开发工具。

1. 下载

MyEclipse 的最新版本(MyEclipse 10.7.1 版本)及其历史版本可从其官方中文网(http://www.myeclipsecn.com/)或 MyEclipse 中文网(http://www.my-eclipse.cn/)下载，其安装文件为 myeclipse-10.7.1-offline-installer-windows.exe。

2. 安装

MyEclipse 的安装过程非常简单，只需双击安装文件 myeclipse-10.7.1-offline-installer-windows.exe，启动 MyEclipse 的安装向导(见图 1.19)，并根据向导完成后续的有关操作即可。在此，指定 MyEclipse 的安装目录为 C:\MyEclipse(见图 1.20)，并选择 All 安装类型(见图 1.21)。

3. 启动与退出

在“开始”菜单中选择“所有程序”→MyEclipse→MyEclipse 10 菜单项，即可启动 MyEclipse，并弹出图 1.22 所示的 Workspace Launcher 对话框。待选定工作区后(在此将工作区选定为 C:\Workspaces\MyEclipse 10)，再单击 OK 按钮，即可打开图 1.23 所示的 MyEclipse Enterprise Workbench 窗口。

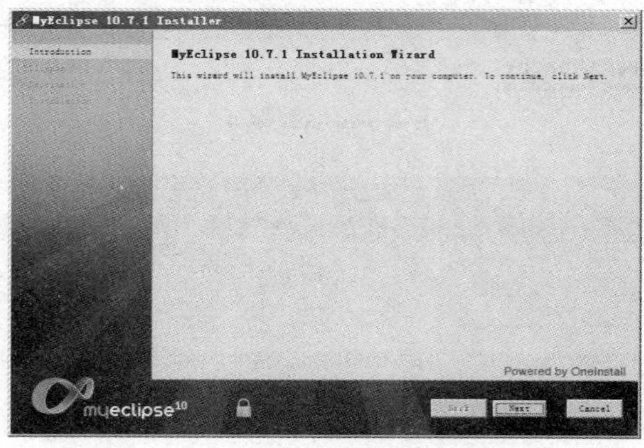

图 1.19　MyEclipse 的安装向导

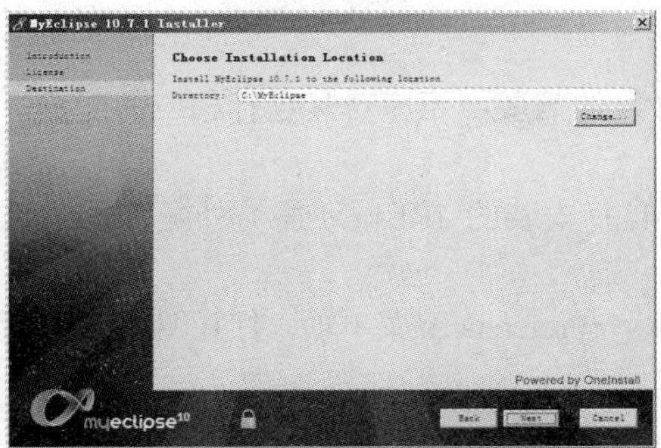

图 1.20　MyEclipse 的安装目录

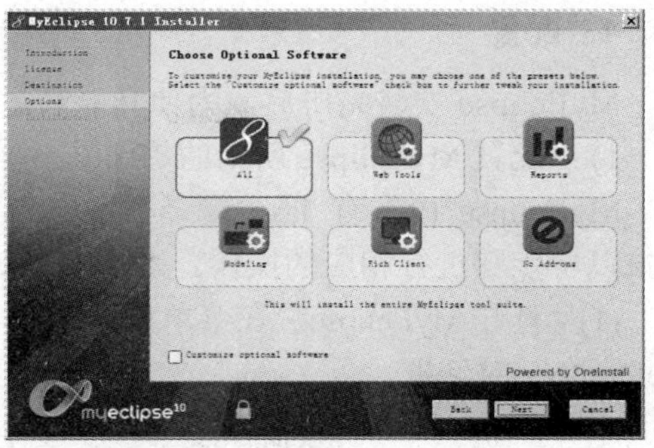

图 1.21　MyEclipse 的安装类型

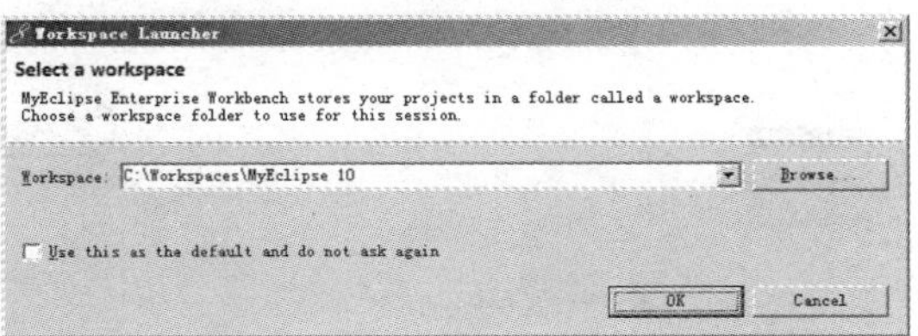

图 1.22　Workspace Launcher 对话框

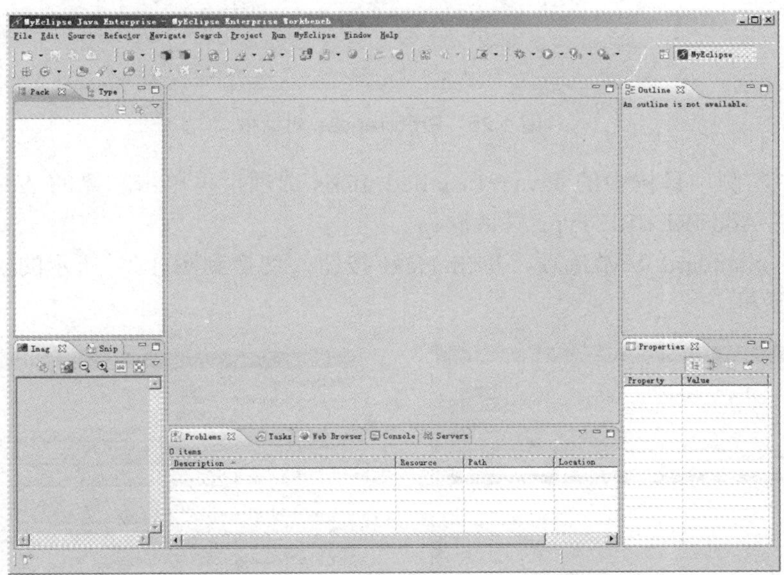

图 1.23　MyEclipse Enterprise Workbench 窗口

单击 MyEclipse Enterprise Workbench 窗口的关闭按钮☒，并在随之弹出的图 1.24 所示的 Confirm Exit 对话框中单击 OK 按钮，即可退出 MyEclipse。

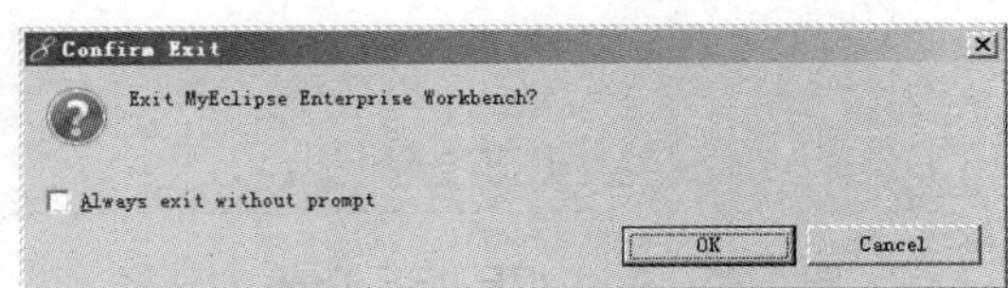

图 1.24　Confirm Exit 对话框

4. 配置

MyEclipse 安装完毕后，为使其能满足应用开发的需要，还应对其进行相应的配置。

1) 配置 MyEclipse 所用的 JRE

MyEclipse 内置有 Java 编译器，但为了使用自行安装的 JRE，还需另外配置。为此，可按以下步骤进行相应的操作。

(1) 打开 MyEclipse，在菜单栏中选择 Windows→Preferences 菜单项，打开图 1.25 所示的 Preferences 对话框。

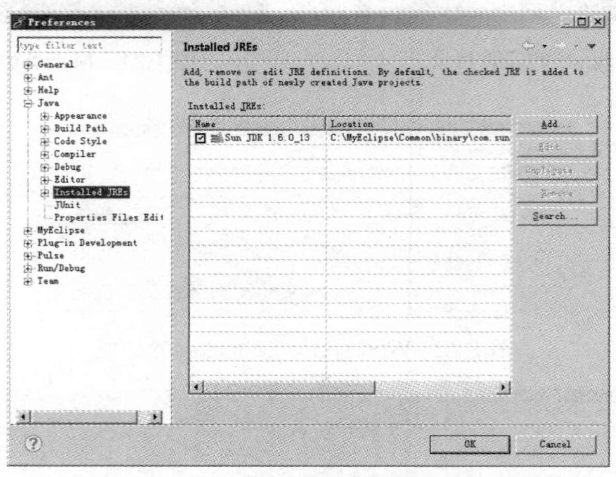

图 1.25 Preferences 对话框

(2) 选中左侧项目树中的 Java→Installed JREs 选项，再单击右侧的 Add 按钮，打开图 1.26 所示的 Add JRE-JRE Type 对话框。

(3) 选中 Standard VM 选项，单击 Next 按钮，切换到图 1.27 所示的 Add JRE-JRE Definition 对话框。

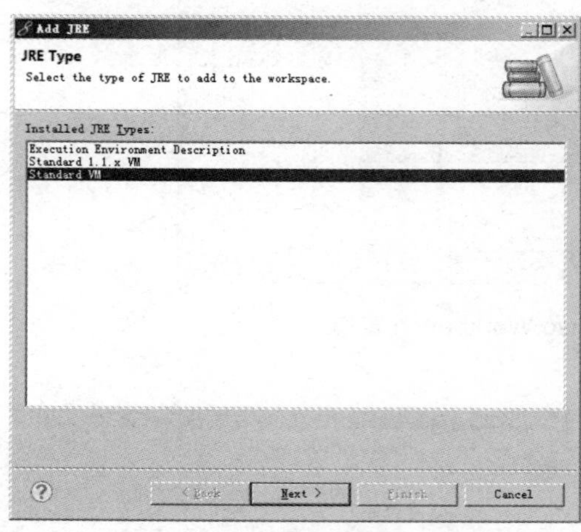

图 1.26 Add JRE-JRE Type 对话框

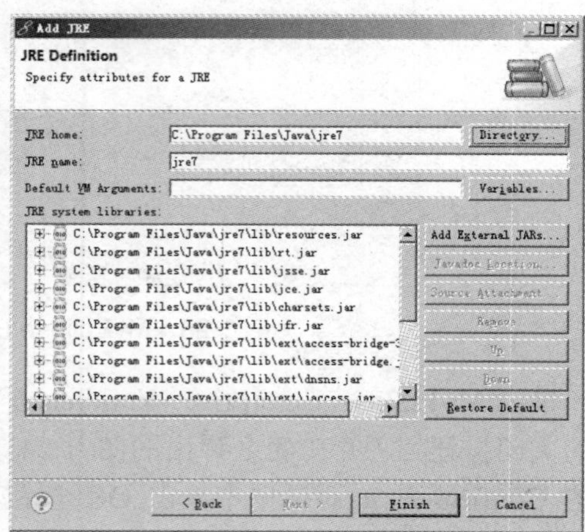

图 1.27 Add JRE-JRE Definition 对话框

(4) 选定相应 JRE 的主目录(在此为 C:\Program Files\Java\jre7)，并指定相应 JRE 名称

(在此为 jre7)，再单击 Finish 按钮，返回 Preferences 对话框(见图 1.28)。

(5) 选中相应的 JRE 复选框(在此为 jre7)，再单击 OK 按钮，关闭 Preferences 对话框。

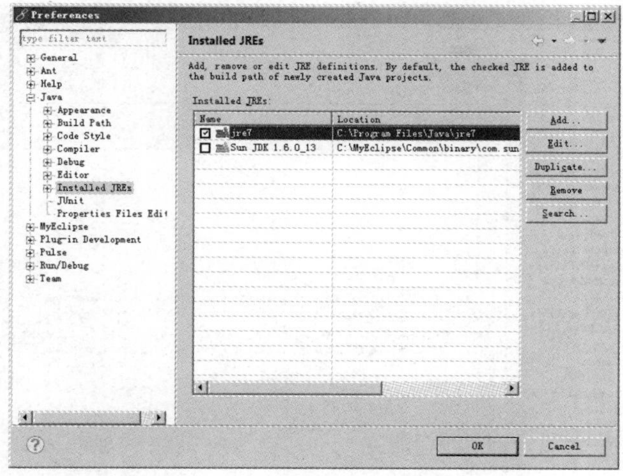

图 1.28　Preferences 对话框

2) 集成 MyEclipse 与 Tomcat

为便于应用的发布与运行，并提高应用的开发效率，可在 MyEclipse 中集成相应的 Tomcat 服务器。为此，可按以下步骤进行相应操作。

(1) 在菜单栏中选择 Windows→Preferences 菜单项，打开 Preferences 对话框。

(2) 选中左侧项目树中的 MyEclipse→Servers→Tomcat→Tomcat 7.x 选项，然后在右侧的 Tomcat home directory 处指定 Tomcat 的安装目录(在此为 C:\Program Files\Apache Software Foundation\Tomcat 7.0)，并选中 Enable 单选按钮激活 Tomcat 服务器，如图 1.29 所示。

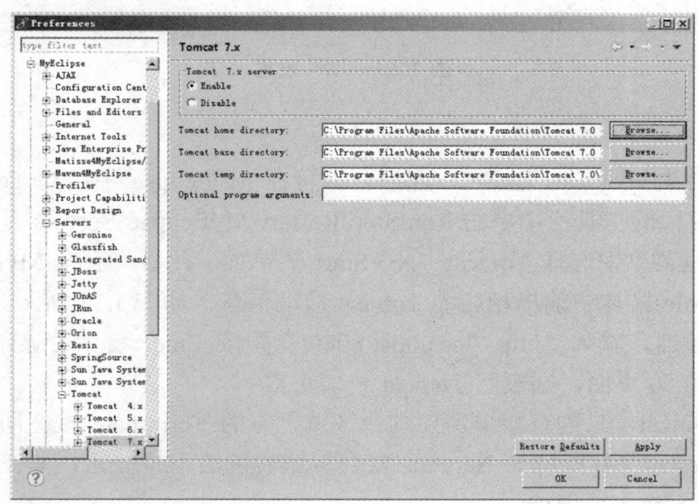

图 1.29　Tomcat 7.x 设置

(3) 选中左侧项目树中 Tomcat 7.x→JDK 选项，然后在右侧的 Tomcat 7.x JDK name 下拉列表框中选中此前刚刚添加的 JRE 所对应的选项(在此为 jre7)，如图 1.30 所示。

(4) 单击 OK 按钮，关闭 Preferences 对话框。

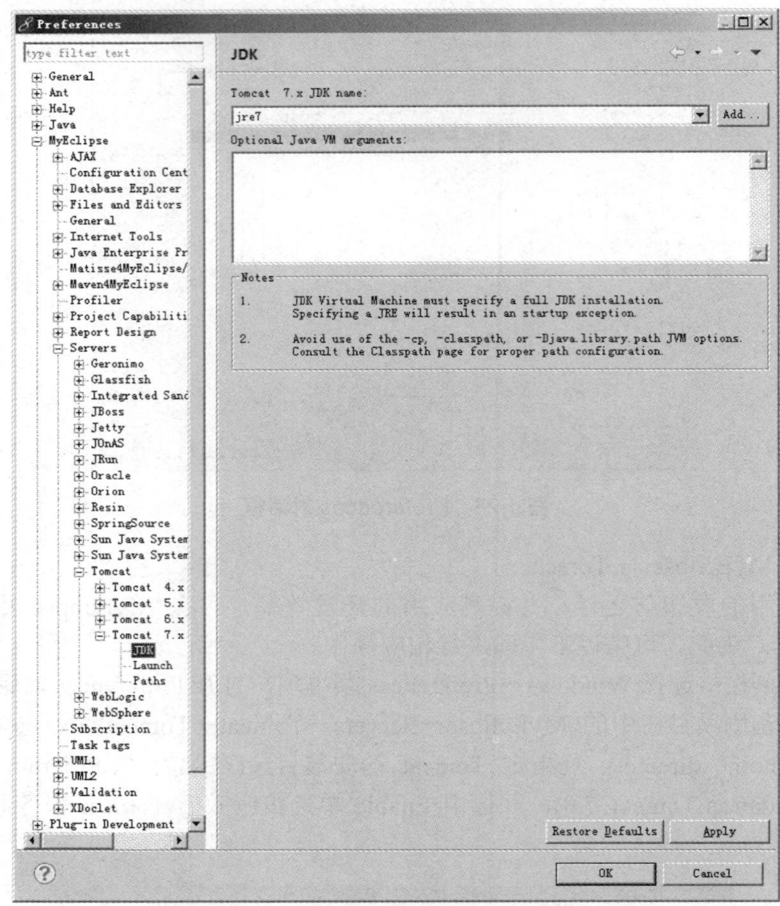

图 1.30　JDK 设置

5. 测试

将 MyEclipse 与 Tomcat 集成在一起后，即可在 MyEclipse 中启动 Tomcat。为此，可在 MyEclipse 主窗口的工具栏中单击 Run/Stop/Restart MyEclipse Servers 复合按钮右侧的下拉按钮，并在下拉菜单中选择 Tomcat 7.x→Start 菜单项。此后，会在 MyEclipse 主窗口下方的控制台(Console)区域中输出相应的 Tomcat 启动信息，如图 1.31 所示。待 Tomcat 启动完毕后，打开浏览器，输入“http://localhost:8080/”并按 Enter 键。若显示图 1.32 所示的 Apache Tomcat/7.0.50 页面，则表明 Tomcat 一切正常。

若要停止 Tomcat，只需直接单击控制台区域上方的 Terminate 按钮即可。此外，也可单击 Run/Stop/Restart MyEclipse Servers 复合按钮右侧的下拉按钮，并在下拉菜单中选择 Tomcat 7.x→Stop Server 菜单项。

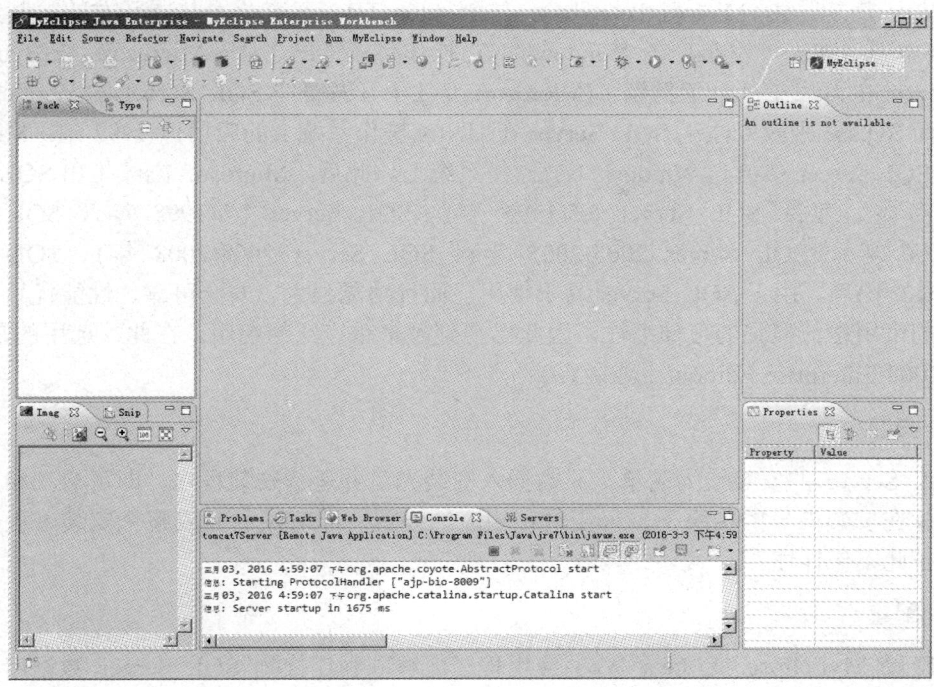

图 1.31　Tomcat 的启动

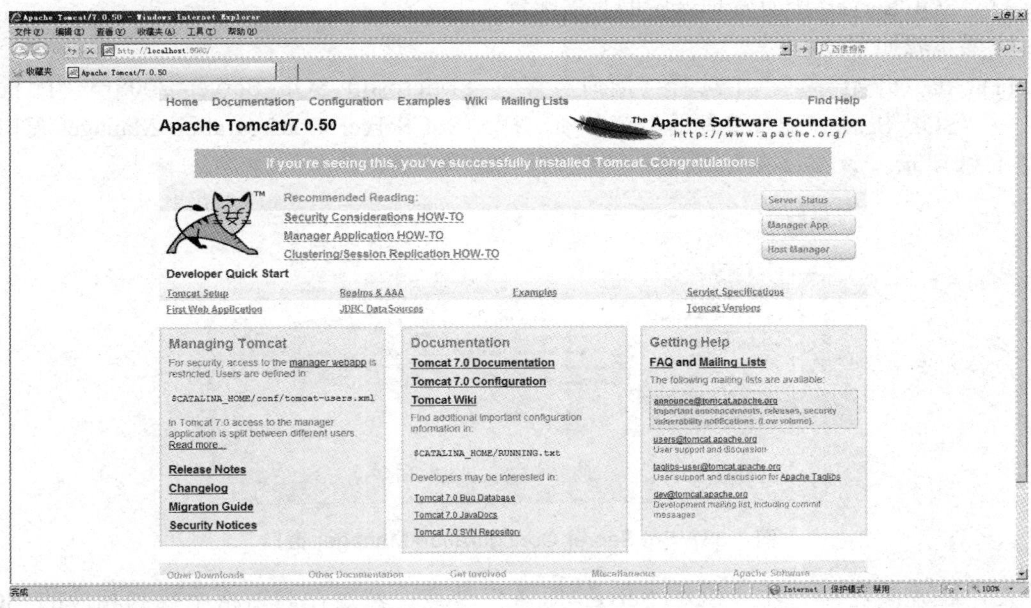

图 1.32　Apache Tomcat/7.0.50 页面

1.3.4　SQL Server 的安装与配置

SQL Server 是一种基于客户机/服务器(C/S)体系结构的大型关系数据库管理系统

(RDBMS)，最初是由 Microsoft、Sybase 与 Ashton-Tate 三家公司共同开发的，其第一个 OS/2 版本于 1988 年发布。在 Windows NT 操作系统推出之后，Microsoft 与 Sybase 在 SQL Server 的开发上就分道扬镳了(Microsoft 专注于开发推广 SQL Server 的 Windows NT 版本，而 Sybase 则较专注于 SQL Server 在 UNIX 操作系统上的应用)。1992 年，Microsoft 成功将 SQL Server 移植到 Windows NT 操作系统上。此后，Microsoft 陆续推出 SQL Server 更高的版本，包括 SQL Server 6.5(1996 年)、SQL Server 7.0(1998 年)、SQL Server 2000(2000 年)、SQL Server 2005(2005 年)、SQL Server 2008(2008 年)、SQL Server 2012(2012 年)等。由于 SQL Server 易于使用，而且功能强大、安全可靠、性能优异，并具有极高的可用性与极强的可伸缩性，因此已得到越来越广泛的应用。在此，选用的是 SQL Server 2008 Enterprise Edition(企业版)。

1. 安装

SQL Server 的安装较为简单，只需插入安装光盘并运行安装程序，即可启动相应的安装向导。在安装向导的指引下，只需进行相应的设置，即可顺利完成整个安装过程。限于篇幅，在此不作详述。

2. 配置

为确保 MyEclipse 或 Java Web 应用程序能够顺利连接到 SQL Server 数据库，应对 SQL Server 数据库服务器的有关配置进行认真检查，并在需要时进行相应的修改。

1) SQL Server 配置管理器中的有关配置

主要步骤如下。

(1) 在“开始”菜单中选择“所有程序”→Microsoft SQL Server 2008→“配置工具”→“SQL Server 配置管理器”菜单项，打开 Sql Server Configuration Manager 窗口，如图 1.33 所示。

图 1.33 Sql Server Configuration Manager 窗口

(2) 在左窗格中选中“SQL Server 服务”节点，然后在右窗格中找到相应的 SQL Server 服务程序(在此为 SQL Server (MSSQLSERVER))，确保其处于“正在运行”状态。为方便起见，建议将该服务的启动模式设置为“自动”。

(3) 在左窗格中选中“SQL Server 网络配置”下的相应协议节点(在此为“MSSQLSERVER 的协议”)，然后在右窗格中找到 TCP/IP 协议，确保其处于“已启用”状态，如图 1.34 所示。若尚未启用，则应双击它，弹出“TCP/IP 属性”对话框，将其中的“协议”选项卡的“已启用”设置为“是”，如图 1.35 所示，同时在“IP 地址”选项卡中将 IPAll 下的“TCP 端口”设置为 1433，如图 1.36 所示。

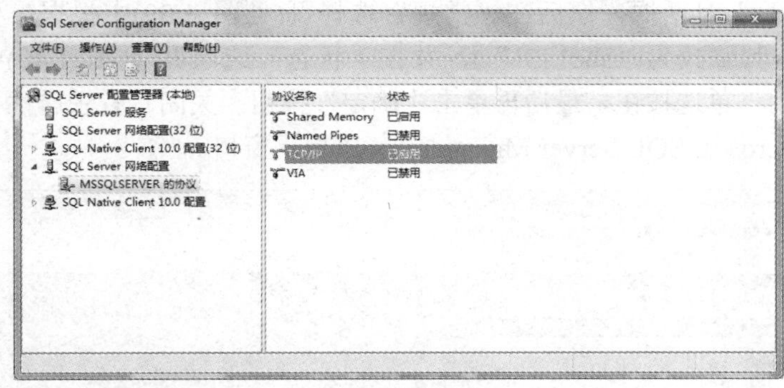

图 1.34　Sql Server Configuration Manager 窗口

图 1.35　“TCP/IP 属性”对话框（“协议”选项卡)

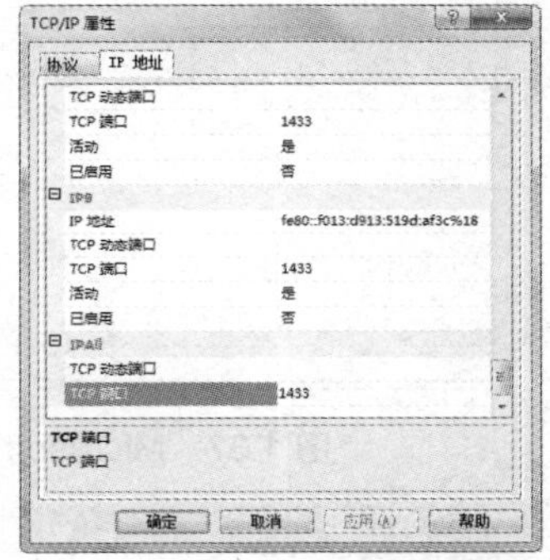

图 1.36　“TCP/IP 属性”对话框（“IP 地址”选项卡)

(4)　重新启动 SQL Server 服务程序(在此为 SQL Server (MSSQLSERVER))，以便有关设置的更改生效，然后关闭 Sql Server Configuration Manager 窗口。

2)　SQL Server Management Studio 中的有关配置

主要步骤如下。

(1)　在“开始”菜单中选择“所有程序”→Microsoft SQL Server 2008→SQL Server Management Studio 菜单项，打开 Microsoft SQL Server Management Studio 窗口，如图 1.37 所示。

(2)　在“对象资源管理器”窗格中，右击 SQL Server 服务器节点，并在其快捷菜单中选择“属性”菜单项，打开“服务器属性”对话框。切换到“安全性”选择页(见图 1.38)，在“服务器身份验证”选项区中选中“SQL Server 和 Windows 身份验证模式”单选按钮，最后单击“确定”按钮，关闭“服务器属性”对话框。

注意： 如果改变了 SQL Server 的身份验证模式，那么应重启 SQL Server 服务，以便身份验证模式的更改生效。

(3) 在“对象资源管理器”窗格中，依次展开 SQL Server 服务器下的“安全性”|“登录名”节点，右击超级管理员登录账号 sa，并在其快捷菜单中选择“属性”菜单项，打开“登录属性-sa”对话框，然后在“常规”选择页(见图 1.39)中根据需要修改登录账号 sa 的密码(在此将其修改为“abc123!”)，并在“状态”选项页(见图 1.40)中选中“授予”单选按钮与“启用”单选按钮，最后再单击“确定”按钮，关闭“登录属性 - sa”对话框。

(4) 关闭 Microsoft SQL Server Management Studio 窗口。

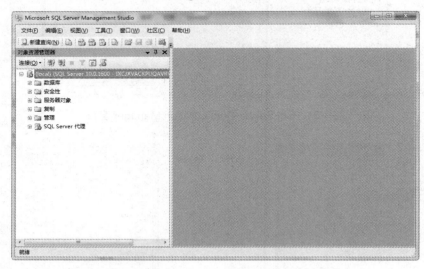

图 1.37 Microsoft SQL Server Management Studio 窗口

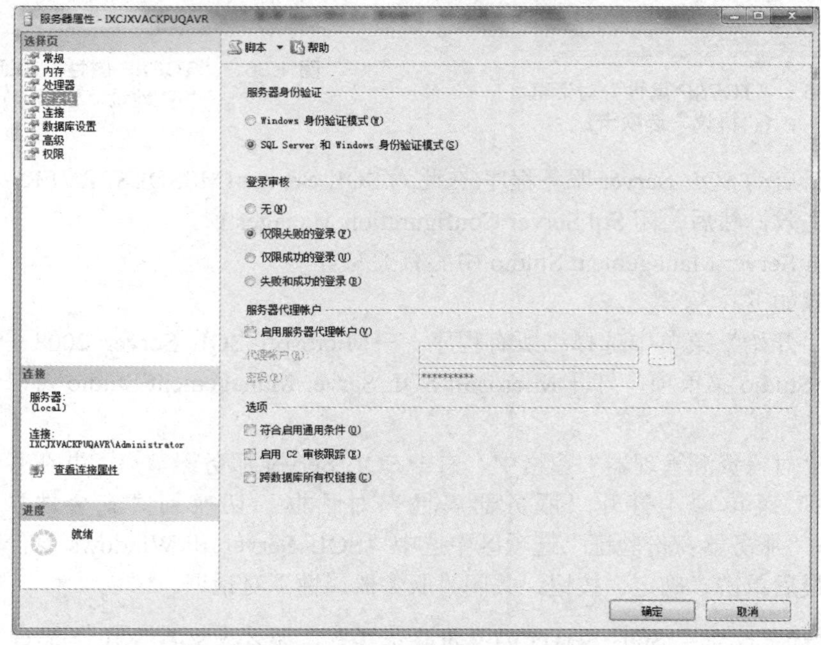

图 1.38 “服务器属性”对话框(“安全性”选择页)

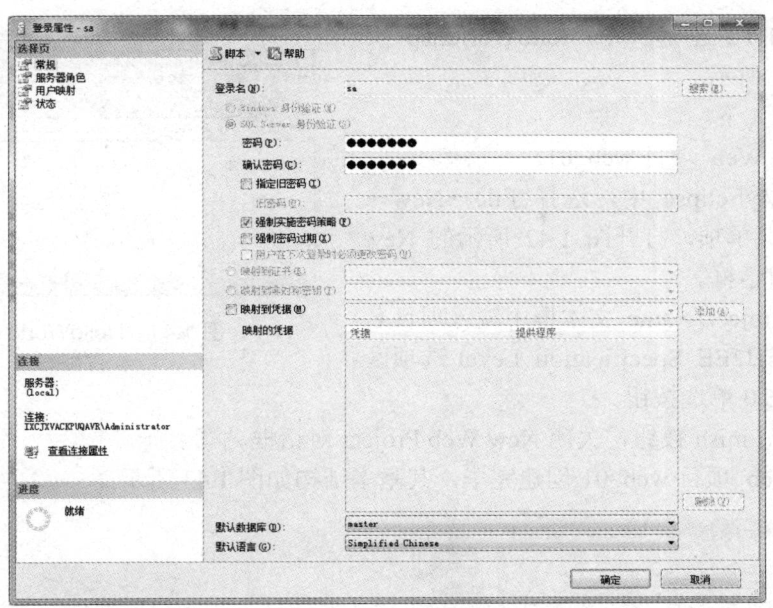

图 1.39　“登录属性-sa”对话框(“常规”选择页)

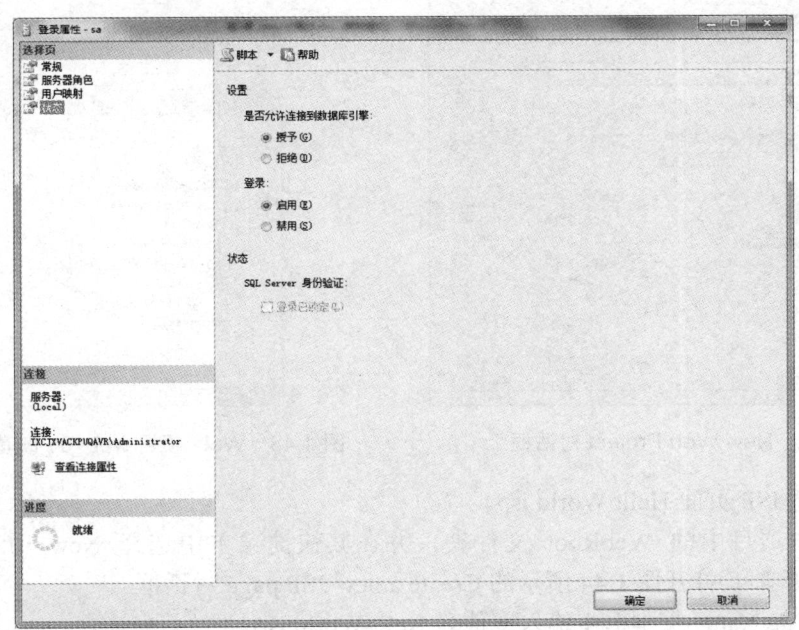

图 1.40　“登录属性-sa”对话框(“状态”选择页)

1.4　Java Web 项目的创建与部署

下面，通过两个简单的实例说明在 MyEclipse 中创建与部署 Java Web 项目的主要步骤与基本方法。

【实例 1-1】创建一个 HelloWorld.jsp 页面，如图 1.41 所示。

主要步骤如下。

(1) 创建 Web 项目 web_01。

① 在 MyEclipse 中，选择 File→New→Web Project 菜单项，打开图 1.42 所示的 New Web Project 对话框。

② 在 Project Name 文本框中输入项目名 web_01，并在 J2EE Specification Level 选项区选中 Java EE 6.0 单选按钮。

图 1.41 HelloWorld.jsp 页面

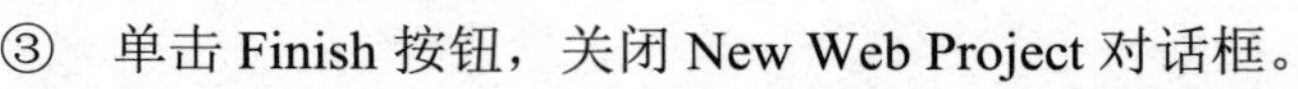

③ 单击 Finish 按钮，关闭 New Web Project 对话框。

至此，Web 项目 web_01 创建完毕，其基本结构如图 1.43 所示。

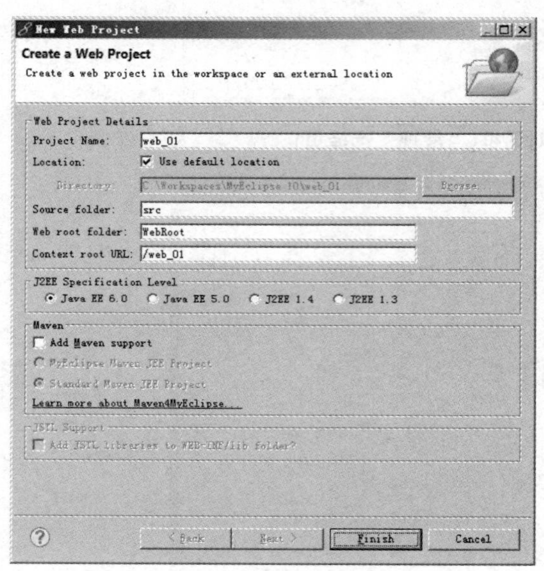

图 1.42 New Web Project 对话框

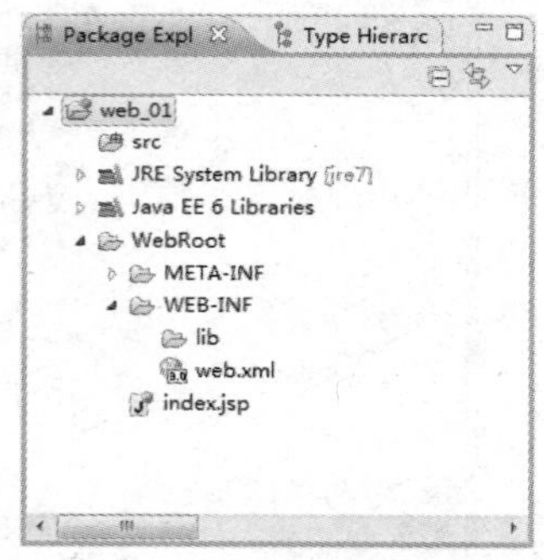

图 1.43 Web 项目 web_01 的基本结构

(2) 创建 JSP 页面 HelloWorld.jsp。

① 右击项目中的 WebRoot 文件夹，并在其快捷菜单中选择 New→JSP(Advanced Templates)菜单项，打开图 1.44 所示的 Create a new JSP page 对话框。

② 在 File Name 文本框中输入文件名 HelloWorld.jsp。

③ 单击 Finish 按钮，关闭 Create a new JSP page 对话框。

④ 删除编辑区中原有代码，然后再输入以下代码。

```
<html>
  <head>
    <title>HelloWorld</title>
    <meta http-equiv="content-type" content="text/html; charset=UTF-8">
  </head>
  <body>
```

```
    Hello,World! <br>
  </body>
</html>
```

⑤　单击工具栏上的 Save 按钮，保存 HelloWorld.jsp 页面。

(3)　修改 web.xml 文件。

双击 WEB-INF 文件夹下的配置文件 web.xml，将其代码修改如下。

```
<?xml version="1.0" encoding="UTF-8"?>
<web-app version="3.0"
    xmlns="http://java.sun.com/xml/ns/javaee"
    xmlns:xsi="http://www.w3.org/2001/XMLSchema-instance"
    xsi:schemaLocation="http://java.sun.com/xml/ns/javaee
    http://java.sun.com/xml/ns/javaee/web-app_3_0.xsd">
  <display-name></display-name>
  <welcome-file-list>
    <welcome-file>HelloWorld.jsp</welcome-file>
  </welcome-file-list>
</web-app>
```

说明： MyEclipse 默认的 Web 项目启动页面为 index.jsp。若要将其改为其他页面(如 HelloWorld.jsp 等)，则要对配置文件 web.xml 进行相应的修改，即修改其中的 <welcome-file>元素。

(4)　部署 Web 项目(将 Web 项目部署到 Tomcat 中)。

①　单击工具栏上的 Deploy MyEclipse J2EE Project to Server 按钮，弹出图 1.45 所示的 Project Deployments 对话框。

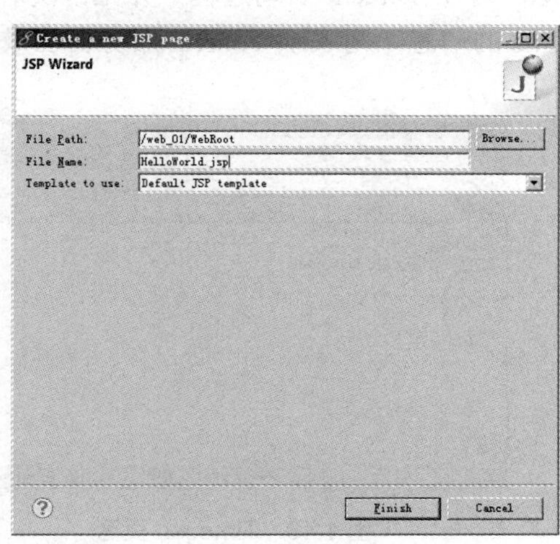

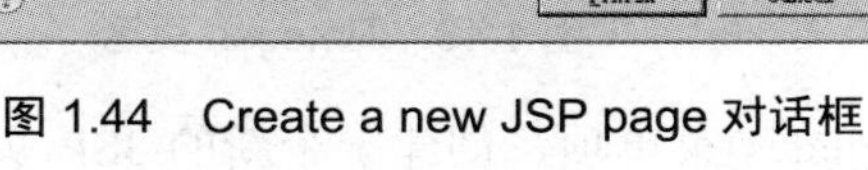

图 1.44　Create a new JSP page 对话框

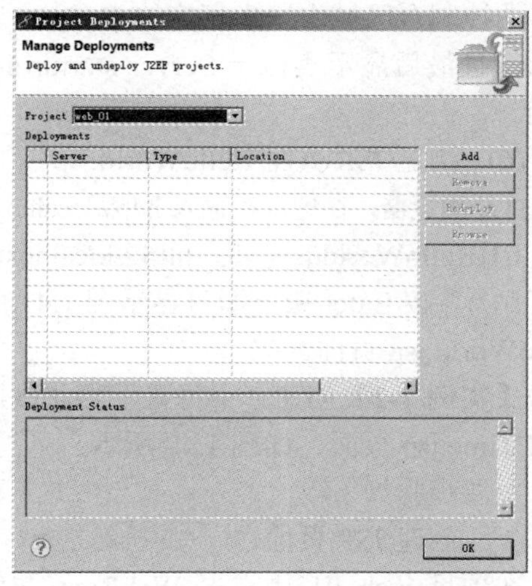

图 1.45　Project Deployments 对话框

②　在 Project 下拉列表框中选中要部署的项目(在此为 web_01)，然后单击 Add 按钮，打开图 1.46 所示的 New Deployment 对话框。

③　在 Server 下拉列表框中选中相应的服务器(在此为 Tomcat 7.x)，然后单击 Finish

按钮，关闭 New Deployment 对话框。

④ 当 Project Deployments 对话框的 Deployment Status 列表框中显示 Successfully deployed 时(见图 1.47)，表明项目已部署成功。此时，应单击 OK 按钮，关闭 Project Deployments 对话框。

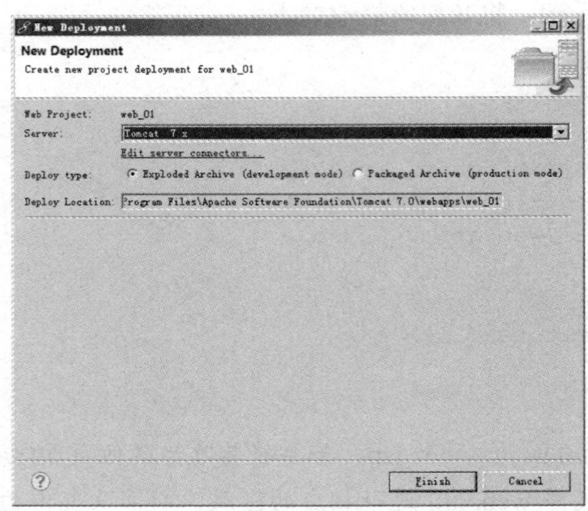

图 1.46　New Deployment 对话框

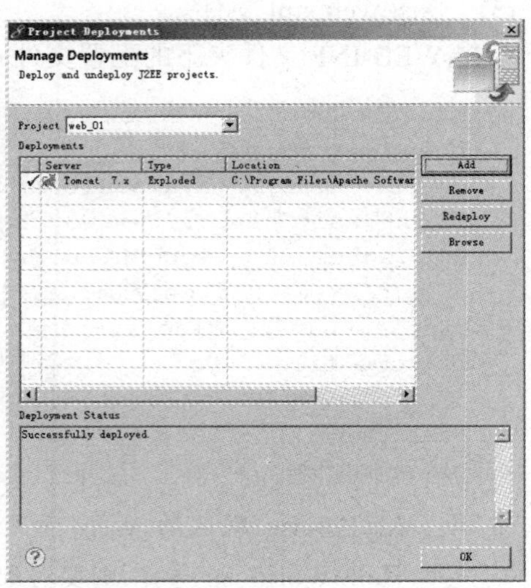

图 1.47　Project Deployments 对话框

(5) 启动 Tomcat。

单击工具栏中的 Run/Stop/Restart MyEclipse Servers 复合按钮右侧的下拉按钮，并在下拉菜单中选择 Tomcat 7.x→Start 菜单项，即可启动 Tomcat。此时，会在 MyEclipse 主窗口下方的控制台区域中输出相应的 Tomcat 启动信息。若在 Tomcat 启动过程中并无错误信息提示，且最后显示类似于 Server startup in 3720 ms 的信息，则表示 Tomcat 服务器已成功启动。

(6) 浏览 JSP 页面 HelloWorld.jsp。

打开浏览器，在地址栏输入 http://localhost:8080/web_01/HelloWorld.jsp(或 http://localhost:8080/web_01/)并按 Enter 键，即可显示图 1.41 所示的 HelloWorld.jsp 页面。

图 1.48　Time.jsp 页面

【实例 1-2】创建一个可显示当前日期与时间的 Time.jsp 页面，如图 1.48 所示。

主要步骤如下。

(1) 创建 JSP 页面。

在项目 web_01 中右击 WebRoot 文件夹，并在其快捷菜单中选择 New→JSP(Advanced Templates)菜单项，创建一个新的 JSP 页面，并将其命名为 Time.jsp。其代码如下。

```
<%@ page language="java" contentType="text/html; charset=UTF-8"%>
<%@ page import="java.util.*" %>
```

```
<html>
  <head>
    <title>HelloWorld</title>
  </head>
  <body>
    <%
    Date d=new Date();
    String s=d.toLocaleString();
    %>
   Hello,World! <br>
   现在的时间是:<%=s%>
  </body>
</html>
```

(2) 重新部署 Web 项目。

① 单击工具栏上的 Deploy MyEclipse J2EE Project to Server 按钮，弹出 Project Deployments 对话框。

② 在 Project 下拉列表框中选中要重新部署的项目(在此为 web_01)，然后单击 Redeploy 按钮。

③ 当 Project Deployments 对话框中显示 Successfully deployed 时，表明项目已重新部署成功。此时，应单击 OK 按钮，关闭 Project Deployments 对话框。

(3) 启动 Tomcat 并浏览 JSP 页面 Time.jsp。

打开浏览器，在地址栏输入 http://localhost:8080/web_01/Time.jsp 并按 Enter 键，即可显示图 1.48 所示的 Time.jsp 页面。

1.5 Java Web 项目的导出、删除与导入

在 MyEclipse 中，所有可编译运行的资源都必须放在项目中，一个项目包括一系列相关的文件与设置。通常，项目目录下的.project、.mymetadata 与.classpath 等文件描述了当前项目的有关信息。对于应用开发人员来说，需要经常对项目进行备份、将项目部署到其他计算机上或者借鉴已有的相关项目，为此，他们就必须掌握一定项目管理方法，包括项目的导出、移除与导入等。下面通过具体实例进行简要说明。

1. 项目的导出

【实例 1-3】导出项目 web_01。

主要步骤如下。

(1) 右击项目名 web_01，并在其快捷菜单中选择 Export 菜单项，打开图 1.49 所示的 Export-Select 对话框。

(2) 在项目树中选择 General→File System(表示将导出的项目保存在本地文件系统中)，然后单击 Next 按钮，打开图 1.50 所示的 Export-File system 对话框。

(3) 单击 Browse 按钮，选定存放路径(在此为 F:\Lu\MyProjects)，然后单击 Finish 按钮，关闭 Export-File system 对话框。

项目导出完毕后，可在指定的路径中(在此为 F:\Lu\MyProjects)找到相应的项目文件夹

(在此为 web_01)。

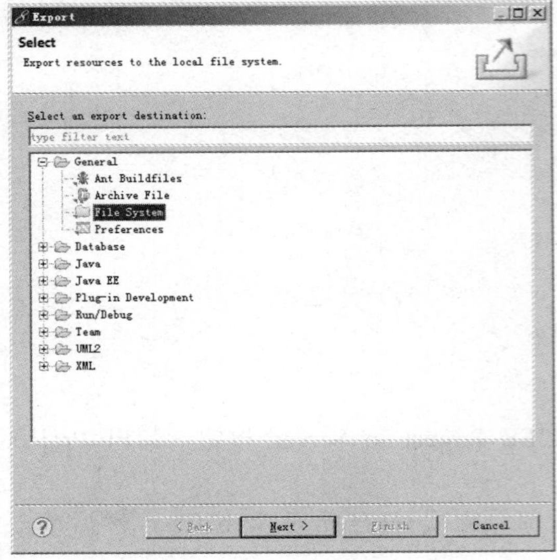

图 1.49　Export-Select 对话框

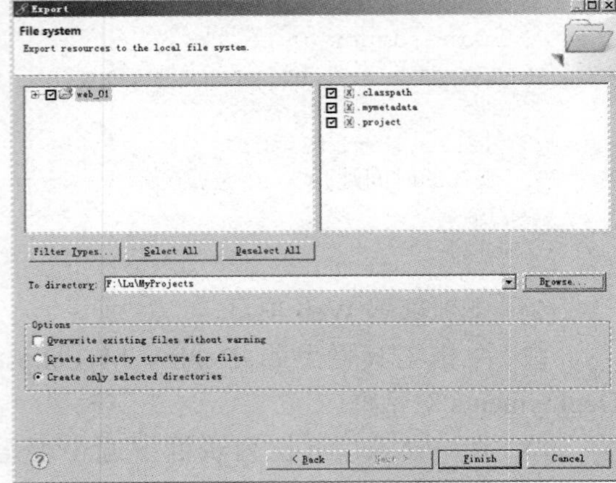

图 1.50　Export-File system 对话框

2．项目的移除

【实例 1-4】移除项目 web_01。

主要步骤如下。

(1) 右击项目名 web_01，并在其快捷菜单中选择 Delete 菜单项，弹出图 1.51 所示的 Delete Resources 对话框。

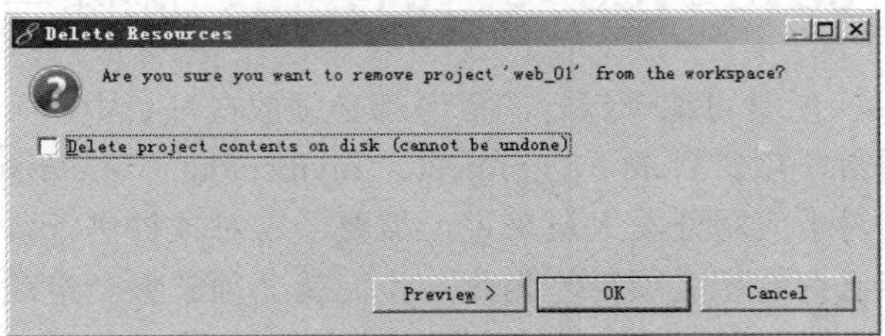

图 1.51　Delete Resources 对话框

(2) 单击 OK 按钮，关闭 Delete Resources 对话框。

项目被移除后，其项目树将不再出现在 MyEclipse 窗口的 Workspace(工作区)窗格中，但其项目文件夹(内含该项目的全部资源文件)仍然存放在 MyEclipse 的工作区目录中(在此为 C:\Workspaces\MyEclipse 10)。如果要彻底删除项目，那么应在 Delete Resources 对话框中选中 Delete project contents on disk (cannot be undone)复选框，然后再单击 OK 按钮。在这种情况下，MyEclipse 会将工作区目录中对应的项目文件夹一起删除，且无法恢复。

3．项目的导入

【实例 1-5】导入项目 web_01。

主要步骤如下。

(1) 将项目文件夹 web_01(内含该项目的全部资源文件)复制到 MyEclipse 的工作区目录中(在此为 C:\Workspaces\MyEclipse 10)。

(2) 在 MyEclipse 中选择 File→Import 菜单项，打开图 1.52 所示的 Import-Select 对话框。

(3) 在项目树中选择 General→Existing Projects into Workspace 选项(表示将导入工作区中已存在的项目)，然后单击 Next 按钮，打开图 1.53 所示的 Import-Import Projects)对话框。

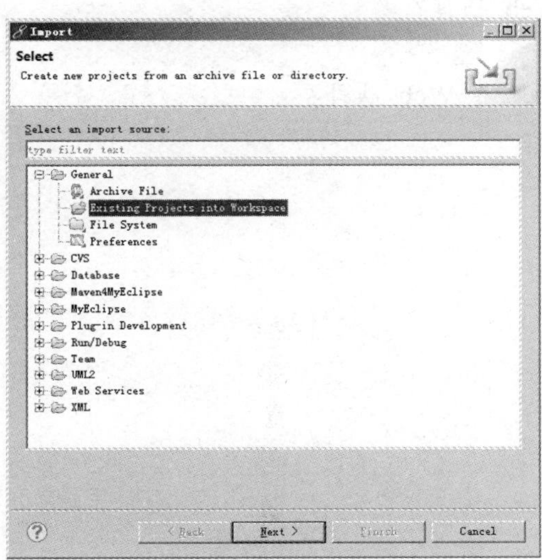
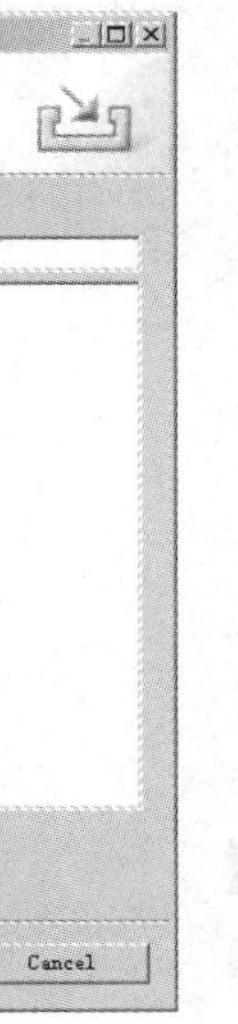

图 1.52　Import-Select 对话框

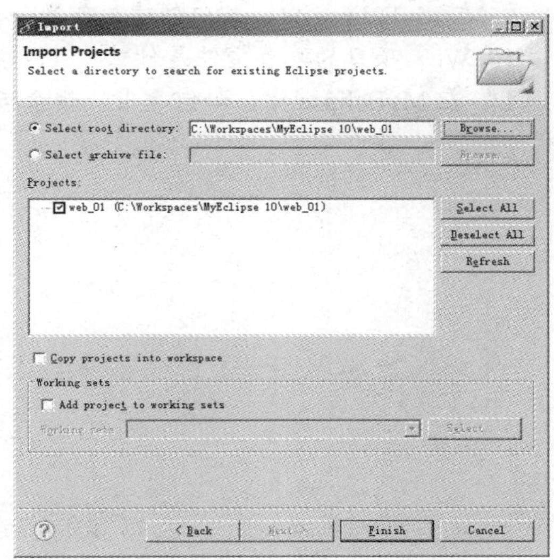

图 1.53　Import-Import Projects 对话框

(4) 单击 Browse 按钮，选中要导入的项目文件夹(在此为 C:\Workspaces\MyEclipse 10\web_01)，然后单击 Finish 按钮，关闭 Import-Import Projects 对话框。

项目导入完毕后，其项目树将再次出现在 MyEclipse 窗口的 Workspace(工作区)窗格中。

本章小结

本章简要地介绍了 Java 平台的分类、Java EE 的概况以及 Java Web 应用开发的主要技术与流行框架，详细讲解了 Java Web 应用开发环境的搭建方法(包括 JDK、Tomcat、MyEclipse、SQL Server 的安装与配置)，并通过具体实例说明了 Java Web 项目的创建、部署与管理方法。通过对本章的学习，读者应熟练掌握 Java Web 应用开发环境的搭建方法、MyEclipse 开发工具的基本用法以及 Java Web 项目的创建、部署与管理方法。

思考题

1. 根据应用领域的不同，Java 平台分为哪几种？其主要应用是什么？

2. Java Web 应用开发的主要技术包括哪些？

3. Java Web 应用开发的流行框架主要有哪些？

4. 请简述 JDK 的配置与测试方法。

5. 请简述 Tomcat 的配置与测试方法。

6. 如何配置 MyEclipse 所用的 JRE?

7. 如何在 MyEclipse 中集成 Tomcat?

8. 对于 Java Web 应用程序来说，SQL Server 数据服务器方面有哪些关键配置？请简述之。

9. 在 MyEclipse 中，如何创建与部署一个 Web 项目？

10. Web 项目的配置文件是什么？

11. 在 MyEclipse 中，如何导出、删除与导入一个 Web 项目？

第 2 章

JSP 基础

JSP 是目前 Web 应用开发或动态网页设计的常用技术之一，也是 Java Web 应用开发的核心技术之一。

本章要点：

- JSP 简介;
- JSP 基本语法;
- JSP 内置对象。

学习目标：

- 了解 JSP 的概况;
- 掌握 JSP 的基本语法(包括各种指令标记与动作标记的基本用法);
- 掌握 JSP 各种内置对象的主要用法。

2.1 JSP 简介

JSP(Java Server Pages)是由 Sun Microsystems 公司倡导、许多公司参与一起建立的一种动态网页技术标准。由于 CGI 自身的缺陷以及 Java 语言的迅速发展，Sun Microsystems 公司于 1997 年推出了 Servlet 1.0 规范。Servlet 的工作原理与 CGI 相似，所完成的功能也与 CGI 相同，但与 CGI 相比，Servlet 具有可移植、易开发、稳健、节省内存与 CPU 资源等优点。Servlet 技术也有一个很大的缺陷，即不擅长编写以显示效果为主的 Web 页面。于是，基于 Servlet 1.0 规范，Sun Microsystems 公司于 1998 年 4 月推出了 JSP 0.90 规范。此后，随着 JSP 的发展，陆续推出了更新的 JSP 规范，包括 JSP 1.1、JSP 1.2、JSP 2.0、JSP 2.1(2006 年 5 月)、JSP 2.2(2009 年 12 月)等。

JSP 技术类似于 ASP 与 PHP 技术，其实就是在 HTML 文档(*.htm、*.html)中嵌入 Java 脚本小程序(Scriptlet)和 JSP 标记(tag)等元素，从而形成 JSP 文件(*.jsp)。使用 JSP 开发的 Web 应用是跨平台的，既能在 Windows 上运行，也能在 UNIX、Linux 等其他操作系统上运行。

JSP 在本质上就是 Servlet。当 JSP 第一次被请求时，Web 服务器上的 JSP 容器(或称为 JSP 引擎)会将其转化为相应的 Servlet 文件，然后再编译为 Servlet 类文件，并且被装载和实例化。此后，每次对此 JSP 的请求均将通过调用已实例化的 Servlet 对象中的方法来产生响应。正因为如此，第一次访问 JSP 页面时响应速度特别慢，而以后就很快了。

JSP 结合了 Servlet 与 JavaBean 技术，充分继承了 Java 的众多优势，包括一次编写随处运行、高效的性能与强大的可扩展性等。如今，JSP 的应用已相当广泛，被认为是当今最有前途的 Web 技术之一，其主要特点如下。

(1) 一次编写，随处运行。

(2) 可重用组件。可通过 JavaBean 等组件技术封装复杂应用，开发人员可共享已开发完成的组件，从而提高 JSP 应用的开发效率与可扩展性。

(3) 标记化页面开发。JSP 技术将许多常用功能封装起来，以 XML 标记的形式展现给开发人员，从而使不熟悉 Java 语言的人员也可以轻松编写 JSP 程序，降低了 JSP 开发的难度。标记化应用也有助于实现“形式和内容相分离”，使得 JSP 页面结构更清晰，更便于维护。

(4) 角色分离。JSP 规范允许将工作量分为两类，即页面的图形内容与页面的动态内容。可先创建好页面的图形内容(HTML 文档)，然后由 Java 程序员向文档中插入 Java 代码，实现动态内容。此特点使得 JSP 的开发和维护更加轻松。

2.2　JSP 基本语法

一个 JSP 页面就是一个以.jsp 为扩展名的程序文件，其组成元素包括 HTML/XHTML 标记、JSP 标记与各种脚本元素。其中，JSP 标记可分为两种，即 JSP 指令标记与 JSP 动作标记。脚本元素则是嵌入 JSP 页面中的 Java 代码，包括声明(Declarations)、表达式(Expressions)与脚本小程序(Scriptlets)等。下面，通过具体实例简要介绍 JSP 的基本语法。

2.2.1　声明

JSP 声明用于定义 JSP 程序所需要的变量、方法与类，其声明方式与 Java 相同，语法格式为：

```
<%! declaration;[ declaration;]… %>
```

其中，declaration 为变量、方法或者类的声明。JSP 声明通常写在脚本小程序的最前面。例如：

```
<%!
Date date;
int sum;
%>
```

注意：在标记符“<%!”和“%>”之间所声明的变量、方法与类在整个 JSP 页面内都是有效的，与其所在的具体位置无关。不过，在方法、类中所定义的变量则只在该方法、类中有效。

2.2.2　表达式

JSP 表达式的值由服务器负责计算，且计算结果将自动转换为字符串并发送到客户端显示。其语法格式为：

```
<%= expression %>
```

其中，expression 为相应的表达式。例如：

```
<%=sum%>
<%=(new java.util.Date()).toLocaleString()%>
```

注意：在“<%=”和“%>”之间所插入的表达式必须要有返回值，且末尾不能加“;”。

2.2.3 脚本小程序

脚本小程序是在“<%”和“%>”之间插入的 Java 程序段(又称程序块或程序片)，其语法格式为：

```
<% Scriptlets %>
```

其中，Scriptlets 为相应的代码序列。在该代码序列中所声明的变量属于 JSP 页面的局部变量。

【实例 2-1】设计一个显示当前日期时间的页面 DisplayDateTime.jsp，如图 2.1 所示。

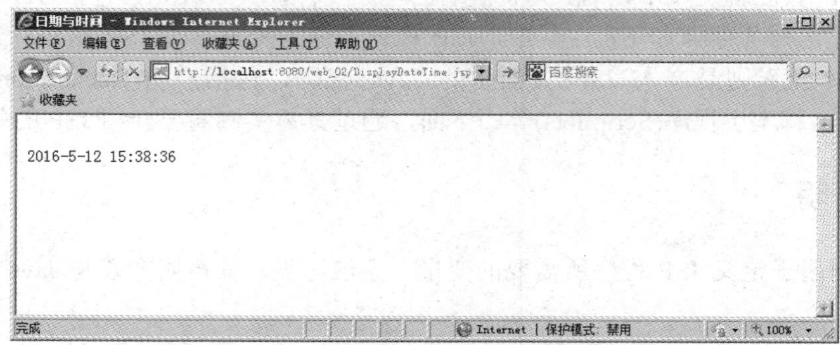

图 2.1　DisplayDateTime.jsp 页面

主要步骤如下。

(1)　新建一个 Web 项目 web_02。

(2)　在项目中添加一个新的 JSP 页面 DisplayDateTime.jsp，其代码如下：

```
<!-- JSP 指令标记 -->
<%@ page contentType="text/html;charset=GB2312"%>
<%!
    //变量声明
    Date date;
%>
<!--HTML 标记 -->
<html>
    <head>
    <title>日期与时间</title>
    </head>
    <body>
    <%
    //java 程序段
    date = new Date();
    out.println("<br>" + date.toLocaleString() + "<br>");
    %>
    </body>
</html>
```

【实例 2-2】设计一个计算 1～100 的全部偶数之和的页面 EvenSum.jsp，如图 2.2

所示。

图 2.2　EvenSum.jsp 页面

在项目 web_02 中添加一个新的 JSP 页面 EvenSum.jsp，其代码如下：

```
<%@ page contentType="text/html;charset=GB2312"%>
<%!
    // 变量声明
    int sum;
    // 方法声明
    public int getEvenSum(int n) {
        for (int i = 1; i <= n; i++) {
            if (i % 2 == 0)
                sum = sum + i;
        }
        return sum;
    }
%>
<html>
    <head>
    <title>偶数和</title>
    </head>
    <body>
    <%
    //java 程序段
    int n=100;
    sum=0;
    %>
    1 ～
    <!-- JSP 表达式 -->
    <%= n %>
    内的全部偶数之和是:
    <!-- JSP 表达式 -->
    <%= getEvenSum(n) %>
    </body>
</html>
```

【实例 2-3】设计一个可显示当前访问者序号的页面 Counter.jsp，如图 2.3 所示。

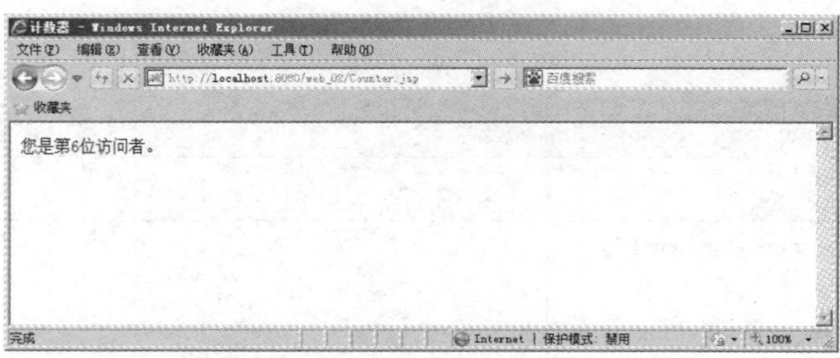

图 2.3 Counter.jsp 页面

在项目 web_02 中添加一个新的 JSP 页面 Counter.jsp，其代码如下：

```
<%@ page contentType="text/html;charset=GB2312"%>
<html>
  <head>
    <title>计数器</title>
  </head>
  <body>
     <%!
     int count = 0; //声明变量
     synchronized void countUser() //声明方法
     {
       count++;
     }
     %>
     <%
       countUser(); //调用方法
     %>
     您是第<%=count%>位访问者。
  </body>
</html>
```

本页面通过声明变量与方法，同时使用脚本小程序与表达式，实现一个简单的计数器功能。

2.2.4 JSP 指令标记

JSP 指令标记是专为 JSP 引擎而设计的，仅用于告知 JSP 引擎如何处理 JSP 页面，而不会直接产生任何可见的输出。指令标记又称为指令元素(directive element)，其语法格式如下：

```
<%@ 指令名 属性="值" 属性="值" … %>
```

JSP 指令标记分为 3 种，即 page 指令、include 指令与 taglib 指令。

1. page 指令

page 指令用于设置整个页面的相关属性与功能，其语法格式为：

```
<%@ page attribute1="value1" attribute2="value2" … %>
```

其中，attribute1、attribute2 等为属性名，value1、value2 等则为属性值。

page 指令的有关属性见表 2.1。

表 2.1　page 指令的有关属性

属　性	说　明
language	声明脚本语言的种类，目前暂时只能用"java"，即 language="java"
import	指明需要导入的作用于程序段、表达式以及声明的 Java 包的列表。该属性的格式为 import="{package.class | package.* }, ..."
contentType	设置 MIME 类型和字符集(字符编码方式)。其中，MIME 类型包括 text/html(默认值)、text/plain、image/gif、image/jpeg 等，字符集包括 ISO-8859-1(默认值)、UTF-8、GB2312、GBK 等。该属性的格式为 contentType="mimeType [;charset=characterSet]" | "text/html; charset=ISO-8859-1"
pageEncoding	设置字符集(字符编码方式)，包括 ISO-8859-1(默认)、UTF-8、GB2312、GBK 等。该属性的格式为 pageEncoding="{characterSet | ISO-8859-1}"
extends	指明 JSP 编译时需要加入的 Java Class 的全名(该属性会限制 JSP 的编译能力，应慎重使用)。该属性的格式为 extends="package.class"
session	设定客户是否需要 HTTP Session。若为 true(默认值)，则 Session 是可用的；若为 false，则 Session 不可用，且不能使用定义了 scope=session 的<jsp:useBean>元素。该属性的格式为 session="true | false"
buffer	指定 buffer 的大小(默认值为 8kb)，该 buffer 由 out 对象用于处理执行后的 JSP 对客户浏览器的输出。该属性的格式为 buffer="none | 8kb | sizekb"
autoFlush	设置 buffer 溢出时是否需要强制输出。若为 true(默认值)，则输出正常；若为 false，则 buffer 溢出时会导致意外错误发生。若将 buffer 设置为 none，则不许将 autoFlush 设置为 false。该属性的格式为 autoFlush="true | false"
isThreadSafe	设置 JSP 文件是否能多线程使用。若为 true(默认值)，则一个 JSP 能够同时处理多个用户的请求；若为 false，则一个 JSP 一次只能处理一个请求。该属性的格式为 isThreadSafe="true | false"
info	指定执行 JSP 时会加入其中的文本(可使用 Servlet.getServletInfo 方法获取)。该属性的格式为 info ="text"
errorPage	设置用于处理异常事件的 JSP 文件。该属性的格式为 errorPage="relativeURL"
isErrorPage	设置当前 JSP 页面是否为出错页面。若为 true，则为出错页面(可使用 exception 对象)；若为 false，则为非出错页面。该属性的格式为 isErrorPage="true | false"

例如：

```
<%@ page contentType="text/html;charset=GBK"%>
<%@ page import="java.util.*, java.lang.*" %>
<%@ page buffer="5kb" autoFlush="false" %>
<%@ page errorPage="error.jsp" %>
<%@ page import="java.util.Date"%>
<%@ page import="java.util.*,java.awt.*"%>
```

注意： 对于 import 属性，可以使用多个 page 指令以指定该属性有多个值。但对于其他属性，则只能使用一次 page 指令以指定该属性的一个值。

2. include 指令

include 指令用于设置 JSP 页面中静态包含一个文件，其语法格式为：

```
<%@ include file="relativeURL" %>
```

其中，file 属性用于指定被包含的文件。被包含文件的路径通常为相对路径。若路径以“/”开头，则该路径为参照 JSP 应用的上下文路径；若路径以文件名或目录名开头，则该路径为正在使用的 JSP 文件的当前路径。例如：

```
<%@ include file="error.jsp" %>
<%@ include file="/include/calendar.jsp" %>
<%@ include file="/templates/header.html" %>
```

include 指令主要用于解决重复性页面问题，其中包含的文件在本页面编译时被引入。

注意： 所谓静态包含，是指 JSP 页面和被包含的文件先合并为一个新的 JSP 页面，然后 JSP 引擎将这个新的 JSP 页面转译为 Java 类文件。其中，被包含的文件应与当前的 JSP 页面处于同一个 Web 项目中，可以是文本文件、HTML/XHTML 文件、JSP 页面或 Java 代码段等，但必须保证合并而成的 JSP 页面要符合 JSP 的语法规则，即能够成为一个合法的 JSP 页面文件。

【实例 2-4】 include 指令示例：设计一个用于显示文本文件 Hello.txt 内容的页面 DisplayText.jsp，如图 2.4 所示。

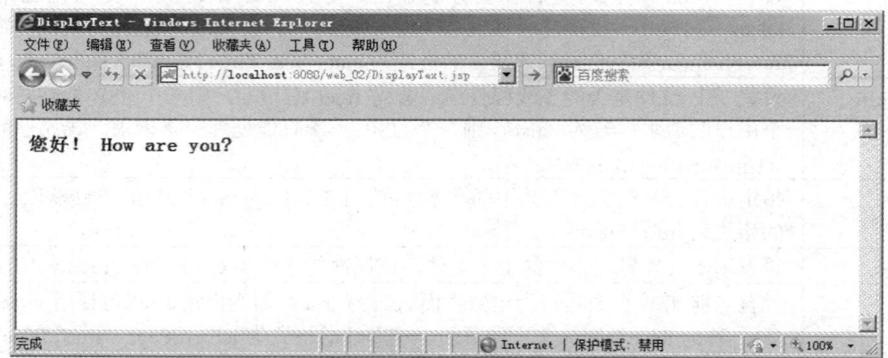

图 2.4　DisplayText.jsp 页面

主要步骤(在项目 web_02 中)如下。

(1) 在 WebRoot 目录中添加一个新的 JSP 页面 DisplayText.jsp，其代码如下：

```
<!-- page 指令 -->
<%@ page contentType="text/html;charset=GB2312" %>
<html>
    <head>
    <title>DisplayText</title>
    </head>
    <body>
        <H3>
        <!-- include 指令 -->
```

```
        <%@ include file="Hello.txt" %>
        </H3>
    </body>
</>
```

(2) 在 WebRoot 目录中添加一个文本文件 Hello.txt，其内容如下：

```
<%@ page contentType="text/html;charset=GB2312" %>
您好！
How are you?
```

3. taglib 指令

taglib 指令用于引用标记库并指定相应的标记前缀，其语法格式为：

```
<%@ taglib uri="tagLibURI" prefix="tagPrefix" %>
```

其中，uri 属性用于指定标记库的路径，prefix 属性用于指定相应的标记前缀，例如：

```
<%@ taglib uri="/struts-tags" prefix="s" %>
```

该 taglib 指令用于引用 Struts 2 的标记库，并将其前缀指定为 s。使用该指令后，即可在当前 JSP 页面中使用<s:form></s:form>、<s:textfield></s:textfield>、<s:password></s:password>等 Struts 2 标记。

2.2.5　JSP 动作标记

JSP 的动作标记又称为动作元素(action element)，共有 7 个，即 param 动作、include 动作、forward 动作、plugin 动作、useBean 动作、getProperty 动作与 getProperty 动作。

1. param 动作

param 动作标记用于以“名称-值”对的形式为其他标记提供附加信息(即参数)，必须与 include、forward 或 plugin 等动作标记一起使用，其语法格式为：

```
<jsp:param name="parameterName" value="{parameterValue | <%=
expression %>}" />
```

其中，name 属性用于指定参数名，value 属性用于指定参数值(可以是 JSP 表达式)，例如：

```
<jsp:param name="username" value="abc"/>
```

该 param 动作标记指定了一个参数 username，其值为 abc。

2. include 动作

include 动作标记用于告知 JSP 页面动态加载一个文件，即在 JSP 页面运行时才将文件引入。其语法格式有两种。

格式 1：

```
<jsp:include page="{relativeURL | <%= expression%>}" flush="true" />
```

格式 2：

```
<jsp:include page="{relativeURL | <%= expression %>}" flush="true" >
<jsp:param name="parameterName" value="{parameterValue | <%=
expression %>}" />
</jsp:include>
```

其中，格式 1 不带 param 子标记，格式 2 带有 param 子标记。在 include 动作标记中，page 属性用于指定要动态加载的文件，而 flush 属性必须设置为 true。

例如：

```
<jsp:include page="scripts/error.jsp" />
<jsp:include page="/copyright.html" />
```

又如：

```
<jsp:include page="scripts/login.jsp">
<jsp:param name="username" value="abc" />
</jsp:include>
```

说明： 所谓动态包含，是指当 JSP 引擎把 JSP 页面转译成 Java 文件时，告诉 Java 解释器，被包含的文件在 JSP 运行时才包含进来。若被包含的文件是普通文本文件，则将文件的内容发送到客户端，由客户端负责显示。若包含的文件是 JSP 文件，则 JSP 引擎执行这个文件，然后将执行的结果发送到客户端，并由客户端负责显示。

注意： include 动作标记与 include 指令标记是不同的。include 动作标记是在执行时才对包含的文件进行处理，因此 JSP 页面和其所包含的文件在逻辑和语法上是独立的。如果对包含的文件进行了修改，那么运行时将看到修改后的结果。而 include 指令标记所包含的文件如果发生了变化，必须重新将 JSP 页面转译成 java 文件(可重新保存该 JSP 页面，然后再访问之，即可转译生成新的 Java 文件)，否则只能看到修改前的文件内容。

【实例 2-5】include 动作标记示例：计算并显示 1 到指定整数内的所有偶数之和，如图 2.5 所示。

图 2.5　EvenSum 页面

主要步骤(在项目 web_02 中)如下。

(1) 在 WebRoot 目录中添加一个新的 JSP 页面 EvenSumSetNumber.jsp，其代码如下：

```
<%@ page contentType="text/html;charset=UTF-8" %>
<html>
    <head>
        <title>EvenSum</title>
    </head>
    <body>
        <jsp:include page="EvenSumByNumber.jsp">
            <jsp:param name="number" value="100" />
        </jsp:include>
    </body>
</html>
```

(2) 在 WebRoot 目录中添加一个新的 JSP 页面 EvenSumByNumber.jsp，其代码如下：

```
<%@ page contentType="text/html;charset=UTF-8" %>
<html>
    <head>
        <title>偶数和</title>
    </head>
    <body>
   <%
        String s=request.getParameter("number"); //获取参数值
    int n=Integer.parseInt(s);
    int sum=0;
    for(int i=1;i<=n;i++){
        if (i % 2 == 0)
            sum=sum+i;
    }
    %>
        从 1 到<%=n%>的偶数和是： <%=sum%>
    </body>
</html>
```

解析：

(1) 在本实例中，EvenSumSetNumber.jsp 页面使用 include 动作动态加载 EvenSumByNumber.jsp 页面，并使用 param 动作将参数值传递到 EvenSumByNumber.jsp 中。

(2) 在 EvenSumByNumber.jsp 页面中，首先获取传递过来的参数值，然后再计算并显示从 1 至该值内的所有偶数之和。

3. forward 动作

forward 动作标记用于跳转至指定的页面。其语法格式有两种。

格式 1：

```
<jsp:forward page={"relativeURL" | "<%= expression %>"} />
```

格式 2：

```
<jsp:forward page={"relativeURL" | "<%= expression %>"} >
<jsp:param name="parameterName" value="{parameterValue | <%=
expression %>}" />
</jsp:forward>
```

其中，格式 1 不带 param 子标记，格式 2 带有 param 子标记。在 forward 动作标记中，page 属性用于指定跳转的目标页面，例如：

```
<jsp:forward page="/servlet/login" />
<jsp:param name="username" value="abc" />
</jsp:forward>
```

说明： 一旦执行到 forward 动作标记，将立即停止当前页面的继续执行，并跳转至该标记中 page 属性所指定的页面。

【实例 2-6】 forward 动作标记示例：计算并显示 1 到指定整数内的所有奇数之和，如图 2.6 所示。

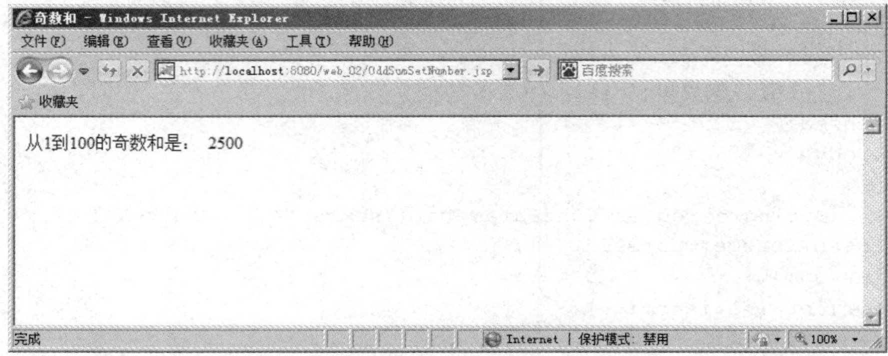

图 2.6 “奇数和”页面

主要步骤(在项目 web_02 中)如下。

(1) 在 WebRoot 目录中添加一个新的 JSP 页面 OddSumSetNumber.jsp，其代码如下：

```
<%@ page contentType="text/html;charset=UTF-8" %>
<html>
    <head>
        <title>EvenSum</title>
    </head>
    <body>
        <jsp:forward page="OddSumByNumber.jsp">
            <jsp:param name="number" value="100" />
        </jsp:forward>
    </body>
</html>
```

(2) 在 WebRoot 目录中添加一个新的 JSP 页面 OddSumByNumber.jsp，其代码如下：

```
<%@ page contentType="text/html;charset=UTF-8" %>
<html>
```

```
    <head>
        <title>奇数和</title>
    </head>
    <body>
   <%
        String s=request.getParameter("number"); //获取参数值
    int n=Integer.parseInt(s);
    int sum=0;
    for(int i=1;i<=n;i++){
        if (i % 2 != 0)
            sum=sum+i;
    }
    %>
        从 1 到<%=n%>的奇数和是： <%=sum%>
    </body>
</html>
```

解析：

(1) 在本实例中，OddSumSetNumber.jsp 页面使用 forward 动作跳转至 OddSumByNumber.jsp 页面，并使用 param 动作将参数值传递到 OddSumByNumber.jsp 中。

(2) 在 OddSumByNumber.jsp 页面中，首先获取传递过来的参数值，然后计算并显示从 1 至该值内的所有奇数之和。

4. plugin 动作

plugin 动作标记用于指示 JSP 页面加载 Java Plugin(插件)，并使用该插件(由客户负责下载)来运行 Java Applet 或 Bean。其语法格式为：

```
<jsp:plugin
type="bean | applet"
code="classFileName"
codebase="classFileDirectoryName"
[ name="instanceName" ]
[ archive="URIToArchive,…" ]
[ align="bottom | top | middle | left | right" ]
[ height="displayPixels" ]
[ width="displayPixels" ]
[ hspace="leftRightPixels" ]
[ vspace="topBottomPixels" ]
[ jreversion="JREVersionNumber | 1.1" ]
[ nspluginurl="URLToPlugin" ]
[ iepluginurl="URLToPlugin" ] >
[ <jsp:params>
[ <jsp:param name="parameterName" value="{parameterValue | <%=
expression %>}" /> ]+
</jsp:params> ]
[ <jsp:fallback> text message for user </jsp:fallback> ]
</jsp:plugin>
```

其中，type 属性用于指定被执行的插件对象的类型(是 Bean 还是 applet)，code 属性用于指定被 Java 插件执行的 Java Class 的文件名(其扩展名.class)，codebase 属性用于指定被执行的 Java Class 文件的目录或路径(若未提供此属性，则使用<jsp:plugin>的 jsp 文件的目录将会被采用)。有关 plugin 动作标记的具体用法请参阅有关手册，在此从略。

5. useBean、getProperty 与 setProperty 动作

useBean 动作标记用于应用一个 JavaBean 组件，而 getProperty 与 setProperty 动作标记分别用于获取与设置 JavaBean 的属性值。这 3 个动作标记与 JavaBean 的使用密切相关，具体用法请参阅后续有关章节。

2.2.6 JSP 注释

JSP 页面中的注释可分为 3 种，即 HTML/XHTML 注释、JSP 注释与 Java 注释。

1. HTML/XHTML 注释

HTML/XHTML 注释指的是在标记符号“<!--”与“-->”之间所加入的内容。其语法格式为：

```
<!--comment|<%=expression%>-->
```

其中，comment 为注释内容，expression 为 JSP 表达式。

对于 HTML/XHTML 注释，JSP 引擎会将其发送到客户端，因此可在浏览器中通过查看源代码的方式查看其内容。因此，HTML/XHTML 注释又称为客户端注释或输出注释。这种注释类似于 HTML 文件中的注释，唯一不同的是前者可在注释中用表达式，以便动态生成不同内容的注释。例如：

```
<!-- 现在时间是：<%=(new java.util.Date()).toLocaleString() %> -->
```

将该代码放在一个 JSP 文件的 body 中运行后，即可以在其源代码中看到相应的注释内容。例如：

```
<!-- 现在时间是：2016-5-11 17:08:26 -->
```

2. JSP 注释

JSP 注释指的是在标记符号“<%--”与“--%>”之间所加入的内容。其语法格式为：

```
<%-- comment --%>
```

其中，comment 为注释内容。

对于 JSP 注释，JSP 引擎在编译 JSP 页面时会自动将其忽略，不会将其发送到客户端。因此，JSP 注释又称为服务器端注释或隐藏注释，仅对服务器端的开发人员可见，对客户端是不可见的。

3. Java 注释

Java 注释只用于注释 JSP 页面中的有关 Java 代码，可分为 3 种情形。

(1) 使用双斜杠“//”进行单行注释，其后至行末的内容均为注释。

(2) 使用“/*”与“*/”进行多行注释，二者之间的内容均为注释。

(3) 使用“/**”与“*/”进行多行注释，二者之间的内容均为注释。这种方式可将所

注释的内容文档化。

【实例 2-7】注释示例：计算并显示 1～100 之间所有的整数之和。

在项目 web_02 中添加一个新的 JSP 页面 CommentExample.jsp，其代码如下：

```
<!-- JSP 页面 <%=(new java.util.Date()).toLocaleString()%> -->
<!-- jsp 指令标记 -->
<%@ page contentType="text/html;charset=GB2312"%>
<%!
    // 变量声明
    int sum;
    // 方法声明
    public int getSum(int n) {
        for (int i = 1; i <= n; i++) {
            sum = sum + i;
        }
        return sum;
    }
%>
<!--html 标记 -->
<%--html 标记 --%>
<html>
    <head>
    <title>累加和</title>
    </head>
    <body>
    <%
    //java 程序段
    int m = 100;
    %>
    1 ～
    <!-- Java 表达式 -->
    <%=m%>
    间的所有整数之和为:
    <%-- Java 表达式 --%>
    <%= getSum(m) %>
    </body>
</html>
```

运行结果如图 2.7 所示，所生成的 HTML 源代码则如图 2.8 所示。

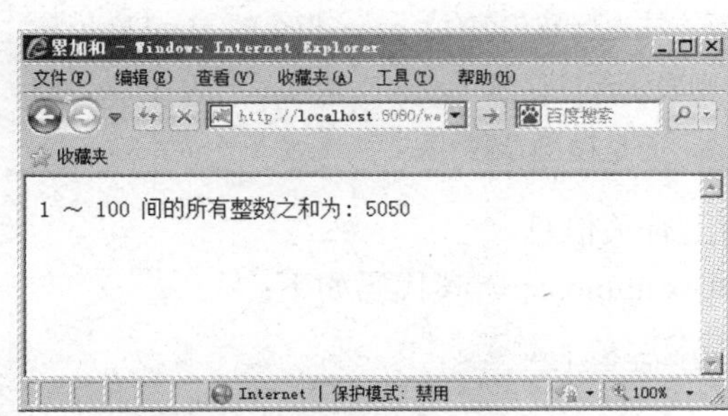

图 2.7　“累加和”页面

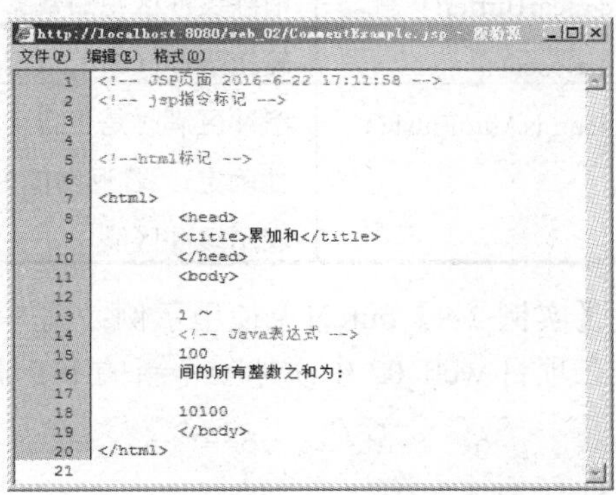

图 2.8　“累加和”页面的 HTML 源代码

2.3 JSP 内置对象

JSP 内置对象是在 JSP 运行环境中已预先定义好的对象，可在 JSP 页面的脚本部分直接使用。在 JSP 中，内置对象共有 9 个，包括 out 对象、request 对象、response 对象、session 对象、application 对象、exception 对象、page 对象、config 对象与 pageContext 对象。

2.3.1 out 对象

out 为输出流对象，主要用于向客户端输出流进行操作，可将有关信息发送到客户端的浏览器。此外，通过 out 对象，还可对输出缓存区与输出流进行相应的控制与管理。

out 是向客户端输出内容时所使用的常用对象，它其实是 javax.servlet.jsp.JspWriter 类的实例，具有 page 作用域。JspWriter 类包含了 java.io.PrintWriter 类中的大部分方法，并新增了一些专为处理缓存而设计的方法，会抛出 IOExceptions 异常(而 PrintWriter 则不会)。根据页面是否使用缓存，JspWriter 类会进行不同的实例化操作。

out 对象的常用方法见表 2.2。

表 2.2　out 对象的常用方法

方　法	说　明
void print(String str)	输出数据
void println(String str)	输出数据并换行
void newLine()	输出一个换行符
int getBufferSize ()	获取缓冲区的大小
int getRemaining()	获取缓冲区剩余空间的大小
void flush()	输出缓冲区中的数据
void clear()	清除缓冲区中的数据，并关闭对客户端的输出流
void clearBuffer()	清除缓冲区中的数据
void close()	输出缓冲区中的数据，并关闭对客户端的输出流
boolean isAutoFlush()	缓冲区满时是否自动清空。该方法返回布尔值(由 page 指令的 autoFlush 属性确定)。若返回值为 true，则表示缓冲区满了会自动清除；若为 false，则表示缓冲区满了不会自动清除，而是抛出异常

【实例 2-8】 out 对象使用示例：向客户端输出有关信息。

在项目 web_02 中添加一个新的 JSP 页面 OutExample.jsp，其代码如下：

```
<%@ page contentType="text/html;charset=GB2312" %>
<html>
    <head>
```

```
        <title>out 对象使用示例</title>
    </head>
    <body>
        <%
        out.println("<br>输出字符串:");
        out.println("Hello,World!");
        out.println("<br>输出字符型数据:");
        out.println('*');
        out.println("<br>输出字符数组数据:");
        out.println(new char[]{'a','b','c'});
        out.println("<br>输出整型数据:");
        out.println(100);
        out.println("<br>输出长整型数据:");
        out.println(123456789123456789L);
        out.println("<br>输出单精度数据:");
        out.println(1.50f);
        out.println("<br>输出双精度数据:");
        out.println(123.50d);
        out.println("<br>输出布尔型数据:");
        out.println(true);
        out.println("<br>输出对象:");
        out.println(new java.util.Date());
        out.println("<br>缓冲区大小:");
        out.println(out.getBufferSize());
        out.println("<br>缓冲区剩余大小:");
        out.println(out.getRemaining());
        out.println("<br>是否自动清除缓冲区:");
        out.println(out.isAutoFlush());
        out.println("<br>调用 out.flush()");
        out.flush();
        out.println("<br>out.flush() OK!");  //不会输出
        out.println("<br>调用 out.clearBuffer()");  //不会输出
        out.clearBuffer();
        out.println("<br>out.clearBuffer() OK!");
        out.println("<br>调用 out.close()");
        out.close();
        out.println("<br>out.close() OK!");  //不会输出
        %>
    </body>
</html>
```

运行结果如图 2.9 所示。

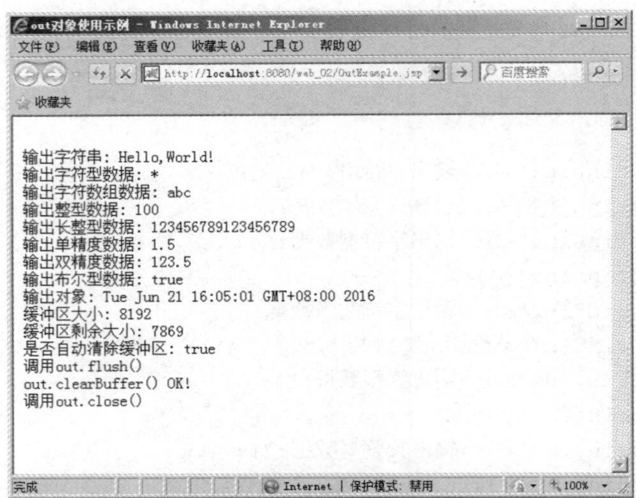

图 2.9 out 对象使用示例页面

2.3.2 request 对象

request 对象为请求对象，其中封装了客户端请求的所有信息，如请求的来源、标头、Cookies 以及与请求相关的参数值等。为获取请求的有关信息(如用户在 form 表单中所填写的数据等)，可调用 request 对象的有关方法。

request 对象是 javax.servlet.http.HttpServletRequest 类的实例，具有 request 作用域，代表的是来自客户端的请求(如 form 表单中写的信息等)，是最常用的对象之一。每当客户端请求一个 JSP 页面时，JSP 引擎就会创建一个新的 request 对象来代表这个请求。

request 对象的常用方法见表 2.3。

表 2.3 request 对象的常用方法

方 法	说 明
String getServerName()	获取接受请求的服务器的主机名
int getServerPort()	获取服务器接受请求所用的端口号
String getRemoteHost()	获取发送请求的客户端的主机名
String getRemoteAddr()	获取发送请求的客户端的 IP 地址
int getRemotePort( )	获取客户端发送请求所用的端口号
String getRemoteUser( )	获取发送请求的客户端的用户名
String getQueryString()	获取查询字符串
String getRequestURI()	获取请求的 URL(不包括查询字符串)
String getRealPath(String path)	获取指定虚拟路径的真实路径
String getParameter(String name)	获取指定参数的值(字符串)
String[] getParameterValue (String name)	获取指定参数的所有值(字符串数组)

续表

方　法	说　明
Enumeration getParameterNames()	获取所有参数名的枚举
Cookie[] getCookies()	获取与请求有关的 Cookie 对象(Cookie 数组)
Map getParameterMap( )	获取请求参数的 Map
Object getAttribute(String name)	获取指定属性的值。若指定属性并不存在，则返回 null
Enumeration getAttributeNames()	获取所有可用属性名的枚举
String getHeader(String name)	获取指定标头的信息
Enumeration getHeaders(String name)	获取指定标头的信息的枚举
int getIntHeader(String name)	获取指定整数类型标头的信息
long getDateHeader(String name)	获取指定日期类型标头的信息
Enumeration getHeaderNames()	获取所有标头名的枚举
String getProtocol()	获取客户端向服务器端传送数据所使用的协议名
String getScheme()	获取请求所使用的协议名
String getMethod( )	获取客户端向服务器端传送数据的方法(如 GET、POST 等)
String getCharacterEncoding()	获取请求的字符编码方式
String getServletPath()	获取客户端所请求的文件的路径
String getContextPath()	获取 Context 路径(即站点名称)
int getContentLength()	获取请求体的长度(字节数)
String getContentType()	获取客户端请求的 MIME 类型。若无法得到该请求的 MIME 类型，则返回-1
ServletInputStream getInputStream()	获取请求的输入流
BufferedReader getReader()	获取解码后的请求体
HttpSession getSession(Boolean create)	获取与当前客户端请求相关联的 HttpSession 对象。若参数 create 为 true，或不指定参数 create，且 session 对象已经存在，则直接返回，否则就创建一个新的 session 对象并返回；若参数 create 为 false，且 session 对象已经存在，则直接返回，否则就返回 null(即不创建新的 session 对象)
String getRequestedSessionId( )	获取 session 对象的 ID 号
void setAttribute(String name, Object obj)	设置指定属性的值
void setCharacterEncoding(String encoding)	设置字符编码方式

【实例 2-9】request 对象使用示例：获取客户端请求的有关信息。

在项目 web_02 中添加一个新的 JSP 页面 RequestExample.jsp，其代码如下：

```
<%@ page contentType="text/html; charset=GB2312" %>
<html>
    <head>
```

```
        <title>request 对象使用示例</title>
    </head>
    <body>
        getServerName: <%= request.getServerName() %> <br>
        getServerPort: <%= request.getServerPort() %> <br>
        getServletPath: <%= request.getServletPath() %> <br>
        getContextPath: <%= request.getContextPath() %> <br>
        getRequestURI : <%= request.getRequestURI() %> <br>
        getRealPath : <%= request.getRealPath(request.getRequestURI()) %>
<br>
        getQueryString : <%= request.getQueryString() %> <br>
    </body>
</html>
```

运行结果如图 2.10 所示。

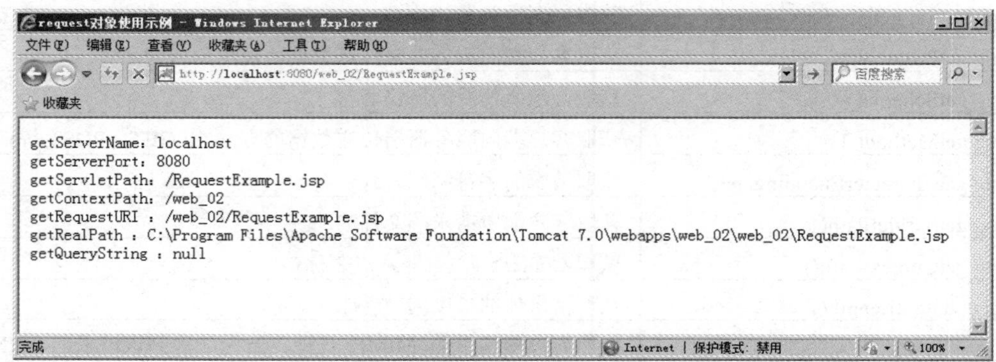

图 2.10　request 对象使用示例页面

【实例 2-10】 request 对象使用示例：用户注册。在图 2.11 所示的用户注册页面中单击“确定”按钮后，即可跳转至图 2.12 所示的注册信息页面，并在其中显示用户所输入的信息。

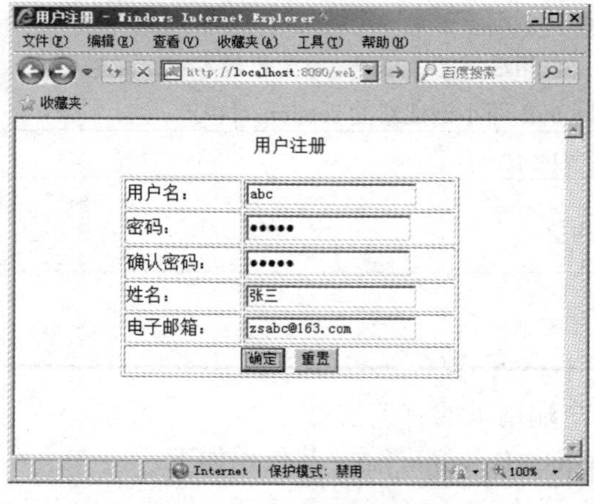

图 2.11　用户注册页面

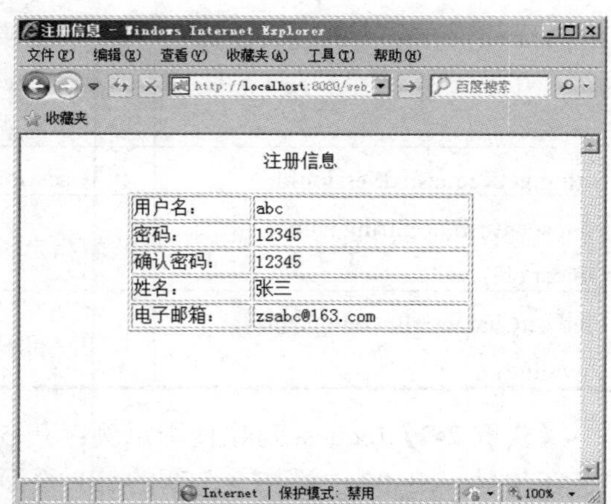

图 2.12　注册信息页面

主要步骤(在项目 web_02 中)如下。

(1) 在 WebRoot 目录中添加一个新的页面 UserRegister.htm，其代码如下：

```
<html>
    <head>
        <title>用户注册</title>
    </head>
    <body>
        <div align="center">
        用户注册
        <form name="form1" id="form1" method="post"
action="UserRegisterResult.jsp">
          <table width="300" border="1">
            <tr>
              <td width="100">用户名：</td>
              <td><input name="username" type="text" id="username" /></td>
            </tr>
            <tr>
              <td>密码：</td>
              <td><input name="password" type="password" id="password"
/></td>
            </tr>
            <tr>
              <td>确认密码：</td>
              <td><input name="confirmpwd" type="password" id="confirmpwd" /></td>
            </tr>
            <tr>
              <td>姓名：</td>
              <td><input name="name" type="text" id="name" /></td>
            </tr>
            <tr>
              <td>电子邮箱：</td>
              <td><input name="email" type="text" id="email" /></td>
            </tr>
            <tr>
              <td colspan="2" valign="" align="center"><input
type="submit" name="Submit" value="确定" />
              <input type="reset" name="Reset" value="重置" /></td>
            </tr>
          </table>
        </form>
        </div>
    </body>
</html>
```

(2) 在 WebRoot 目录中添加一个新的 JSP 页面 UserRegisterResult.jsp，其代码如下：

```
<%@ page contentType="text/html; charset=gb2312" %>
<html>
```

```
    <head>
        <title>注册信息</title>
    </head>
    <body>
        <%
        request.setCharacterEncoding("gb2312");
        String id=request.getParameter("username");
        String password=request.getParameter("password");
        String confirmpwd=request.getParameter("confirmpwd");
        String name=request.getParameter("name");
        String email=request.getParameter("email");
        %>
        <div align="center">
        <p>注册信息</p>
        <table width="300" border="1">
            <tr>
              <td width="100">用户名：</td>
              <td><%=id%></td>
            </tr>
            <tr>
              <td>密码：</td>
              <td><%=password%></td>
            </tr>
            <tr>
              <td>确认密码：</td>
              <td><%=confirmpwd%></td>
            </tr>
            <tr>
              <td>姓名：</td>
              <td><%=name%></td>
            </tr>
              <tr>
              <td>电子邮箱：</td>
              <td><%=email%></td>
            </tr>
          </table>
        </div>
    </body>
</html>
```

解析：

(1) UserRegister.htm 是一个静态页面，内含一个供用户输入注册信息的表单。该表单被提交后，将由 JSP 页面 UserRegisterResult.jsp 进行处理。

(2) 在 UserRegisterResult.jsp 页面中，调用 request 对象的 getParameter()方法获取通过表单提交过来的数据。为确保能够正确处理汉字，避免出现乱码，在获取数据前，先调用 request 对象的 setCharacterEncoding()方法设置好相应的汉字编码。

2.3.3 response 对象

response 对象为响应对象，用于对客户端的请求进行动态响应，可向客户端发送数据，如 Cookie、时间戳、HTTP 标头信息、HTTP 状态码等。在实际应用中，response 对象主要用于将 JSP 数据处理后的结果传回到客户端。

response 对象是 javax.servlet.http.HttpServletResponse 类的实例，具有 page 作用域。当服务器创建 request 对象时，会同时创建用于响应该客户端的 response 对象。

response 对象的常用方法见表 2.4。

表 2.4 response 对象的常用方法

方 法	说 明
viod addHeader(String name,String value)	添加指定的标头。若指定标头已存在，则覆盖其值
viod setHeader(String name,String value)	设置指定标头的值
boolean containsHeader(String name)	判断指定的标头是否存在
void sendRedirect(String url)	重定向(跳转)到指定的页面(URL)
String encodeRedirectURL(String url)	对用于重定向的 URL 进行编码
String encodeURL(String url)	对 URL 进行编码
void setCharacterEncoding(String encoding)	设置响应的字符编码方式
String getCharacterEncoding()	获取响应的字符编码方式
void setContentType(String type)	设置响应的 MIME 类型
String getContentType()	获取响应的 MIME 类型
void addCookie(Cookie cookie)	添加指定的 Cookie 对象(用于保存客户端的用户信息)
int getBuffersize()	获取缓冲区的大小(KB)
viod setBuffersize(int size)	设置缓冲区的大小(KB)
viod flushBuffer()	将当前缓冲区的内容强制发送到客户端
void reset()	清空缓冲区中的所有内容
void resetBuffer()	清空缓冲区中除了标头与状态信息以外的所有内容
ServletOutputStream getOutputStream()	获取客户端的输出流对象
PrintWriter getWriter()	获取输出流对应的 Writer 对象
void setContentLength(int length)	设置响应的 BODY 长度
void setStatus(int sc)	设置状态码(status code)。常用的状态码有：404(指示网页找不到的错误)、505(指示服务器内部错误)等
void sendError(int sc)	发送状态码(status code)
void sendError(int sc, String msg)	发送状态码与状态信息
void addDateHeader(String name, long date)	添加指定的日期类型标头
void addHeader(String name, String value)	添加指定的字符串类型标头

续表

方 法	说 明
void addIntHeader(String name, int value)	添加指定的整数类型标头
void setDateHeader(String name, long date)	设置指定日期类型标头的值
void setHeader(String name, String value)	设置指定字符串类型标头的值
void setIntHeader(String name, int value)	设置指定整数类型标头的值

【实例 2-11】response 对象使用示例：自动刷新页面。图 2.13 所示为“页面自动刷新”页面，其中的数字每隔 1s 会自动增加 1(从 0 开始)。

在项目 web_02 中添加一个新的 JSP 页面 AutoRefresh.jsp，其代码如下：

```
<%@page contentType="text/html;charset=gb2312"%>
<HTML>
    <HEAD>
        <TITLE>页面自动刷新</TITLE>
    </HEAD>
    <BODY>
        <%!int i = 0;%>
        <%
            response.setHeader("refresh", "1");
        %>
        <h1><%=i++%></h1>
    </BODY>
</HTML>
```

在本页面中，调用 response 对象的 setHeader()方法发送一个值为 1 的 refresh 标头，让页面每隔 1s 便自动刷新一次，从而更新所要显示的数字。

【实例 2-12】response 对象使用示例：重定向页面(友情链接)。图 2.14 为“友情链接”页面，在“友情链接”下拉列表框中选择某个选项(在此为“百度”)后，即可打开相应的目标页面(如图 2.15 所示)。

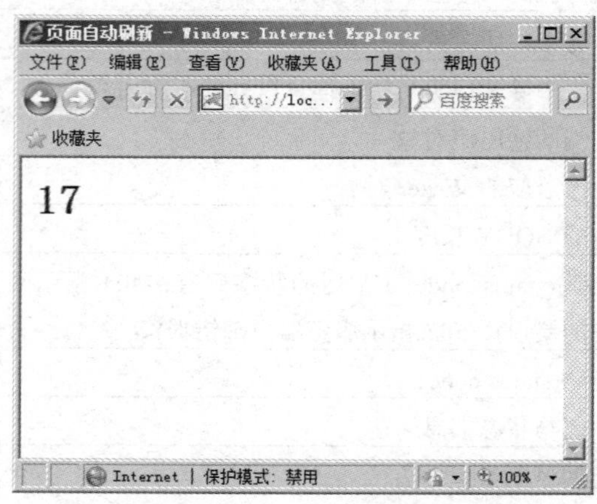

图 2.13 “页面自动刷新”页面

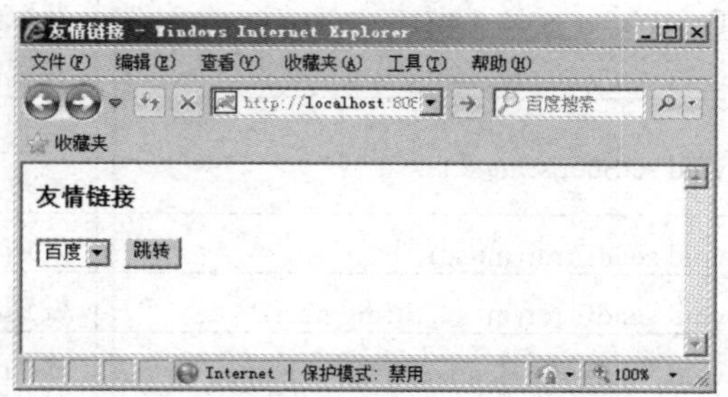

图 2.14 “友情链接”页面

图 2.15　“百度”页面

主要步骤(在项目 web_02 中)如下。

(1) 在 WebRoot 目录中添加一个新的页面 FriendLinks.htm，其代码如下：

```
<html>
    <head>
        <title>友情链接</title>
    </head>
    <body>
        <b>友情链接</b><br>
        <form action="FriendLinksResult.jsp" method="get">
        <select name="where">
            <option value="baidu" selected>百度
            <option value="sogou">搜狗
            <option value="youdao">友道
        </select>
        <input type="submit" value="跳转">
        </form>
    </body>
</html>
```

(2) 在 WebRoot 目录中添加一个新的 JSP 页面 FriendLinksResult.jsp，其代码如下：

```
<%@ page contentType="text/html;charset=GB2312"%>
<html>
    <head>
        <title>response 对象的 sendRedirect 方法使用示例</title>
    </head>
    <body>
        <%
            String address = request.getParameter("where");
            if (address != null) {
                if (address.equals("baidu"))
                    response.sendRedirect("http://www.baidu.com");
                else if (address.equals("sogou"))
                    response.sendRedirect("http://www.sogou.com/");
                else if (address.equals("youdao"))
                    response.sendRedirect("http://www.youdao.com/");
```

```
            }
        %>
    </body>
</html>
```

解析：

(1) FriendLinks.htm 是一个静态页面，内含一个供用户选择搜索引擎的表单。该表单被提交后，将由 JSP 页面 FriendLinksResult.jsp 进行处理。

(2) 在 FriendLinksResult.jsp 页面中，先调用 request 对象的 getParameter()方法获取通过表单提交过来的值，然后对其进行判断，再调用 response 对象的 sendRedirect()方法将页面重定向到指定的 URL 地址。

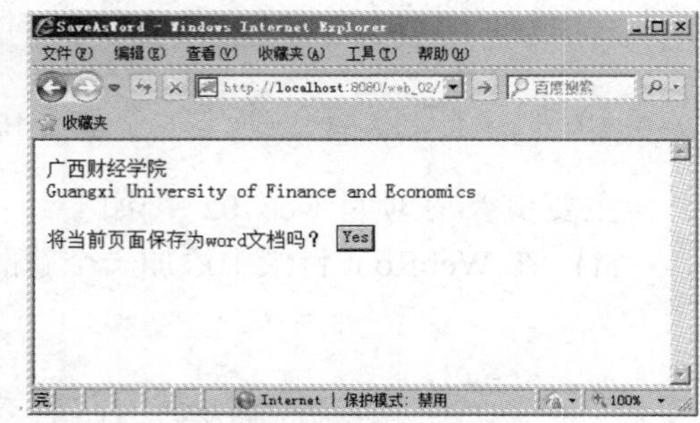

图 2.16　SaveAsWord 页面

【实例 2-13】response 对象使用示例：将当前页面保存为 Word 文档。图 2.16 所示为 SaveAsWord 页面，单击 Yes 按钮后，可弹出图 2.17 所示的“文件下载”对话框。单击其中的“保存”按钮后，即可将当前页面保存为相应的 Word 文档(默认为 SaveAsWord.doc，其内容如图 2.18 所示)。

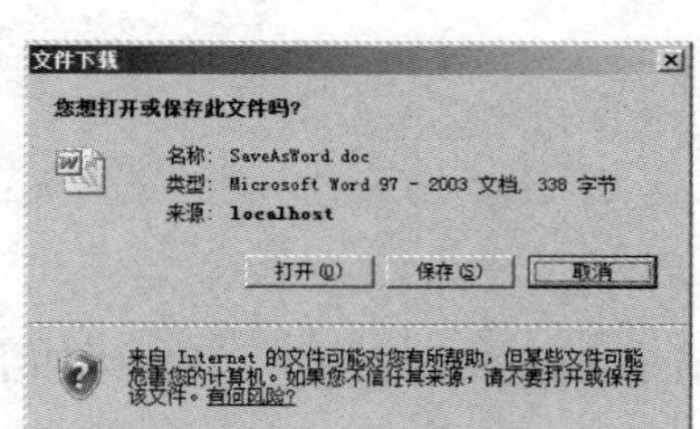

图 2.17　“文件下载”对话框

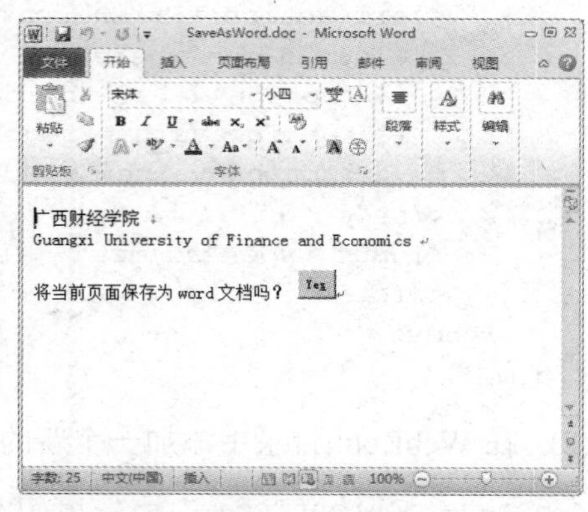

图 2.18　SaveAsWord.doc 的内容

在项目 web_02 中添加一个新的 JSP 页面 SaveAsWord.jsp，其代码如下：

```
<%@ page contentType="text/html;charset=GB2312"%>
<HTML>
    <HEAD>
        <TITLE>SaveAsWord</TITLE>
    </HEAD>
    <BODY>
        <FONT size=3>
            <P>
                广西财经学院
                <BR>
                Guangxi University of Finance and Economics
```

```
            <P>
        </FONT>
        <FORM action="" method="get">
            将当前页面保存为 word 文档吗?
            <INPUT type="submit" value="Yes" name="submit">
        </FORM>
<%
    String submit = request.getParameter("submit");
    if (submit == null)
        submit = "";
    if (submit.equals("Yes"))
        response.setContentType("application/msword;charset=GB2312");
 %>
    </BODY>
</HTML>
```

解析：

(1) 在本页面中，由于<FORM>标记的 action 属性为空，因此在单击 Yes 按钮提交表单后，将由当前页面自身进行处理。

(2) 为将当前页面保存为 Word 文档，只需调用 response 对象的 setContentType()方法将响应的 MIME 类型设置为“application/msword”即可。

【实例 2-14】 Cookies 使用示例：用户登录。图 2.19 所示为“用户登录”页面，单击“登录”按钮后，可打开图 2.20 所示的“登录结果”页面，以显示用户所输入的用户名与密码。单击其中的[OK]链接后，将打开图 2.21 所示的“登录信息”页面，以显示当前用户的有关信息。

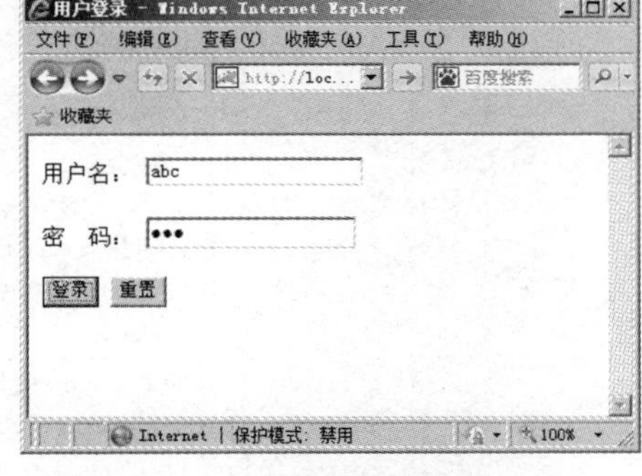

图 2.19　“用户登录”页面

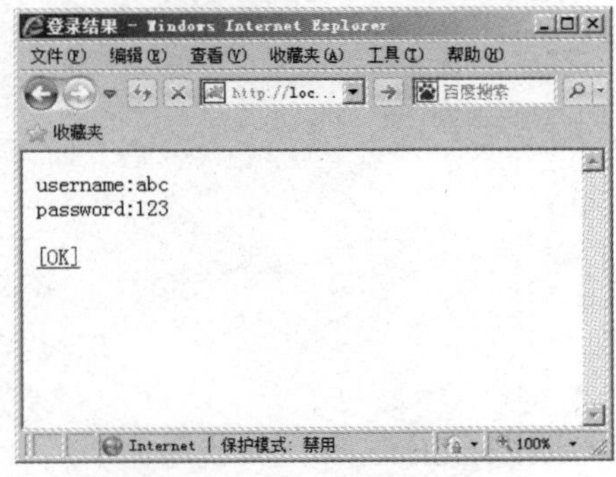

图 2.20　“登录结果”页面

图 2.21　“登录信息”页面

主要步骤(在项目 web_02 中)如下。

(1) 在 WebRoot 目录中添加一个新的页面 UserLogin.htm，其代码如下：

```
<html>
    <head>
```

```
        <title>用户登录</title>
    </head>
    <body>
        <form method="post" action="UserLoginResult.jsp">
            <p>
                用户名:
                <input type="text" name="username" size="20">
            </p>
            <p>
                密  码:
                <input type="password" name="password" size="20">
            </p>
            <p>
                <input type="submit" value="登录" name="ok">
                <input type="reset" value="重置" name="cancel">
            </p>
        </form>
    </body>
</html>
```

(2) 在 WebRoot 目录中添加一个新的 JSP 页面 UserLoginResult.jsp，其代码如下：

```
<%@ page contentType="text/html;charset=GB2312"%>
<html>
    <head>
        <title>登录结果</title>
    </head>
    <%
        String username = request.getParameter("username");
        String password = request.getParameter("password");
        out.println("username:"+username+"<br>");
        out.println("password:"+password+"<br>");
        Cookie cookieUsername=new Cookie("username",username);
        Cookie cookiePassword=new Cookie("password",password);
        //cookieUsername.setMaxAge(30);
        //cookiePassword.setMaxAge(30);
        response.addCookie(cookieUsername);
        response.addCookie(cookiePassword);
    %>
    <br>
    <a href="UserLoginInfo.jsp">[OK]</a>
</html>
```

(3) 在 WebRoot 目录中添加一个新的 JSP 页面 UserLoginInfo.jsp，其代码如下：

```
<%@ page contentType="text/html;charset=GB2312"%>
<html>
    <head>
        <title>登录信息</title>
    </head>
    <%
        Cookie[] cookies=request.getCookies();
```

```
        if (cookies!=null){
            for(int i=0;i<cookies.length;i++){

    out.println(cookies[i].getName()+":"+cookies[i].getValue()+"<br>");
            }
        }
    %>
</html>
```

解析：

(1) Userlogin.htm 是一个静态页面，内含一个供用户输入用户名与密码的表单。该表单被提交后，将由 JSP 页面 UserLoginResult.jsp 进行处理。

(2) 在 UserLoginResult.jsp 页面中，先获取并显示用户所输入的用户名与密码，然后调用 response 对象的 addCookie()方法添加相应的用户名与密码 Cookie，最后再生成一个目标页面为 UserLoginInfo.jsp 的[OK]链接。

(3) 在 UserLoginInfo.jsp 页面中，通过调用 request 对象的 getCookies()方法获取相应的 Cookie 数组，然后再对其进行遍历，输出各个 Cookie 的名称与值。

说明： 必要时，可为各个 Cookie 分别设置相应的生存期，即关闭浏览器后 Cookie 的有效期。为此，只需分别调用 Cookie 对象的 setMaxAge()方法即可。该方法的基本格式为：

```
void setMaxAge(int expiry)
```

其中，参数 expiry 用于指定当前 Cookie 的最大生存期(以秒为单位)。通常，expiry 的值应为正整数，以指定当前 Cookie 在关闭浏览器后还可保存多长时间；若 expiry 值为负，则表示当前 Cookie 在关闭浏览器后将立即被删除掉；若 expiry 值为 0，则表示删除当前 Cookie。

2.3.4　session 对象

session 对象为会话对象，该对象封装了当前用户会话的有关信息。借助于 session 对象，可对各个客户端请求期间的会话进行跟踪。在实际应用中，通常用 session 对象存储用户在访问各个页面期间所产生的有关信息，并在页面之间进行共享。

session 对象是 javax.servlet.http.HttpSession 类的实例(类似于 Servlet 中的 session 对象)，具有 session 作用域。当一个用户首次访问服务器上的一个 JSP 页面时，JSP 引擎就会产生一个 session 对象，同时为该 session 对象分配一个 String 类型的 ID 号，并将该 ID 号发送到客户端，存放在用户的 Cookie 中。当该用户再次访问连接该服务器的其他页面时，或从该服务器连接到其他服务器再返回到该服务器时，JSP 引擎将继续使用此前所创建的同一个 session 对象。待用户关闭浏览器(即终止与服务器端的会话)后，服务器端才将该用户的 session 对象销毁掉。可见，每个用户都对应一个 session 对象，可专门用来存放与该用户有关的信息。

session 对象的常用方法见表 2.5。

表 2.5　session 对象的常用方法

方　法	说　明
String getId()	获取 session 对象的 ID 号
boolean isNew()	判断是否为新的 session 对象。新的 session 对象是指该 session 对象已由服务器产生，但尚未被客户端使用过
void setMaxInactiveInterval (int interval)	设置 session 对象的有效时间或生存时间(单位为秒)，即会话期间客户端两次请求的最长时间间隔。超过此时间，session 对象将会失效。若为 0 或负值，则表示该 session 对象永远不会过期
int getMaxInactiveInterval ()	获取 session 对象的有效时间或生存时间(单位为秒)。若为 0 或负值，则表示该 session 对象永远不会过期
void setAttribute (String name,Object obj)	在 session 中设置指定的属性及其值
Object getAttribute(String name)	获取 session 中指定的属性值。若该属性不存在，则返回 null
Enumeration getAttributeNames()	获取 session 中所有属性名的枚举
void removeAttribute(String name)	删除 session 中指定的属性
void invalidate()	注销当前的 session 对象，并删除其中的所有属性
long getCreationTime()	获取 session 对象的创建时间，单位为毫秒(ms)，由 1970 年 1 月 1 日零时算起)
long getLastAccessedTime()	返回当前会话中客户端最后一次发出请求的时间(单位为毫秒(ms)，由 1970 年 1 月 1 日零时算起)

【实例 2-15】 session 对象使用示例：站点计数器。图 2.22 所示为“站点计数器”页面，其主要功能为显示当前用户是访问本站点的第几个用户。

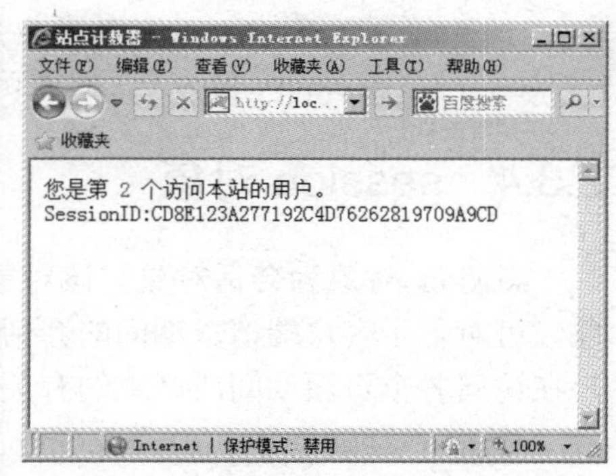

图 2.22　“站点计数器”页面

在项目 web_02 中添加一个新的 JSP 页面 SiteCounter.jsp，其代码如下：

```
<%@ page
contentType="text/html;charset=GB2312"%
>
<html>
    <head>
        <title>站点计数器</title>
    </head>
    <body>
    <%!
    int counter=0;
    synchronized void countPeople(){
        counter=counter+1;
    }
    %>
    <%
    if (session.isNew()) {
```

```
        countPeople();
        session.setAttribute("counter", String.valueOf(counter));
    }
    %>
    <P>
    您是第
    <font color="red">
<%=(String)session.getAttribute("counter")%></font>
    个访问本站的用户。
    <br>
    SessionID:<%=session.getId()%><br>
    </body>
</html>
```

解析：

(1) 本页面通过调用 session 对象的 isNew()方法来判断当前用户是否开始一个新的会话(刷新页面时不会开始一个新的会话，即 session 对象的 ID 号是不会改变的)。只有在开始一个新的会话时，才会增加计数，并将相应的计数结果作为 session 对象的 counter 属性保存。反之，显示的计算结果是从 session 对象的 counter 属性获取的。因此，本页面具有禁止用户通过刷新页面增加计数的功能。

(2) 本页面同时显示 session 的 ID 号。在刷新页面时，session 的 ID 号是不会改变的，这表明，刷新页面时并不会开始一个新的会话。

2.3.5 application 对象

application 对象为应用对象，负责提供 Web 应用程序在服务器运行期间的某些全局性信息。与 session 对象不同，application 对象针对 Web 应用程序的所有用户，并由所有用户所共享(session 对象只针对各个不同的用户，是由各个用户所独享的)。

application 对象是 javax.servlet.ServletContext 类的实例(其实是直接包装了 Servlet 的 ServletContext 类对象)，具有 application 作用域。当 Web 服务器启动一个 Web 应用程序时，就为其产生一个 application 对象。当关闭 Web 服务器或停止 Web 应用程序时，该 application 对象才会被销毁掉。各个 Web 应用程序的 application 对象是互不相同的。

application 对象的常用方法见表 2.6。

表 2.6 application 对象的常用方法

方 法	说 明
void setAttribute(String name,Object obj)	在 application 中设置指定的属性及其值
Object getAttribute(String name)	获取 application 中指定的属性值。若该属性不存在，则返回 null
Enumeration getAttributeNames()	获取 application 中所有属性名的枚举
void removeAttribute(String name)	删除 application 中指定的属性

续表

方 法	说 明
Object getInitParameter(String name)	获取 application 中指定的属性的初始值。若该属性不存在，则返回 null
String getServerInfo()	获取 JSP(Servlet)引擎的名称及版本号
int getMajorVersion()	获取服务器支持的 Servlet API 的主要版本号
int getMinorVersion()	获取服务器支持的 Servlet API 的次要版本号
String getRealPath(String path)	获取指定虚拟路径的真实路径(绝对路径)
ServletContext getContext(String uripath)	获取指定 Web Application 的 application 对象
String getMimeType(String file)	获取指定文件的 MIME 类型
URL getResource(String path)	获取指定资源(文件或目录)的 URL 路径
InputStream getResourceAsStream(String path)	获取指定资源的输入流
RequestDispatcher getRequestDispatcher(String uripath)	获取指定资源的 RequestDispatcher 对象
Servlet getServlet(String name)	获取指定名称的 Servlet
Enumeration getServlets()	获取所有 Servlet 的枚举
Enumeration getServletNames()	获取所有 Servlet 名称的枚举
void log(String msg)	将指定的信息写入 log(日志)文件中
void log(String msg,Throwable throwable)	将 stack trace(栈轨迹)及所产生的 Throwable 异常信息写入 log(日志)文件中
void log(Exception exception,String msg)	将指定异常的 stack trace(栈轨迹)及错误信息写入 log(日志)文件中

【实例 2-16】 application 对象使用示例：站点计数器。图 2.23 所示为“站点计数器”页面，其功能为显示当前用户是访问本站点的第几个用户。

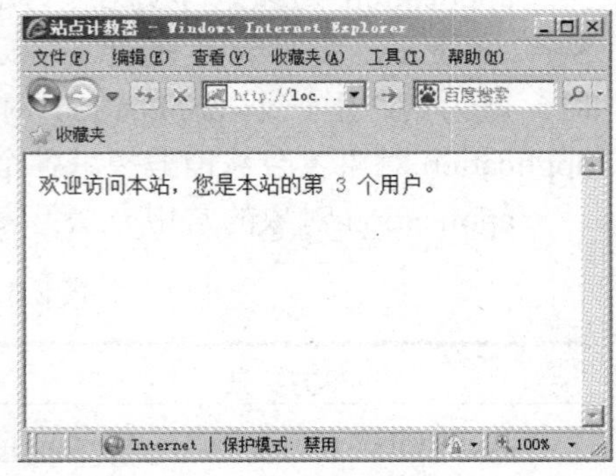

图 2.23 “站点计数器”页面

在项目 web_02 中添加一个新的 JSP 页面 WebSiteCounter.jsp，其代码如下：

```
<%@ page language="java"
contentType="text/html;charset=GB2312"%>
<html>
    <head>
        <title>站点计数器</title>
    </head>
    <body>
    <%!
    synchronized void countPeople() //声明方法
    {
        ServletContext application = getServletContext();
        Integer counter = (Integer) application.getAttribute("counter");
```

```
        if (counter == null) {
            counter=1;
            application.setAttribute("counter", counter);
        } else {
            counter=counter+1;
            application.setAttribute("counter", counter);
        }
    }
    %>
    <%
        Integer allCounter = (Integer)
application.getAttribute("counter");
        if (session.isNew() || allCounter == null) {
            countPeople();
        }
        Integer myCounter = (Integer)
application.getAttribute("counter");
    %>
    <P>
        欢迎访问本站，您是本站的第
        <font color="red"> <%=myCounter%></font>
        个用户。
        <br>
    </body>
</html>
```

解析：

(1) 本页面将站点的访问计数结果保存到 application 对象的 counter 属性中。这样，站点的各个用户均可对其进行访问，并在需要时递增其值。

(2) 本页面根据当前会话是否为一个新的会话来决定是否递增站点的访问计数，因此具有防止用户通过刷新页面来增加计数的功能。

【实例 2-17】request、session 与 application 对象使用示例。

主要步骤(在项目 web_02 中)如下。

(1) 在 WebRoot 目录中添加一个新的 JSP 页面 MyPage1.jsp，其代码如下：

```
<%@ page language="java" pageEncoding="gb2312"%>
<html>
    <head>
        <title>MyPage1</title>
    </head>
    <body>
        <%
    request.setAttribute("request","I am in Request.");
    session.setAttribute("session","I am in Session.");
    application.setAttribute("application","I am in Application.");
        %>
        <jsp:forward page="MyPage2.jsp"></jsp:forward>
    </body>
</html>
```

(2) 在 WebRoot 目录中添加一个新的 JSP 页面 MyPage2.jsp，其代码如下：

```
<%@ page language="java" pageEncoding="gb2312"%>
<html>
    <head>
        <title>MyPage2</title>
    </head>
    <body>
        <%
        out.println("request:
"+(String)request.getAttribute("request")+"<br>");
        out.println("session:
"+(String)session.getAttribute("session")+"<br>");
        out.print("application:
"+(String)application.getAttribute("application")+"<br>");
        %>
    </body>
</html>
```

解析：

(1) MyPage1.jsp 页面分别在 request、session 与 application 对象中设置相应的属性，而 MyPage2 页面则分别获取并显示 request、session 与 application 对象的相应属性值。

(2) 打开浏览器，输入地址“http://localhost:8080/web_02/MyPage1.jsp”并按 Enter 键，结果如图 2.24 所示。

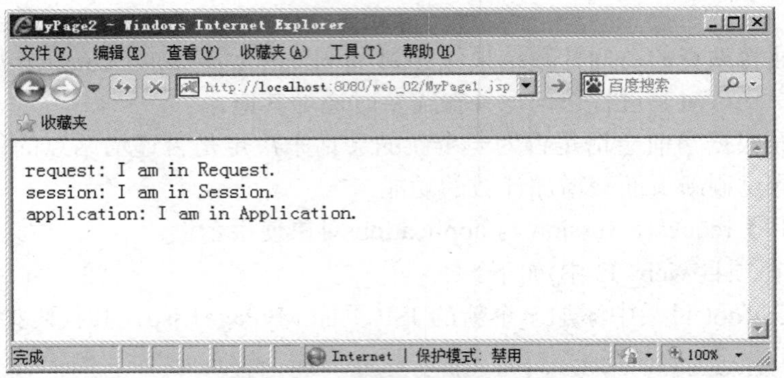

图 2.24 MyPage2 页面 1

在 MyPage1.jsp 中，由于使用 forward 动作标记将页面跳转到 MyPage2.jsp，因此浏览器中的地址不会改变，也就是说请求并没有改变。在同一个请求中，request、session 与 application 对象都是有效的。因此，MyPage2.jsp 页面可顺利获取并显示这三个对象的有关属性值。

(3) 若直接将浏览器中的地址“http://localhost:8080/web_02/MyPage1.jsp”改为“http://localhost:8080/web_02/MyPage2.jsp”并按 Enter 键，则结果如图 2.25 所示。

地址改变了，请求也就不同了，于是此前的 request 对象就失效了，因此 MyPage2.jsp 页面无法访问到 request 对象的相应属性。但由于浏览器并未关闭，依然处于同一个会话中，session 与 application 对象仍然有效，因此 MyPage2.jsp 页面可顺利获取并显示这两个对象中的有关属性值。

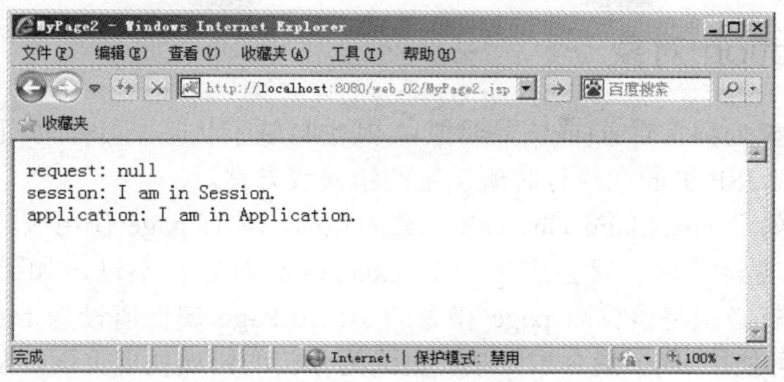

图 2.25　MyPage2 页面 2

(4) 若重新打开一个浏览器，然后输入地址“http://localhost:8080/web_02/MyPage2.jsp”并按 Enter 键，则结果如图 2.26 所示。

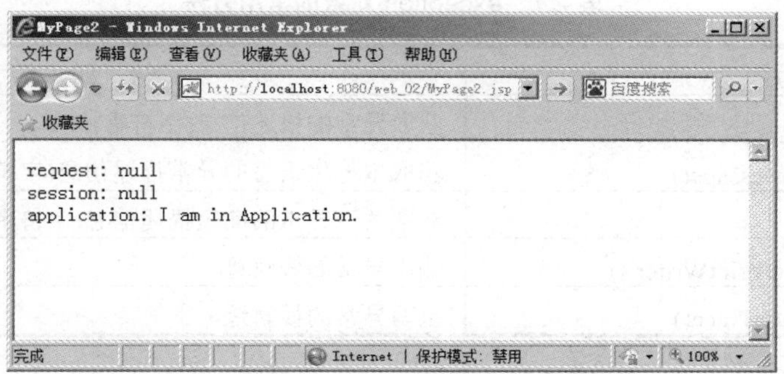

图 2.26　MyPage2 页面 3

由于开始了一个新的会话，此前的 request 与 session 对象就失效了，因此 MyPage2.jsp 页面无法访问到其中的相应属性。不过，application 对象依然有效，因此 MyPage2.jsp 页面可顺利获取并显示该对象中的有关属性值。

(5) 若停止并重启 Tomcat 服务器，再重新打开一个浏览器，然后直接输入地址“http://localhost:8080/web_02/MyPage2.jsp”并按 Enter 键，则结果如图 2.27 所示。

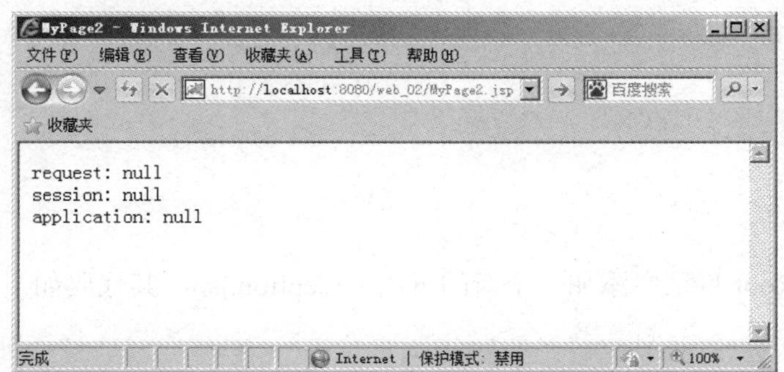

图 2.27　MyPage2 页面 4

由于停止并重启了 Tomcat 服务器，此前的 request、session 与 application 对象就全部失效了，因此 MyPage2.jsp 页面无法访问到其中的相应属性。

2.3.6 exception 对象

exception 对象为异常对象(或例外对象)，其中封装了从某个 JSP 页面中所抛出的异常信息，常用于处理 JSP 页面在执行时所发生的错误或异常。

exception 对象是 javax.lang.Throwable 类的实例，具有 page 作用域。当一个 JSP 页面在运行过程中出现异常时，就会产生一个 exception 对象。不过，如果一个页面要使用 exception 对象，就必须将该页面 page 指令的 isErrorPage 属性值设为 true，否则将无法进行编译。通常，可使用 page 指令指定某一页面为专门的错误处理页面，从而将有关页面的异常或错误都集中到该页面中进行处理。这样，可让整个系统变得更加健壮，并使系统的执行流程变得更加清晰。

exception 对象的常用方法见表 2.7。

表 2.7 exception 对象的常用方法

方 法	说 明
String getMessage()	获取异常的描述信息(字符串)
String getLocalizedMessage()	获取本地化语言的异常描述信息(字符串)
String toString()	获取关于异常的简短描述信息(字符串)
void printStackTrace(PrintWriter s)	输出异常的栈轨迹
Throwable FillInStackTrace()	重写异常的栈轨迹

【实例 2-18】exception 对象使用示例。

主要步骤(在项目 web_02 中)如下。

(1) 在 WebRoot 目录中添加一个新的页面 OnePage.jsp，其代码如下：

```
<%@ page language="java" contentType="text/html;charset=GB2312"%>
<%@ page errorPage="Exception.jsp" %>
<html>
    <head>
        <title>exception 对象使用示例</title>
    </head>
    <body>
        <%
        int i= 100/0;
        %>
    </body>
</html>
```

(2) 在 WebRoot 目录中添加一个新的页面 Exception.jsp，其代码如下：

```
<%@ page language="java" contentType="text/html;charset=GB2312"%>
<%@ page isErrorPage="true" import="java.io.*" %>
<html>
```

```
    <head>
        <title>exception 对象使用示例</title>
    </head>
    <body>
        <font color="red">
        <%=exception.toString() %>
        <br>
        <%
        exception.printStackTrace(new PrintWriter(out));
        %>
        </font>
    </body>
</html>
```

运行结果：在浏览器中输入地址“http://localhost:8080/web_02/OnePage.jsp”并按 Enter 键，结果如图 2.28 所示。

解析：

(1) 在 OnePage.jsp 页面中，执行“int i= 100/0;”语句时将产生一个异常。由于该页面包含有一条“<%@ page errorPage="Exception.jsp" %>”指令，因此在产生异常时将自动跳转到 Exception.jsp 页面，并传入相应的 exception 对象。

(2) 在 Exception.jsp 页面，通过 exception 对象即可获取异常的有关信息。不过，为访问 exception 对象，需使用 page 指令将 isErrorPage 属性设置为 true。

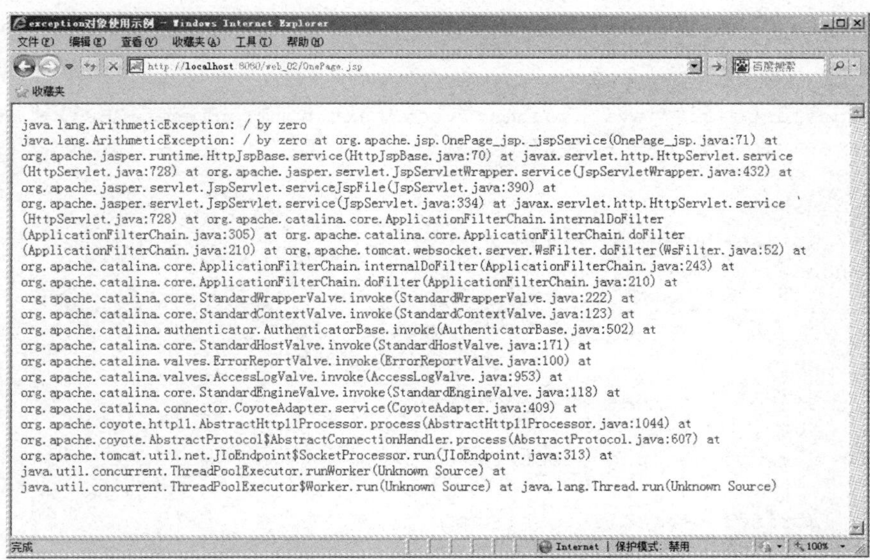

图 2.28　“exception 对象使用示例”页面

2.3.7　page 对象

page 对象为页面对象，是页面实例的引用，代表 JSP 页面本身，即 JSP 页面被转译后的 Servlet。

page 对象是 java.lang.Object 类的实例，具有 page 作用域。从本质上看，page 对象是一个包含当前 Servlet 接口引用的变量，相当于 this 变量的一个别名。通过 page 对象，可调用 Servlet 类所定义的方法。不过，在 JSP 中 page 对象其实很少使用。

page 对象的常用方法见表 2.8。

表 2.8　page 对象的常用方法

方　法	说　明
class getClass()	获取当前对象的类
int hashCode()	获取当前对象的哈希(Hash)代码
boolean equals(Object obj)	判断当前对象是否与指定的对象相等
void copy(Object obj)	把当前对象复制到指定的对象中
Object clone()	克隆当前对象
String toString()	获取表示当前对象的一个字符串
void notify()	唤醒一个等待的线程
void notifyAll()	唤醒所有等待的线程
void wait(int timeout)	使一个线程处于等待状态，直到指定的超时时间结束或被唤醒
void wait()	使一个线程处于等待状态，直到被唤醒

【实例 2-19】 page 对象使用示例。

在项目 web_02 的 WebRoot 目录中添加一个新的 JSP 页面 PageExample.jsp，其代码如下：

```
<%@ page language="java" contentType="text/html;charset=GB2312"%>
<%@ page info="My JSP Page."%>
<html>
    <head>
        <title>page 对象使用示例</title>
    </head>
    <body>
        Class: <%= page.getClass() %>
        <br>
        Class: <%= this.getClass() %>
        <br>
        String: <%= page.toString() %>
        <br>
        String: <%= this.toString() %>
        <br>
        Page Info: <%=
((javax.servlet.jsp.HttpJspPage)page).getServletInfo() %>
        <!--
        <br>
        Page Info: <%= this.getServletInfo() %>
        -->
    </body>
</html>
```

运行结果如图 2.29 所示。

解析：

(1) 在本页面中，直接调用 page 对象的 getClass()方法与 toString()方法获取当前页面对象的类以及表示该对象的字符串。通常，在代码中可用 this 代替 page。

(2) page 对象的类型为 java.lang.Object，将其强制类型转换为 javax.servlet.jsp.HttpJspPage，即可调用 getServletInfo()函数，获取页面中 page 指令的 info 属性值。

图 2.29　"page 对象使用示例"页面

2.3.8　config 对象

config 对象为配置对象，主要用于获取 Servlet 或者 JSP 引擎的初始化参数。

config 对象是 javax.servlet.ServletConfig 类的实例(其实是直接包装了 Servlet 的 ServletConfig 类对象)，具有 page 作用域。config 对象包含了初始化参数以及一些实用方法，可为使用 web.xml 文件的服务器程序与 JSP 页面在其环境中设置初始化参数。

config 对象的常用方法见表 2.9。

表 2.9　config 对象的常用方法

方　法	说　明
String getInitParameter(String name)	获取指定 Servlet 初始化参数的值
Enumeration getInitParameterNames()	获取所有 Servlet 初始化参数的枚举
Sring getServletName()	获取 Servlet 的名称
ServletContext getServletContext()	获取 Servlet 上下文(ServletContext)

【实例 2-20】config 对象使用示例。

主要步骤(在项目 web_02 中)如下。

(1) 在项目的配置文件 web.xml 中添加一个 Servlet 的配置，其具体代码如下：

```
<?xml version="1.0" encoding="UTF-8"?>
<web-app version="3.0"
   xmlns="http://java.sun.com/xml/ns/javaee"
   xmlns:xsi="http://www.w3.org/2001/XMLSchema-instance"
   xsi:schemaLocation="http://java.sun.com/xml/ns/javaee
   http://java.sun.com/xml/ns/javaee/web-app_3_0.xsd">
  <display-name></display-name>
  <welcome-file-list>
    <welcome-file>index.jsp</welcome-file>
  </welcome-file-list>
  <servlet>
    <servlet-name>ConfigExample</servlet-name>
```

```
    <jsp-file>/ConfigExample.jsp</jsp-file>
     <init-param>
     <param-name>www</param-name>
         <param-value>http://www.gxufe.cn</param-value>
     </init-param>
    <init-param>
        <param-name>port</param-name>
        <param-value>80</param-value>
    </init-param>
  </servlet>
  <servlet-mapping>
    <servlet-name>ConfigExample</servlet-name>
    <url-pattern>/ConfigExample.jsp</url-pattern>
  </servlet-mapping>
</web-app>
```

(2) 在项目的 WebRoot 文件夹中添加一个新的 JSP 页面 ConfigExample.jsp，其代码如下：

```
<%@ page language="java" contentType="text/html;charset=GB2312"%>
<html>
    <head>
        <title>config 对象使用示例</title>
    </head>
    <body>
        <%
        String www = (String)config.getInitParameter("www");
        String port = (String)config.getInitParameter("port");
        %>
        www: <%= www %><br>
        port: <%= port %><br>
    </body>
</html>
```

运行结果如图 2.30 所示。

解析：在本页面中，通过调用 config 对象的 getInitParameter()方法，获取配置文件 web.xml 中有关 Servlet 的初始化参数值。

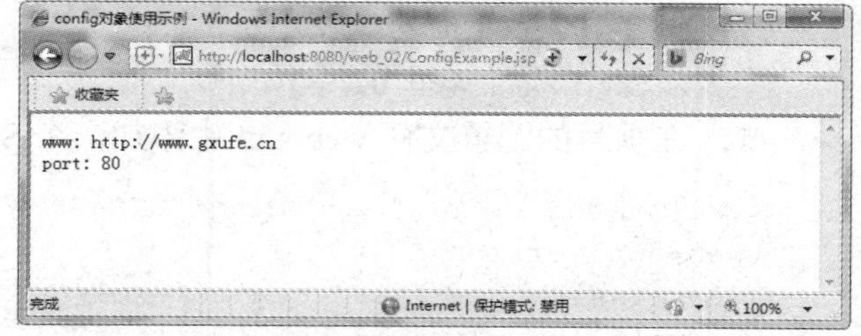

图 2.30 “config 对象使用示例”页面

2.3.9 pageContext 对象

pageContext 对象为页面上下文对象，主要用于访问页面的有关信息。其实，pageContext 对象是整个 JSP 页面的代表，相当于页面中所有功能的集大成者，可实现对页面内所有对象的访问。

pageContext 对象是 javax.servlet.jsp.PageContext 类的实例，具有 page 作用域，其创建与初始化均由容器完成。PageContext 类定义了一些范围常量，包括 PAGE_SCOPE，REQUEST_SCOPE，SESSION_SCOPE 与 APPLICATION_SCOPE，分别表示 Page 范围

(即 pageContext 对象的属性范围)、Request 范围(即 request 对象的属性范围)、Session 范围(即 session 对象的属性范围)与 Application 范围(即 application 对象的属性范围)。

在 pageContext 对象中，包含了传递给 JSP 页面的指令信息，存储了 request 对象与 response 对象的引用。此外，out 对象、session 对象、application 对象与 config 对象也可以从 pageContext 对象中导出。可见，通过 pageContext 对象可以存取其他内置对象。

pageContext 对象的常用方法见表 2.10。

表 2.10　pageContext 对象的常用方法

方　法	说　明
Exception getException()	获取当前页(该页应为 Error Page)的 exception 对象
JspWriter getOut()	获取当前页的 out 对象
Object getPage()	获取当前页的 page 对象
ServletRequest getRequest()	获取当前页的 request 对象
ServletResponse getResponse()	获取当前页的 response 对象
ServletConfig getServletConfig()	获取当前页的 config 对象
ServletContext getServletContext()	获取当前页的 application 对象
HttpSession getSession()	获取当前页的 session 对象
void setAttribute(String name,Object obj)	设置指定的属性及其值(在 Page 范围内)
void setAttribute(String name,Object obj, int scope)	在指定范围内设置指定的属性及其值
public Object getAttribute(String name)	获取指定属性的值(在 Page 范围内)。若无指定的属性，则返回 null
Object findAttribute(String name)	按顺序在 Page、Request、Session 与 Application 范围内查找并返回指定属性的值。若无指定的属性，则返回 null
public Object getAttribute(String name, int scope)	在指定范围内获取指定属性的值。若无指定的属性，则返回 null
void removeAttribute(String name)	在所有范围内删除指定的属性
void removeAttribute(String name,int scope)	在指定范围内删除指定的属性
int getAttributeScope(String name)	获取指定属性的作用范围
Enumeration getAttributeNamesInScope (int scope)	获取指定范围内的属性名枚举
void release()	释放 pageContext 对象所占用的资源
void forward(String relativeUrlPath)	将页面重定向到指定的地址
void include(String relativeUrlPath)	在当前位置包含指定的文件

【实例 2-21】 pageContext 对象使用示例。

在项目 web_02 中添加一个新的 JSP 页面 PageContextExample.jsp，其代码如下：

```
<%@ page language="java" pageEncoding="gb2312"%>
<html>
    <head>
        <title>pageContext 对象使用示例</title>
```

```
    </head>
    <body>
        <%
        ServletRequest myRequest=pageContext.getRequest();
    myRequest.setAttribute("request","I am in Request.");
        HttpSession mySession=pageContext.getSession();
        mySession.setAttribute("session","I am in Session.");
    ServletContext myApplication=pageContext.getServletContext();
    myApplication.setAttribute("application","I am in Application.");
    out.println(pageContext.getAttribute("request",
    PageContext.REQUEST_SCOPE) + "<br>");
    out.println(pageContext.getAttribute("session",
    PageContext.SESSION_SCOPE) + "<br>");
    out.println(pageContext.getAttribute("application",
    PageContext.APPLICATION_SCOPE) + "<br>");
        %>
    </body>
</html>
```

运行结果如图 2.31 所示。

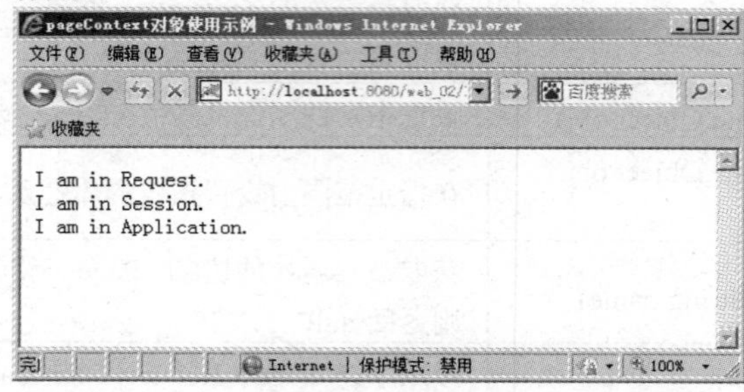

图 2.31 “pageContext 对象使用示例”页面

2.4 JSP 应用案例

2.4.1 系统登录

系统登录是各类应用系统中用于确保系统安全性的一项至关重要的功能，其主要作用是验证用户的身份，并确定其操作权限。下面采用 JSP 技术实现 Web 应用系统中的系统登录功能(在此暂时假定合法用户的用户名与密码分别为 abc 与 123)。

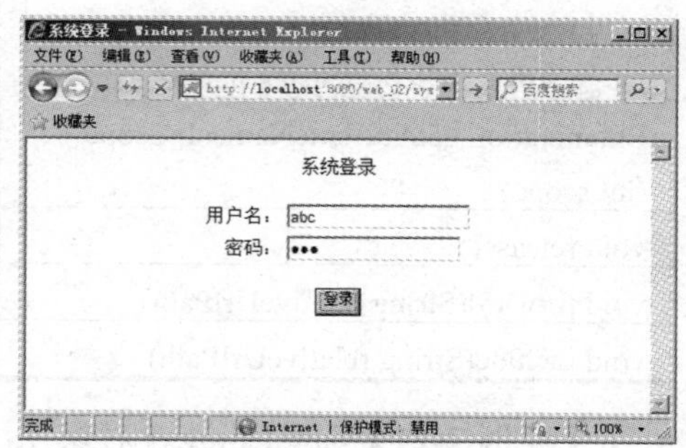

图 2.32 “系统登录”页面

图 2.32 所示为“系统登录”页面。在此页面输入用户名与密码后，再单击“登录”按钮，即可提交信息至服务器对用户的身份进行验证。若该用户为系统的合法用户，则跳转至“登录成功”页面(见图 2.33)；反之，则跳转至“登录失败”页面(见图 2.34)。

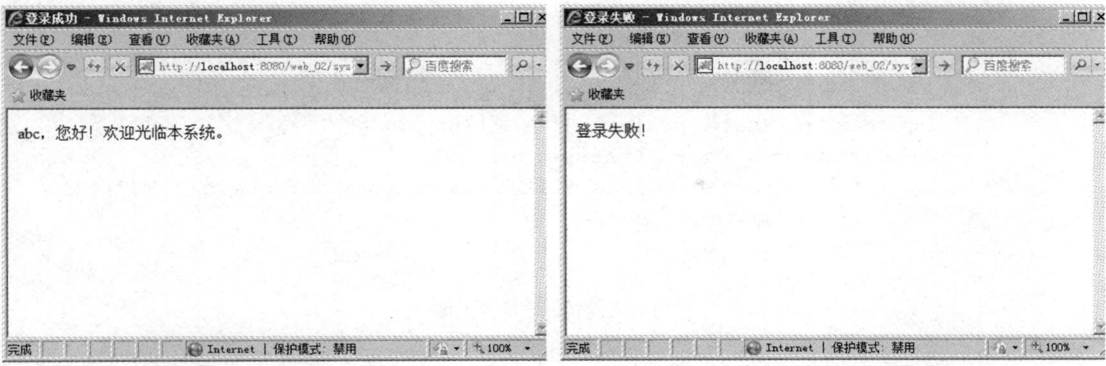

图 2.33　“登录成功”页面　　　　图 2.34　“登录失败”页面

实现步骤(在项目 web_03 中)如下。

(1) 在项目的 WebRoot 文件夹中新建一个子文件夹 syslogin。

(2) 在子文件夹 syslogin 中添加一个新的 JSP 页面 login.jsp，其代码如下：

```
<%@ page language="java" pageEncoding="utf-8" %>
<html>
    <head>
    <title>系统登录</title>
    </head>
    <body>
        <div align="center">
        <form action="validate.jsp" method="post">
            系统登录<br><br>
            <table>
            <tr><td align="right">用户名：</td><td><input type="text"
name="username"></td></tr>
            <tr><td align="right">密码：</td><td><input type="password"
name="password"></td></tr>
            </table>
            <br>
            <input type="submit" value="登录">
        </form>
        </div>
    </body>
</html>
```

(3) 在子文件夹 syslogin 中添加一个新的 JSP 页面 validate.jsp，其代码如下：

```
<%@ page language="java" pageEncoding="utf-8" %>
<%request.setCharacterEncoding("utf-8"); %>
<html>
    <head>
        <title>登录验证</title>
        <meta http-equiv="Content-Type" content="text/html;charset=utf-
8">
    </head>
```

```
    <body>
        <%
        String username0=request.getParameter("username");  //获取提交的姓名
        String password0=request.getParameter("password");  //获取提交的密码
        boolean validated=false;  //验证标识
        if (username0!=null&&password0!=null){
            if (username0.equals("abc") && password0.equals("123")) {
                validated=true;
            }
        }
        if(validated)  //验证成功跳转到成功页面
        {
        %>
            <jsp:forward page="welcome.jsp"></jsp:forward>
        <%
        }
        else  //验证失败跳转到失败页面
        {
        %>
            <jsp:forward page="error.jsp"></jsp:forward>
        <%
        }
        %>
    </body>
</html>
```

(4) 在子文件夹 syslogin 中添加一个新的 JSP 页面 welcome.jsp，其代码如下：

```
<%@ page language="java" pageEncoding="utf-8" %>
<%request.setCharacterEncoding("utf-8"); %>
<html>
    <head>
        <title>登录成功</title>
    </head>
    <body>
        <%out.print(request.getParameter("username")); %>，您好！欢迎光临本
系统。
    </body>
</html>
```

(5) 在子文件夹 syslogin 中添加一个新的 JSP 页面 error.jsp，其代码如下：

```
<%@ page language="java" pageEncoding="utf-8" %>
<html>
    <head>
        <title>登录失败</title>
    </head>
    <body>
        登录失败！
    </body>
</html>
```

打开浏览器，输入地址“http://localhost:8080/web_02/syslogin/login.jsp”并按 Enter 键，即可打开图 2.32 所示的“系统登录”页面。

2.4.2　简易聊天室

聊天室是目前较为常见的一种网络应用，下面综合应用 JSP 的有关内置对象，设计并实现一个简易的聊天室。

图 2.35 所示为“用户登录”页面。输入用户名与密码，再单击“登录”按钮后，即可打开图 2.36 所示的 ChatRoom 页面。在该页面中，可显示所有用户的聊天内容，当前用户也可输入并发送自己的言论。

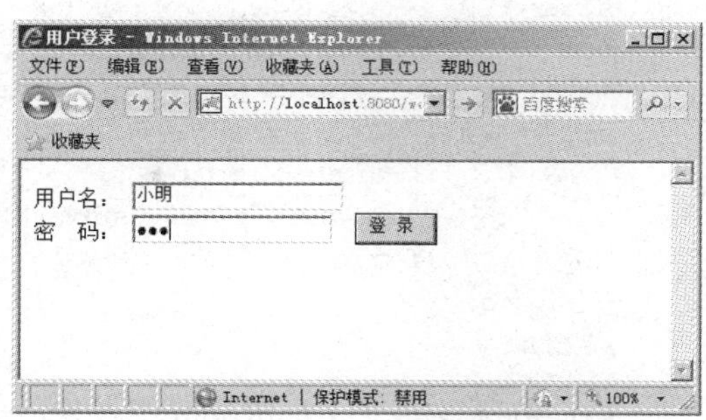

图 2.35　“用户登录”页面

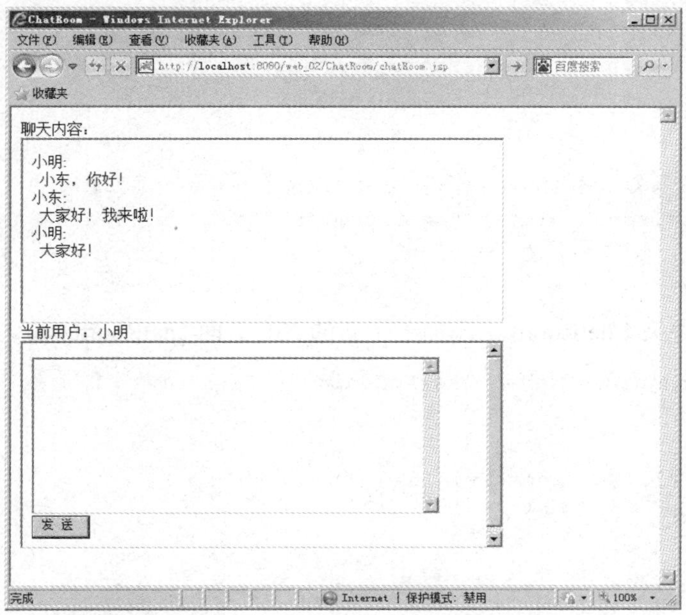

图 2.36　ChatRoom 页面

实现步骤(在项目 web_03 中)如下。

(1) 在项目的 WebRoot 文件夹中新建一个子文件夹 ChatRoom。

(2) 在子文件夹 ChatRoom 中添加一个新的页面 login.htm，其代码如下：

```
<html>
    <head>
        <meta http-equiv="Content-Type" content="text/html;charset=UTF-8">
        <title>用户登录</title>
    </head>
    <body>
        <form method="post" action="loginCheck.jsp">
```

```
            用户名:
            <input name="username" type="text" id="username"
size="20"><br>
            密    码:
            <input name="password" type="password" id="password"
size="20">
              
            <input type="submit" name="Submit" value="登 录 ">
        </form>
    </body>
</html>
```

(3) 在子文件夹 ChatRoom 中添加一个新的 JSP 页面 loginCheck.jsp，其代码如下：

```
<%@ page contentType="text/html;charset=UTF-8" %>
<% request.setCharacterEncoding("UTF-8"); %>
<%
String username=request.getParameter("username");
String password=request.getParameter("password");
if(username == null || password == null){
    response.sendRedirect("login.htm");
}
else{
    session.setAttribute("username",username);
    response.sendRedirect("chatRoom.jsp");
}
%>
```

(4) 在子文件夹 ChatRoom 中添加一个新的 JSP 页面 chatRoom.jsp，其代码如下：

```
<%@ page language="java" pageEncoding="UTF-8" %>
<html>
    <head>
        <title>ChatRoom</title>
    </head>
    <body>
        <div>
            聊天内容:<br>
            <iframe name="displayFrame" width="500" height=200
src="displayText.jsp">
            </iframe>
            <br>当前用户:<%=(String)
session.getAttribute("username")%><br>
            <iframe name="sendFrame" width=500 height=220
src="sendText.htm">
            </iframe>
        </div>
    </BODY>
</HTML>
```

(5) 在子文件夹 ChatRoom 中添加一个新的 JSP 页面 displayText.jsp，其代码如下：

```
<%@ page contentType="text/html;charset=UTF-8" %>
<%response.setHeader("refresh", "3");%>
<%
  String allChatText = (String)application.getAttribute("allChatText");
  if ( allChatText != null)
  {
   out.print(allChatText);
  }
  else
  {
    out.print("[聊天记录]");
  }
%>
```

(6) 在子文件夹 ChatRoom 中添加一个新的页面 sendText.htm，其代码如下：

```
<html>
    <head>
        <meta http-equiv="content-type" content="text/html; charset=UTF-
8">
        <title></title>
    </head>
    <body>
    <form action="sendText.jsp" method="post">
        <textarea name="chatText" rows=10 cols=50></textarea>
        <br>
        <input type="submit" value="发 送 "/>
    </form>
</body>
</html>
```

(7) 在子文件夹 ChatRoom 中添加一个新的 JSP 页面 sendText.jsp，其代码如下：

```
<%@ page contentType="text/html;charset=UTF-8" %>
<% request.setCharacterEncoding("UTF-8"); %>
<%
    String userName = (String)session.getAttribute("username");
    //获得提交聊天内容的用户名
    String chatText = request.getParameter("chatText");//获得聊天内容
    chatText = userName + ":<br>  " + chatText + "<br>";
    String allChatText = (String)application.getAttribute
    ("allChatText");//获得历史聊天记录
    if (allChatText == null)
    {
      allChatText = chatText;
    }
```

```
    else
    {
        allChatText = chatText + allChatText;
    }
    application.setAttribute("allChatText",allChatText);
    response.sendRedirect("sendText.htm");
%>
```

打开浏览器，输入地址“http://localhost:8080/web_02/ChatRoom/login.htm”并按 Enter 键，即可打开图 2.35 所示的“用户登录”页面。

本章小结

本章首先介绍了 JSP 的概况，然后通过具体实例讲解了 JSP 的基本语法(包括各种指令标记与动作标记的基本用法)以及 JSP 各种内置对象的主要用法，最后再通过具体案例说明了 JSP 的综合应用技术。通过本章的学习，读者应熟练掌握 JSP 的基本语法以及 JSP 各种指令标记、动作标记与内置对象的主要用法。

思考题

1. JSP 声明、表达式与脚本小程序的基本语法格式是什么？
2. JSP 指令标记的基本语法格式是什么？
3. JSP 指令标记分为哪几种？
4. page 指令的作用是什么？其基本语法格式是什么？
5. include 指令的作用是什么？其基本语法格式是什么？
6. taglib 指令的作用是什么？其基本语法格式是什么？
7. JSP 的动作标记有哪些？
8. param 动作标记的作用是什么？其基本语法格式是什么？
9. include 动作标记的作用是什么？其基本语法格式是什么？
10. forward 动作标记的作用是什么？其基本语法格式是什么？
11. plugin 动作标记的作用是什么？
12. useBean、getProperty 与 setProperty 动作标记的作用是什么？
13. JSP 页面中的注释分为哪几种？其基本语法格式是什么？
14. JSP 的内置对象有哪些？
15. out 对象的主要作用是什么？有哪些常用方法？
16. request 对象的主要作用是什么？有哪些常用方法？
17. response 对象的主要作用是什么？有哪些常用方法？
18. 如何创建 Cookie？如何读取 Cookie 的信息？
19. session 对象的主要作用是什么？有哪些常用方法？
20. 如何在 session 对象中设置属性？如何获取 session 对象中的属性？

21. application 对象的主要作用是什么？有哪些常用方法？
22. 如何在 application 对象中设置属性？如何获取 application 对象中的属性？
23. exception 对象的主要作用是什么？有哪些常用方法？
24. page 对象的主要作用是什么？有哪些常用方法？
25. config 对象的主要作用是什么？有哪些常用方法？
26. pageContext 对象的主要作用是什么？有哪些常用方法？

第 3 章

JDBC 技术

JDBC 是 Java 语言中通用的一种数据库访问技术，其实质是 Java 与数据库间的一套接口规范。在各类 Java 应用中，JDBC 的使用是相当普遍的。

本章要点：

- JDBC 简介;
- JDBC 的核心类与接口;
- JDBC 的基本应用。

学习目标：

- 了解 JDBC 的概况;
- 掌握 JDBC 核心类与接口的基本用法;
- 掌握 JDBC 的数据库编程技术;
- 掌握 Web 应用系统开发的 JSP+JDBC 模式。

3.1 JDBC 简介

JDBC(Java Data Base Connectivity，Java 数据库连接)是一种用于执行 SQL 语句的 Java API(Application Programming Interface，应用编程接口或应用程序编程接口)，由一组用 Java 语言编写的类和接口组成，可让开发人员以纯 Java 的方式连接数据库，并完成相应的操作。

从结构上看，JDBC 是 Java 语言访问数据库的一套接口集合。而从本质上看，JDBC 则是调用者(开发人员)与执行者(数据库厂商)之间的一种协议。JDBC 的实现由数据库厂商以驱动程序的形式提供。对于不同类型的数据库管理系统(DBMS)来说，其 JDBC 驱动程序是不同的。而对于某些数据库管理系统，若其版本不同，则所用的 JDBC 驱动程序也可能有所不同。

JDBC 为各种数据库的访问提供了一致的方式，从而便于开发人员使用 Java 语言编写数据库应用程序。使用 JDBC，应用程序能自动地将 SQL 语句传送给相应的数据库管理系统，因此，在开发数据库应用时，Java 与 JDBC 的结合可让开发人员真正实现“一次编写，处处运行”。

在 JSP 中，可使用 JDBC 技术实现对数据库的访问，并完成有关表中记录的查询、增加、修改、删除等操作。

3.2 JDBC 的核心类与接口

JDBC 的类库包含在两个包中，即 java.sql 包 与 javax.sql 包。

java.sql 包提供了 JDBC 的一些基本功能，主要针对基本的数据库编程服务(如生成连接、执行语句以及准备语句与运行批处理查询等)，同时也有一些高级的处理(如批处理更新、事务隔离与可滚动结果集等)。JDBC 的核心类库均包含在 java.sql 包中。java.sql 包所包含的重要类与接口主要有 java.sql.DriverManager 类、java.sql.Driver 接口、

java.sql.Connection 接口、java.sql.Statement 接口、java.sql.PreparedStatement 接口、java.sql.CallableStatement 接口、java.sql.ResultSet 接口等。

javax.sql 包最初是作为 JDBC 2.0 的可选包而引入的，提供了 JDBC 的一些扩展功能，主要是为数据库方面的高级操作提供相应的接口与类。例如，为连接管理、分布式事务与旧有的连接提供更好的抽象，引入了容器管理的连接池、分布式事务与行集等。javax.sql 包所包含的重要的接口有 javax.sql.DataSource、javax.sql.RowSet 等。

3.2.1 DriverManager 类

DriverManager 类相当于驱动程序的管理器，负责注册 JDBC 驱动程序，建立与指定数据库(或数据源)的连接。该类的常用方法见表 3.1。

表 3.1 DriverManager 类的常用方法

方 法	说 明
Connection getConnection(String url,String user,String password)	建立与指定数据库的连接，并返回一个数据库连接对象
void setLoginTimeout(int seconds)	设置连接数据库时驱动程序等待的时间(以秒为单位)
void registerDriver(Driver driver)	注册指定的驱动程序
Enumeration getDrivers()	返回当前加载的驱动程序的枚举
Driver getDriver(String url)	在已向 DriverManager 注册的驱动程序中查找并返回一个能打开指定数据库的驱动程序

对于 DriverManager 类来说，最常用的方法应该是 getConnection()。该方法的常用方式为：

```
Static Connection getConnection(String url,String user,String password)
throw SQLException
```

其中，参数 user 为连接数据库时所使用的用户名，password 为相应用户的密码，url 则为所要连接的数据库(或数据源)的 URL(相当于连接字符串)。数据库(或数据源)不同，其 URL 的格式往往是不同的，如以下示例：

```
    // ODBC 数据源
    String url="jdbc:odbc:<dataSourceName>";
    //SQL Server 2005/2008 数据库
String url="jdbc:sqlserver://localhost:1433;databaseName=<databaseName>";
//Mysql 数据库
String url="jdbc:mysql://localhost:3306/<databaseName>";
//Oracle 数据库
String url="jdbc:oracle:thin:@localhost:1521:<databaseName>";
其中，<dataSourceName>表示数据源名，<databaseName>表示数据库名。
```

3.2.2 Driver 接口

Driver 接口是每个 JDBC 驱动程序都要实现的接口。换言之，每个 JDBC 驱动程序都要提供一个实现 Driver 接口的类。DriverManager 会尝试加载尽可能多的可以找到的驱动程序，并让每个驱动程序依次尝试连接到指定的目标 URL。在加载某个 Driver 类时，会创建相应的实例并向 DriverManager 注册。

要使用 JDBC 访问数据库，首先要加载相应的驱动程序。驱动程序的加载与注册可通过调用 Class 类(在 java.lang 包中)的静态方法 forName()实现。该方法的常用方式为：

```
Class.forName(String driverClassName);
```

其中，参数 driverClassName 用于指定 JDBC 驱动程序的类名。

说明： Class.forName()实例化相应的驱动程序类并自动向 DriverManager 注册，因此无须再显式地调用 DriverManager.registerDriver()方法进行注册。

对于 Java 应用来说，目前常用的数据库访问方式主要有两种，即 JDBC-ODBC 桥驱动方式与直接驱动方式。

要通过 JDBC-ODBC 桥驱动方式访问数据库，加载驱动程序的代码如下：

```
Class.forName("sun.jdbc.odbc.JdbcOdbcDriver");
```

要通过直接驱动方式访问数据库，则针对不同的 DBMS(数据库管理系统)，所要加载的驱动程序是不同的。例如：

```
//SQL Server 数据库
Class.forName("com.microsoft.jdbc.sqlserver.SQLServerDriver");
//Mysql 数据库
Class.forName("com.mysql.jdbc.Driver");
//Oracle 数据库
Class.forName("Oracle.jdbc.driver.OracleDriver");
```

3.2.3 Connection 接口

Connection 接口代表与数据库的连接，用于执行 SQL 语句并返回相应的结果，也可为数据库的事务处理提供支持(如提供提交、回滚等方法)。该接口的常用方法见表 3.2。

表 3.2 Connection 接口的常用方法

方　法	说　明
Statement createStatement()	创建并返回一个 Statement 对象。该对象通常用于执行不带参数的 SQL 语句
PreparedStatement prepareStatement(String sql)	创建一个 PreparedStatement 对象。该对象通常用于执行带有参数的 SQL 语句，并对 SQL 语句进行编译预处理
CallableStatement prepareCall(String sql)	创建一个 CallableStatement 对象。该对象通常用于调用数据库的存储过程
void close()	关闭与数据库的连接(即关闭当前的 Connection 对象，并释放其所占用的资源
boolean isClosed()	判断当前 Connection 对象是否已被关闭。若已关闭，则返回 true，否则返回 false

续表

方　法	说　明
void setReadOnly(boolean readOnly)	开启(readOnly 为 true 时)或关闭(readOnly 为 false 时)当前 Connection 对象的只读模式(默认为非只读模式)。不能在事务中执行此操作，否则将抛出异常
boolean getReadOnly() boolean isReadOnly()	判断当前 Connection 对象是否为只读模式。若为只读模式，则返回 true，否则返回 false
void setAutoCommit(boolean autoCommit)	开启(autoCommit 为 true 时)或关闭(autoCommit 为 false 时)当前 Connection 对象的自动提交模式
boolean getAutoCommit()	判断当前 Connection 对象是否为自动提交模式。若为自动提交模式，则返回 true，否则返回 false
void commit()	提交事务，也就是将从上一次提交或回滚以来所进行的所有更改保存到数据库
void rollback()	回滚事务，也就是取消当前事务中的所有更改。该方法只能在非自动提交模式下调用，否则将抛出异常。此外，该方法的一种重载形式以 SavePoint 实例为参数，可用于取消指定 SavePoint 实例之后的所有更改
SavePoint setSavepoint()	在当前事务中创建并返回一个 SavePoint 实例。要求当前 Connection 对象为非自动提交模式，否则将抛出异常
void releaseSavePoint(SavePoint savepoint)	从当前事务中删除指定的 SavePoint 实例

要创建一个 Connection 对象，只需调用 DriverManager 类的 getConnection()方法即可。如以下示例：

```
Class.forName("com.microsoft.sqlserver.jdbc.SQLServerDriver");
String url="jdbc:sqlserver://localhost:1433;DatabaseName=rsgl";
String user="sa";
String password="abc123!";
Connection conn=DriverManager.getConnection(url,user,password);
```

在此，创建了一个 Connection 对象 conn。

一个 Connection 对象表示与特定的数据库的连接，主要用于执行 SQL 语句并得到执行的结果。默认情况下，Connection 对象处于自动提交模式，即每条 SQL 语句在执行后都会自动进行提交。若禁用了自动提交模式，则必须显式调用其 commit()方法以提交对数据库的更改(否则无法将更改保存到数据库中)。

3.2.4　Statement 接口

Statement 接口用于执行不带参数的 SQL 语句(即静态 SQL 语句)，并返回相应的执行结果。该接口的常用方法见表 3.3。

表 3.3　Statement 接口的常用方法

方　法	说　明
ResultSet executeQuery(String sql)	执行指定的查询类 SQL 语句(通常为 Select 语句)，并返回一个 ResultSet 对象
int executeUpdate(String sql)	执行指定的更新类 SQL 语句(通常为 Insert、Update 或 Delete 语句)，并返回受影响的行数
boolean execue(String sql)	执行指定的 SQL 语句(该语句可能返回多个结果)。若该语句返回的第一个结果是 ResultSet 对象，则返回 true，否则(第一个结果是受影响的行数或没有结果)返回 false
void addBatch(String sql)	将指定的 SQL 语句(通常为静态的更新类语句)添加到 Batch(批)中。若驱动程序不支持批处理，将抛出异常
void clearBatch()	清除 Batch 中的所有 SQL 语句。若驱动程序不支持批处理，将抛出异常
int[] executeBatch()	执行 Batch 中的所有 SQL 语句。若全部执行成功，则返回相应的整数数组。若驱动程序不支持批处理，或未能全部执行成功，将抛出异常
void close()	关闭 Statement 对象(实例)，释放其所占用的资源

为创建一个 Statement 对象，只需调用 Connection 对象的 createStatement()方法即可。如以下示例：

```
...
Connection conn=DriverManager.getConnection(url,user,password);
Statement stmt=conn.createStatement();
```

在此，创建了一个 Statement 对象 stmt。

对于 Connection 接口来说，createStatement()是其最为常用的方法之一。其实，该方法还有带参数的形式，所创建的 Statement 对象可用于生成具有指定特性的 ResultSet 对象。其具体的重载形式如下。

(1) Statement createStatement(int resultSetType，int resultSetConcurrency) throws SQLException。

(2) Statement createStatement(int resultSetType，int resultSetConcurrency,int resultSetHoldability) throws SQLException。

其中，第一种形式用于创建一个可生成具有给定结果集类型与并发性的 ResultSet 对象的 Statement 对象，第二种形式用于创建一个可生成具有给定结果集类型、并发性与可保持性的 ResultSet 对象的 Statement 对象。各参数的说明如下。

(1) resultSetType 用于指定结果集游标(Cursor)的滚动方式，其取值为以下 ReasultSet 常量之一。

① ResultSet.TYPE_FORWORD_ONLY，默认的游标类型，仅支持结果集的 forward 操作，不支持 backforward 、random 、last、first 等操作。

② ResultSet.TYPE_SCROLL_INSENSITIVE，支持结果集的 forward、backforward、 random、last、first 等操作，对其他 session 对数据库中数据做出的更改是不敏感的。

③ ResultSet.TYPE_SCROLLL_SENSITIVE，支持结果集的 forward、backforward、random、last、first 等操作，对其他 session 对数据库中数据做出的更改是敏感的，即其他 session 修改了数据库中的数据，会反映到本结果集中。

(2) resultSetConcurrency 用于指定是否可以用结果集更新数据库，其取值为以下 ReasultSet 常量之一。

① ResultSet.CONCUR_READ_ONLY，ResultSet 中的数据记录是只读的，不允许修改。

② ResultSet.CONCUR_UPDATABLE，ResultSet 中的数据记录可以任意修改，然后更新到数据库中，即可以进行记录的插入、修改或删除。

(3) resultSetHoldability 用于指定结果集的保持性，其取值为以下 ResultSet 常量之一。

① ResultSet.HOLD_CURSORS_OVER_COMMIT，在事务 Commit(提交)或 Rollback (回滚)后，ResultSet 仍然可用。

② ResultSet.CLOSE_CURSORS_AT_COMMIT，在事务 Commit 或 Rollback 后，ResultSet 被关闭。

如以下示例：

```
…
Connection conn=DriverManager.getConnection(url,user,password);
stmt=conn.createStatement(ResultSet.TYPE_SCROLL_INSENSITIVE,
ResultSet.CONCUR_READ_ONLY);
```

在此，同样创建了一个 Statement 对象 stmt，但该 stmt 对象所创建的结果集可具有指定的特性，使用起来将更加灵活。

3.2.5 PreparedStatement 接口

PreparedStatement 接口继承自 Statement 接口，是 Statement 接口的扩展，用于执行预编译的动态 SQL 语句，即包含有参数的 SQL 语句(参数占位符为问号“?”)。该接口的常用方法见表 3.4。

表 3.4　PreparedStatement 接口的常用方法

方　法	说　明
ResultSet executeQuery()	执行预编译的 Select 语句，并返回一个 ResultSet 对象
int executeUpdate()	执行预编译的 Insert、Update 或 Delete 语句，并返回受影响的行数
void setXxx(int parameterIndex, Xxx x)	将 SQL 语句中的第 parameterIndex 个参数的值设置为 x(Xxx 表示数据类型)。参数的编号从 1 开始，用来设置参数值的 setter 方法(如 setInt、setShort、setString 等)必须与相应参数的 SQL 类型兼容。例如，若参数具有 SQL 类型 int，则应该使用 setInt()方法设置其值

续表

方　法	说　明
void clearParameters()	清除当前所有的参数值
void close()	关闭 PreparedStatement 对象(实例)，释放其所占用的资源

为创建一个 PreparedStatement 对象，只需调用 Connection 对象的 preparedStatement() 方法即可。如以下示例：

```
…
Connection conn=DriverManager.getConnection(url,user,password);
String sql="insert into zgb(bh,xm,xb,bm,csrq,jbgz,gwjt)
values(?,?,?,?,?,?,?)";
PreparedStatement stmt=conn.prepareStatement(sql);
stmt.setString(1,bh);
stmt.setString(2,xm);
stmt.setString(3,xb);
stmt.setString(4,bm);
stmt.setDate(5,Date.valueOf(csrq));
stmt.setFloat(6,Float.valueOf(jbgz));
stmt.setFloat(7,Float.valueOf(gwjt));
int n = stmt.executeUpdate();
…
```

在此，创建了一个 PreparedStatement 对象 stmt，并利用该对象向职工表 zgb 插入了一条职工记录。

PreparedStatement 对象表示预编译的 SQL 语句对象，相应的 SQL 语句被预编译并存储在该对象之中，因此通过该对象可高效地重复执行相应的 SQL 语句。

3.2.6　CallableStatement 接口

CallableStatement 接口继承自 PreparedStatement 接口，是 PreparedStatement 接口的扩展，用于执行数据库的存储过程。该接口的常用方法见表 3.5。

表 3.5　CallableStatement 接口的常用方法

方　法	说　明
ResultSet executeQuery()	执行查询操作，并返回一个 ResultSet 对象
int executeUpdate()	执行更新操作，并返回受影响的行数
void setXxx(int parameterIndex, Xxx x)	将第 parameterIndex 个参数的值设置为 x(参数的编号从 1 开始，Xxx 表示数据类型)
void registerOutParameter (int parameterIndex, JDBCType)	将第 parameterIndex 个参数注册为相应的 JDBC 类型(参数的编号从 1 开始，JDBCType 表示 JDBC 类型，如 java.sql.Types.TINYINT、java.sql.Types.DECIMAL 等)。该方法用于 OUT 参数与 INOUT 参数

续表

方　法	说　明
Xxx getXxx(int parameterIndex)	获取第 parameterIndex 个参数的值(参数的编号从 1 开始，Xxx 表示数据类型)。该方法用于 OUT 参数与 INOUT 参数
void clearParameters()	清除当前所有的参数值
void close()	关闭 PreparedStatement 对象(实例)，释放其所占用的资源

CallableStatement 对象为所有的 DBMS 提供了一种以标准形式调用存储过程的方法。这种调用方法使用换码语法(或转义语法)，分为带结果参数与不带结果参数两种形式(结果参数表示存储过程的返回值)，其语法格式如下：

```
{call 存储过程名[(?, ?, …)]}
{? = call 存储过程名[(?, ?, …)]}
```

其中，第一种形式为不返回结果参数的存储过程的调用语法，第二种形式为返回结果参数的存储过程的调用语法。在这两种语法格式中，方括号表示可选项，问号则作为参数的占位符使用(第一个参数的编号为 1，依次类推)。其实，存储过程的调用形式是根据其具体定义来确定的。若存储过程并无返回值，则调用时无须使用结果参数，反之则应在前面添加一个问号；若存储过程并无参数，则调用时只需指定其名称即可，反之则应在其名称后加上圆括号，并在其中添加相应数量的问号。存储过程的参数分为 3 种，即输入参数(IN 参数)、输出参数(OUT 参数)与输入输出参数(INOUT 参数)。至于结果参数，则属于输出参数。

为创建一个 CallableStatement 对象，只需调用 Connection 对象的 prepareCall()方法即可。如以下示例：

```
…
Connection conn=DriverManager.getConnection(url,user,password);
CallableStatement stmt=conn.prepareCall("{call getZgInfo(?, ? , ?)}");
…
```

在此，创建了一个 CallableStatement 对象 stmt。其中，getZgInfo 为存储过程名，其后各参数的类型取决于该存储过程的定义。

3.2.7　ResultSet 接口

ResultSet 接口表示查询返回的结果集，类似于一个数据表。该接口的常用方法见表 3.6。

表 3.6　ResultSet 接口的常用方法

方　法	说　明
void first()	将指针移到结果集第一行。若结果集为空，则返回 false，否则返回 true。若结果集类型为 TYPE_FORWORD_ONLY，将抛出异常

续表

方　法	说　明
void last()	将指针移到结果集最后一行。若结果集为空，则返回 false，否则返回 true。若结果集类型为 TYPE_FORWORD_ONLY，将抛出异常
boolean previous()	将指针移到上一行。若有上一行，则返回 true，否则返回 false。若结果集类型为 TYPE_FORWORD_ONLY，将抛出异常
boolean next()	将指针移到下一行(指针最初位于第一行之前)。若有下一行，则返回 true，否则返回 false
void beforeFirst()	将指针移到第一行之前(即结果集的开头)。若结果集类型为 TYPE_FORWORD_ONLY，将抛出异常
void afterLast()	将指针移到最后一行之后(即结果集的末尾)。若结果集类型为 TYPE_FORWORD_ONLY，将抛出异常
boolean absolute(int row)	将指针移到指定行(若 row 为正整数，则表示从前向后编号；若 row 为负整数，则表示从后向前编号)。若存在指定行，则返回 true，否则返回 false。若结果集类型为 TYPE_FORWORD_ONLY，将抛出异常
boolean relative(int row)	将指针移到相对于当前行的指定行(若 row 为正整数，则表示向后移动；若 row 为负整数，则表示向前移动；若 row 为 0，则表示当前行)。若存在指定行，则返回 true，否则返回 false。若结果集类型为 TYPE_FORWORD_ONLY，将抛出异常
int getRow()	获取当前行的行号或索引号(从 1 开始)。未处于有效行时，将返回 0
Xxx getXxx(int columnIndex)	返回当前行中指定列的值(通过索引号指定列)
Xxx getXxx(String columnName)	返回当前行中指定列的值(通过名称指定列)
void close()	关闭 ResultSet 对象(实例)，释放其所占用的资源
int findColumn(String columnName)	返回指定列的索引号。若无指定列，将抛出异常
boolean isFirst()	判断指针是否位于第一行。如果是，则返回 true，否则返回 false
boolean isLast()	判断指针是否位于最后一行。如果是，则返回 true，否则返回 false
boolean isBeforeFirst()	判断指针是否位于第一行之前(即结果集的开头)。如果是，则返回 true，否则返回 false
boolean isAfterLast()	判断指针是否位于最后一行之后(即结果集的末尾)。如果是，则返回 true，否则返回 false

ResultSet 对象(即结果集)是通过执行相应的查询语句或存储过程产生的。如以下示例：

```
…
Connection conn=DriverManager.getConnection(url,user,password);
Statement stmt=conn.createStatement();
```

```
String sql = "select * from bmb order by bmbh";
ResultSet rs = stmt.executeQuery(sql);
…
```

在此，创建了一个 ResultSet 对象 rs，内含部门表 bmb 的所有记录(按部门编号的升序排列)。

ResultSet 对象主要用于获取检索结果以及相应数据表的有关信息。每个 ResultSet 对象都具有一个指向其当前数据行的指针，且该指针最初均被置于第一行之前(即结果集的开头)。通过调用 ResultSet 对象的 next()方法，可将指针移到下一行。若无下一行，则 next()方法返回 false。因此，可在循环中使用 next()方法来对结果集进行迭代。

默认的结果集(即 ResultSet 对象)是不可更新的，且只能向前移动指针，因此只能迭代一次，且只能按从第一行到最后一行的顺序依次进行。必要时，可专门生成可滚动和/或可更新的结果集，这样，即可更加灵活地满足相关应用的需要。如以下示例：

```
…
Connection conn=DriverManager.getConnection(url,user,password);
Statement stmt= conn.createStatement(ResultSet.TYPE_SCROLL_INSENSITIVE,
ResultSet.CONCUR_UPDATABLE);
String sql = "select * from bmb order by bmbh";
ResultSet rs = stmt.executeQuery(sql);
…
```

在此，所创建的结果集(即 ResultSet 对象)rs 是可滚动、可更新的，而且不受其他更新的影响。

3.3　JDBC 基本应用

通过使用 JDBC，可轻松实现对数据库的访问，并完成有关操作。JDBC 编程的基本步骤如下。

(1) 加载驱动程序。通过调用 Class.forName(driverClassName)方法，即可加载相应的驱动程序。

(2) 建立与数据库的连接。通过调用 DriverManager.getConnection()方法，即可建立与指定数据库的连接，并获得相应的连接对象 conn。

(3) 执行 SQL 语句。首先，通过调用连接对象 conn 的 createStatement()方法生成相应的语句对象 stmt；然后，调用语句对象 stmt 的 executeQuery()或 executeUpdate()方法执行指定的查询类语句或更新类语句，并返回相应的结果。其中，executeQuery()方法返回一个代表查询结果的结果集 rs，executeUpdate()方法返回一个表示受影响记录数的整数值 n。

(4) 处理返回的结果。通过返回的结果集 rs 或整数值 n，判断相应语句的执行情况。特别地，对于返回的非空结果集 rs，可通过 while(rs.next())等循环结构进行迭代输出。

(5) 关闭与数据库的连接。通过调用连接对象 conn 的 close()方法关闭与数据库的连接。在此之前，也可先调用结果集 rs 与语句对象 stmt 的 close()方法关闭。

下面，通过一些具体实例说明 JDBC 的应用方法。在此，所要访问的数据库为在 SQL Server 2005/2008 中所创建的人事管理数据库 rsgl。该数据库中，共有 3 个表，即部门表

bmb、职工表 zgb 与用户表 users。各表的结构见表 3.7~表 3.9，各表中所包含的记录见表 3.10～表 3.12。

表 3.7　部门表 bmb 的结构

列　名	类　型	约　束	说　明
bmbh	char(2)	主键	部门编号
bmmc	varchar(20)		部门名称

表 3.8　职工表 zgb 的结构

列　名	类　型	约　束	说　明
bh	char(7)	主键	编号
xm	char(10)		姓名
xb	char(2)		性别
bm	char(2)		所在部门(编号)
csrq	datetime		出生日期
jbgz	decimal(7,2)		基本工资
gwjt	decimal(7,2)		岗位津贴

表 3.9　用户表 users 的结构

列　名	类　型	约　束	说　明
username	char(10)	主键	用户名
password	varchar(20)		用户密码
usertype	varchar(10)		用户类型

表 3.10　部门记录

部门编号	部门名称
01	计信系
02	会计系
03	经济系
04	财政系
05	金融系

表 3.11　职工记录

编　号	姓名	性别	所在部门	出生日期	基本工资(元)	岗位津贴(元)
1992001	张三	男	01	1969-06-12	1500.00	1000.00
1992002	李四	男	01	1968-12-15	1600.00	1100.00
1993001	王五	男	02	1970-01-25	1300.00	800.00
1993002	赵一	女	03	1970-03-15	1300.00	800.00
1993003	赵二	女	01	1971-01-01	1200.00	700.00

表 3.12　用户记录

用 户 名	用户密码	用户类型
abc	123	普通用户
abcabc	123	普通用户
admin	12345	系统管理员
system	12345	系统管理员

【实例 3-1】应用实例：连接 SQL Server 2005/2008 数据库 rsgl，然后再关闭与该数据库的连接(假定 SQL Server 2005/2008 安装在本地计算机上，其超级管理员登录账号 sa 的密码为“abc123!”)。

主要步骤如下。

(1) 新建一个 Web 项目 web_03。

(2) 将 SQL Server 2005/2008 的 JDBC 驱动程序 sqljdbc4.jar 添加到项目中。为此，只需先复制 sqljdbc4.jar，然后再右击项目的 WebRoot\WEB-INF\lib 文件夹，并在其快捷菜单中选择 Paste 菜单项即可。添加了驱动程序 sqljdbc4.jar 后的项目结构如图 3.1 所示。

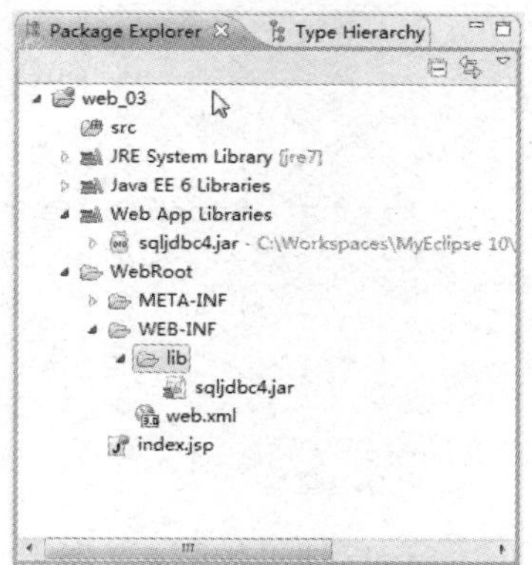

图 3.1　项目的结构

注意： 目前常用的 SQL Server JDBC 驱动程序主要有两种，即 Microsoft SQL Server 2005 JDBC Driver(sqljdbc.jar)与 Microsoft JDBC Driver 4.0 for SQL Server(sqljdbc4.jar)。其中，前者为与 JDBC 3.0 兼容的驱动程序，可提供对 SQL Server 2000 与 SQL Server 2005 数据库的可靠访问；后者为与 JDBC 4.0 兼容的驱动程序，可提供对 SQL Server 2005 与 SQL Server 2008 数据库的可靠访问。

(3) 在项目的 WebRoot 文件夹中添加一个新的 JSP 页面 ConnectDB.jsp，其代码如下：

```
<%@ page contentType="text/html;charset=GB2312" language="java" %>
<%@ page import="java.sql.*"%>
<html>
<head><title>SQL Server</title></head>
<body>
<%
    Connection conn=null;
    try{
        Class.forName("com.microsoft.sqlserver.jdbc.SQLServerDriver");
        String url="jdbc:sqlserver://localhost:1433;DatabaseName=rsgl";
       //String  url="jdbc:sqlserver://127.0.0.1:1433;DatabaseName=rsgl";
        String user="sa";
        String password="abc123!";
        conn=DriverManager.getConnection(url,user,password);
```

```
        out.println("数据库连接成功!<br>");
    }
    catch(ClassNotFoundException e){
        out.println("!"+e.getMessage());
    }
    catch(SQLException e){
        out.println(e.getMessage());
    }
    finally{
        try{
            if (conn!=null){
                conn.close();
                out.println("数据库连接关闭成功!<br>");
            }
        }
        catch(Exception e){
            out.println(e.getMessage());
        }
    }
 %>
</body>
</html>
```

运行结果如图 3.2 所示。

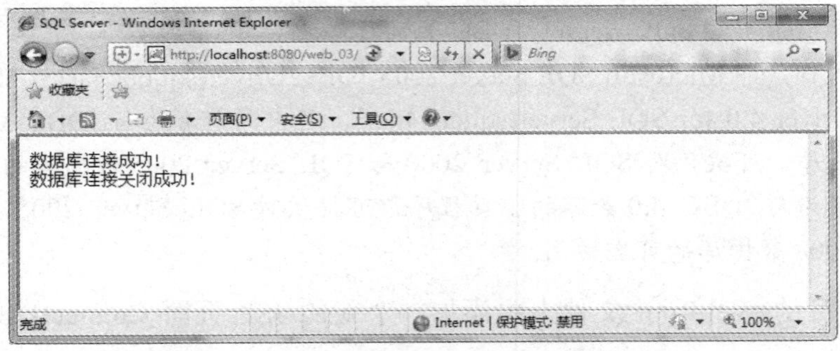

图 3.2 页面 ConnectDB.jsp 的运行结果

解析：

(1) 在 ConnectDB.jsp 页面中，先调用 Class.forName()显式地加载指定的驱动程序类，然后再调用 DriverManager.getConnection()方法建立与指定数据库的连接，并获取相应的连接对象 conn，最后再调用 conn 的 close()方法关闭与数据库的连接。

(2) 在 ConnectDB.jsp 页面中，为妥善处理可能出现的各种异常，使用了相应的异常处理语句。

【实例 3-2】应用实例：部门增加(增加一个部门，其编号为“10”，名称为“学生办”)。

主要步骤如下。

在项目 web_03 的 WebRoot 文件夹中添加一个新的 JSP 页面 BmZj.jsp，其代码如下：

```
<%@ page contentType="text/html;charset=GB2312" language="java" %>
<%@ page import="java.sql.*"%>
<%request.setCharacterEncoding("gb2312"); %>
<html>
<head><title></title></head>
<body>
<%
    try {
        Class.forName("com.microsoft.sqlserver.jdbc.SQLServerDriver");
        String url="jdbc:sqlserver://localhost:1433;DatabaseName=rsgl";
        String user="sa";
        String password="abc123!";
        Connection conn=DriverManager.getConnection(url,user,password);
        String sql = "insert into bmb values('10','学生办')";
       Statement stmt=conn.createStatement();
       int n = stmt.executeUpdate(sql);
       if (n==1)
            out.print("部门增加成功!<br>");
        else
            out.print("部门增加失败!<br>");
       stmt.close();
        conn.close();
    }
    catch (Exception e) {
        out.print(e.toString());
    }
%>
</body>
</html>
```

运行结果如图 3.3 所示。

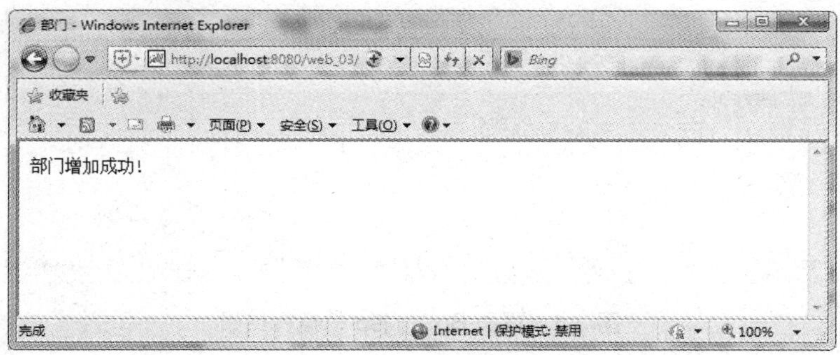

图 3.3　页面 BmZj.jsp 的运行结果

【实例 3-3】应用实例：部门修改(将编号为“10”的部门的名称修改为“学工部”)。主要步骤如下。

在项目 web_03 的 WebRoot 文件夹中添加一个新的 JSP 页面 BmXg.jsp，其代码如下：

```
<%@ page contentType="text/html;charset=GB2312" language="java" %>
<%@ page import="java.sql.*"%>
```

```
<%request.setCharacterEncoding("gb2312"); %>
<html>
<head><title>部门</title></head>
<body>
<%
    try {
        Class.forName("com.microsoft.sqlserver.jdbc.SQLServerDriver");
        String url="jdbc:sqlserver://localhost:1433;DatabaseName=rsgl";
        String user="sa";
        String password="abc123!";
        Connection conn=DriverManager.getConnection(url,user,password);
        String sql = "update bmb set bmmc='学工部' where bmbh='10'";
       Statement stmt=conn.createStatement();
       int n = stmt.executeUpdate(sql);
       if (n==1)
            out.print("部门修改成功!<br>");
        else
            out.print("部门修改失败!<br>");
       stmt.close();
        conn.close();
    }
    catch (Exception e) {
        out.print(e.toString());
    }
%>
</body>
</html>
```

运行结果如图 3.4 所示。

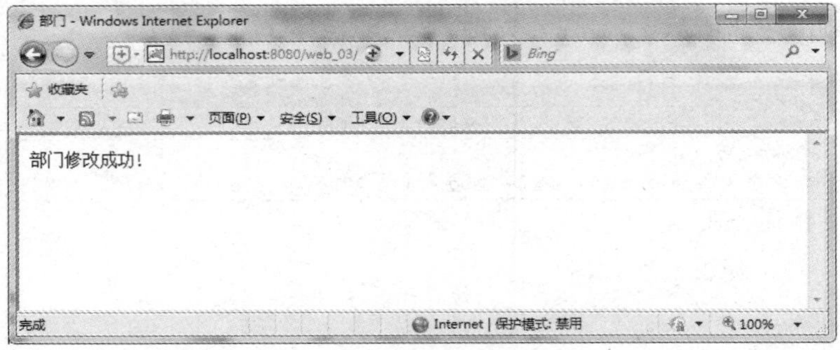

图 3.4　页面 BmXg.jsp 的运行结果

【实例 3-4】应用实例：部门删除(将编号为“10”的部门删除掉)。

主要步骤如下。

在项目 web_03 的 WebRoot 文件夹中添加一个新的 JSP 页面 BmSc.jsp，其代码如下：

```
<%@ page contentType="text/html;charset=GB2312" language="java" %>
<%@ page import="java.sql.*"%>
<%request.setCharacterEncoding("gb2312"); %>
<html>
```

```
<head><title>部门</title></head>
<body>
<%
    try {
        Class.forName("com.microsoft.sqlserver.jdbc.SQLServerDriver");
        String url="jdbc:sqlserver://localhost:1433;DatabaseName=rsgl";
        String user="sa";
        String password="abc123!";
        Connection conn=DriverManager.getConnection(url,user,password);
        String sql = "delete from bmb where bmbh='10'";
       Statement stmt=conn.createStatement();
       int n = stmt.executeUpdate(sql);
       if (n==1)
            out.print("部门删除成功!<br>");
        else
            out.print("部门删除失败!<br>");
       stmt.close();
        conn.close();
    }
    catch (Exception e) {
        out.print(e.toString());
    }
%>
</body>
</html>
```

运行结果如图 3.5 所示。

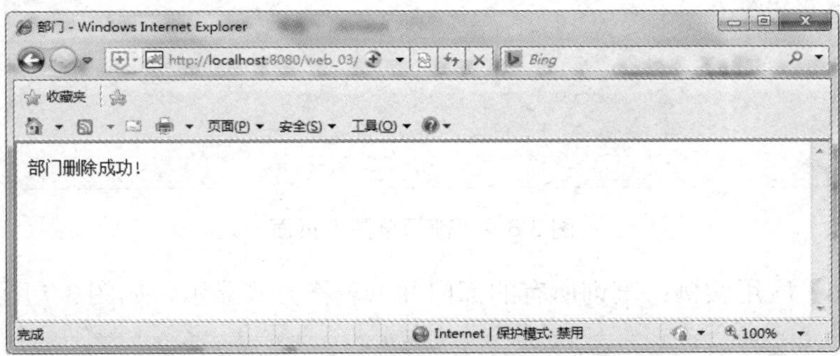

图 3.5　页面 BmSc.jsp 的运行结果

【实例 3-5】应用实例：查询并显示部门的总数。

主要步骤如下。

在项目 web_03 的 WebRoot 文件夹中添加一个新的 JSP 页面 BmZs.jsp，其代码如下：

```
<%@ page contentType="text/html;charset=GB2312" language="java" %>
<%@ page import="java.sql.*"%>
<%request.setCharacterEncoding("gb2312"); %>
<html>
<head><title>部门总数</title></head>
<body>
<%
```

```
    try {
        Class.forName("com.microsoft.sqlserver.jdbc.SQLServerDriver");
        String url="jdbc:sqlserver://localhost:1433;DatabaseName=rsgl";
        String user="sa";
        String password="abc123!";
        Connection conn=DriverManager.getConnection(url,user,password);
        String sql = "select count(*) from bmb";
       Statement stmt=conn.createStatement();
       ResultSet rs = stmt.executeQuery(sql);
       rs.next();
       int n=rs.getInt(1);
       out.print("部门总数："+n);
       rs.close();
       stmt.close();
        conn.close();
    }
    catch (Exception e) {
        out.print(e.toString());
    }
%>
</body>
</html>
```

运行结果如图 3.6 所示。

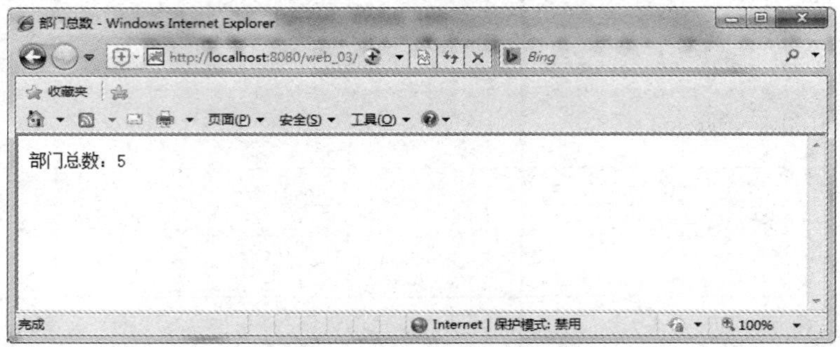

图 3.6 “部门总数”页面

【**实例 3-6**】应用实例：查询所有的部门并以表格方式显示，如图 3.7 所示。

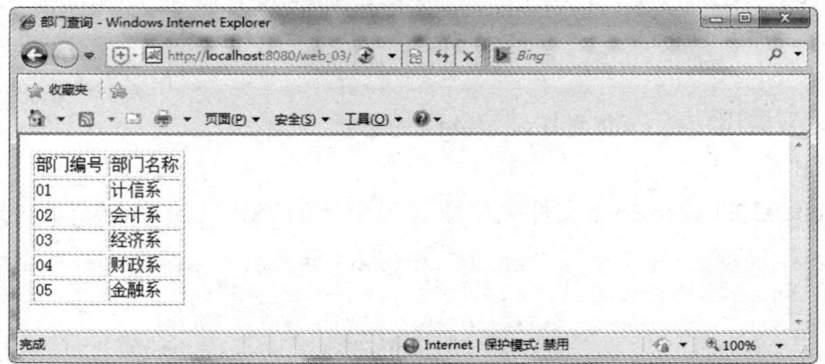

图 3.7 “部门查询”页面

主要步骤如下。

在项目 web_03 的 WebRoot 文件夹中添加一个新的 JSP 页面 BmCx.jsp，其代码如下：

```
<%@ page contentType="text/html;charset=GB2312" language="java" %>
<%@ page import="java.sql.*"%>
<%request.setCharacterEncoding("gb2312"); %>
<html>
<head><title>部门查询</title></head>
<body>
<%
    try {
        Class.forName("com.microsoft.sqlserver.jdbc.SQLServerDriver");
        String url="jdbc:sqlserver://localhost:1433;DatabaseName=rsgl";
        String user="sa";
        String password="abc123!";
        Connection conn=DriverManager.getConnection(url,user,password);
        String sql = "select * from bmb order by bmbh";
       Statement stmt=conn.createStatement();
       ResultSet rs = stmt.executeQuery(sql);
%>
    <table border="1">
    <tr><td>部门编号</td><td>部门名称</td></tr>
<%
       while(rs.next())
       {
        String bmbh=rs.getString("bmbh");
        String bmmc=rs.getString("bmmc");
%>
    <tr><td><%=bmbh%></td><td><%=bmmc%></td></tr>
<%
       }
%>
    </table>
<%
       rs.close();
       stmt.close();
        conn.close();
    }
    catch (Exception e) {
        out.print(e.toString());
    }
%>
</body>
</html>
```

【实例 3-7】应用实例：图 3.8 所示为“部门选择”页面。通过部门下拉列表框选择部门后，再单击“确定”按钮，即可显示出所选部门的编号，如图 3.9 所示。

图 3.8 “部门选择”页面

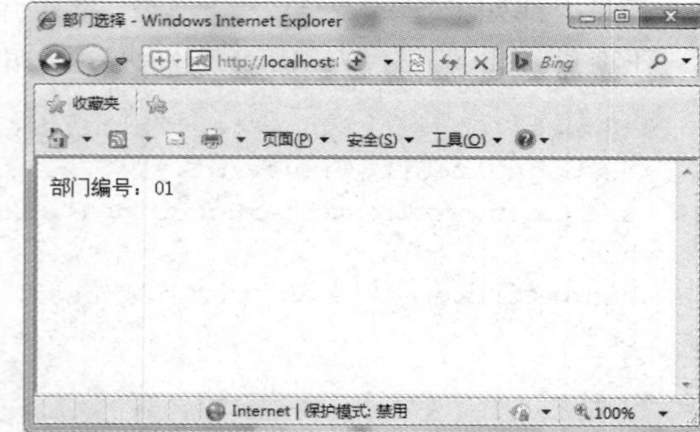

图 3.9 部门选择结果页面

主要步骤(在项目 web_03 中)如下。

(1) 在项目的 WebRoot 文件夹中添加一个新的 JSP 页面 BmXz.jsp，其代码如下：

```
<%@ page contentType="text/html;charset=GB2312" language="java" %>
<%@ page import="java.sql.*"%>
<%request.setCharacterEncoding("gb2312"); %>
<html>
<head><title>部门选择</title></head>
<body>
<%
    try {
        Class.forName("com.microsoft.sqlserver.jdbc.SQLServerDriver");
        String url="jdbc:sqlserver://localhost:1433;DatabaseName=rsgl";
        String user="sa";
        String password="abc123!";
        Connection conn=DriverManager.getConnection(url,user,password);
        String sql = "select * from bmb order by bmbh";
       Statement stmt=conn.createStatement();
       ResultSet rs = stmt.executeQuery(sql);
%>
    <form method="post" action="BmXz0.jsp">
    部门：
    <select name="bmbh">
<%
       while(rs.next())
       {
        String bmbh=rs.getString("bmbh");
        String bmmc=rs.getString("bmmc");
%>
    <option value="<%=bmbh%>"><%=bmmc%></option>
<%
       }
%>
    </select>
    <input name="Submit" type="submit" value="确定" />
    </form>
<%
```

```
        rs.close();
        stmt.close();
         conn.close();
    }
    catch (Exception e) {
        out.print(e.toString());
    }
%>
</body>
</html>
```

(2) 在项目的 WebRoot 文件夹中添加一个新的 JSP 页面 BmXz0.jsp，其代码如下：

```
<%@ page contentType="text/html;charset=GB2312" language="java" %>
<%@ page import="java.sql.*"%>
<%request.setCharacterEncoding("gb2312"); %>
<html>
<head><title>部门选择</title></head>
<body>
    <%
    String bmbh=request.getParameter("bmbh");
    %>
    部门编号：<%=bmbh%>
</body>
</html>
```

解析：

(1) 在本实例中，BmXz.jsp 页面其实是一个表单(其处理页面为 BmXz0.jsp)，表单内包含有一个部门下拉列表框。该部门下拉列表框中的各个选项是根据部门表动态生成的，每个选项的显示文本为部门的名称，而相应的取值则为部门的编号。

(2) 在 BmXz0.jsp 页面中，通过调用 request 对象的 getParameter()方法获取提交表单后传递过来的部门编号，然后显示。

3.4　JDBC 应用案例

3.4.1　系统登录

下面，以 JSP+JDBC 模式实现 Web 应用系统中的系统登录功能(基于 SQL Server 2005/2008 数据库 rsgl 中的用户表 users)。

图 3.10 所示为“系统登录”页面。在此页面输入用户名与密码后，再单击“登录”按钮，即可提交至服务器对用户的身份进行验证。若该用户为系统的合法用户，则跳转至“登录成功”页面(见图 3.11)；反之，则跳转至“登录失败”页面(见图 3.12)。

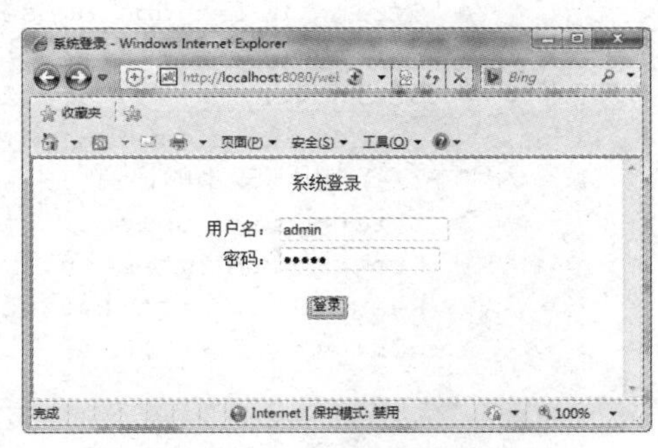

图 3.10　“系统登录”页面

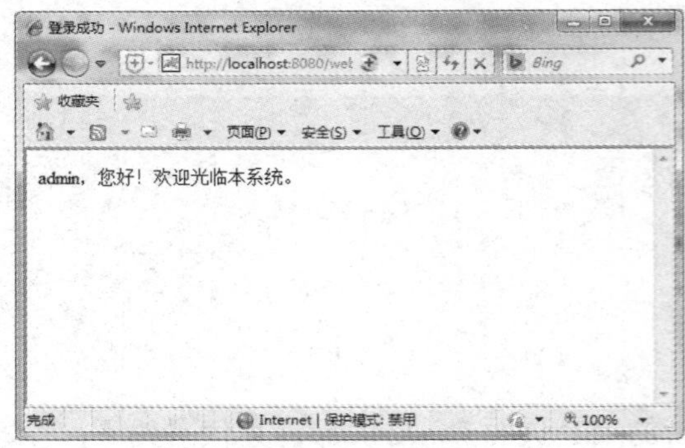

图 3.11 “登录成功”页面

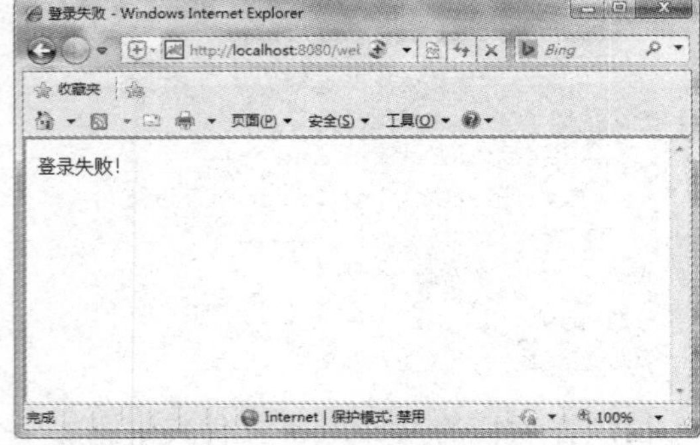

图 3.12 “登录失败”页面

实现步骤(在项目 web_03 中)如下。

(1) 在项目 web_03 的 WebRoot 文件夹中新建一个子文件夹 syslogin。

(2) 在子文件夹 syslogin 中添加一个新的 JSP 页面 login.jsp。其代码与第 2 章 2.4.1 小节系统登录案例中的相同。

(3) 在子文件夹 syslogin 中添加一个新的 JSP 页面 validate.jsp，其代码如下：

```
<%@ page language="java" pageEncoding="utf-8" import="java.sql.*" %>
<%request.setCharacterEncoding("utf-8"); %>
<html>
    <head>
        <title>登录验证</title>
        <meta http-equiv="Content-Type" content="text/html;charset=utf-
8">
    </head>
    <body>
        <%
        String username0=request.getParameter("username"); //获取提交的姓名
        String password0=request.getParameter("password"); //获取提交的密码
        boolean validated=false;  //验证标识

    Class.forName("com.microsoft.sqlserver.jdbc.SQLServerDriver").newIns
tance();
        String url="jdbc:sqlserver://localhost:1433;databaseName=rsgl";
        String user="sa";
        String password="abc123!";
       Connection conn=DriverManager.getConnection(url, user, password);
        //查询用户表中的记录
        String sql="select * from users";
        Statement stmt=conn.createStatement();
        ResultSet rs=stmt.executeQuery(sql);  //获取结果集
        while(rs.next())
        {

    if((rs.getString("username").trim().compareTo(username0)==0)&&
              (rs.getString("password").trim().compareTo(password0)==0))
            {
```

```
                validated=true;
            }
        }
        rs.close();
        stmt.close();
        conn.close();
        if(validated)  //验证成功跳转到成功页面
        {
        %>
            <jsp:forward page="welcome.jsp"></jsp:forward>
        <%
        }
        else  //验证失败跳转到失败页面
        {
        %>
            <jsp:forward page="error.jsp"></jsp:forward>
        <%
        }
        %>
    </body>
</html>
```

(4) 在子文件夹 syslogin 中添加一个新的 JSP 页面 welcome.jsp。其代码与第 2 章 2.4.1 小节系统登录案例中的相同。

(5) 在子文件夹 syslogin 中添加一个新的 JSP 页面 error.jsp。其代码其代码与第 2 章 2.4.1 小节系统登录案例中的相同。

3.4.2 数据添加

Web 应用系统的一大功能就是数据的添加(或增加)。在此，以职工增加为例进行简要说明，并通过 JSP+JDBC 模式加以实现(基于 SQL Server 2005/2008 数据库 rsgl 中的职工表 zgb 与部门表 bmb)。

图 3.13 所示为“职工增加”页面。在其中的表单输入职工的各项信息后，再单击“确定”按钮提交表单，即可将该职工记录添加到职工表中。

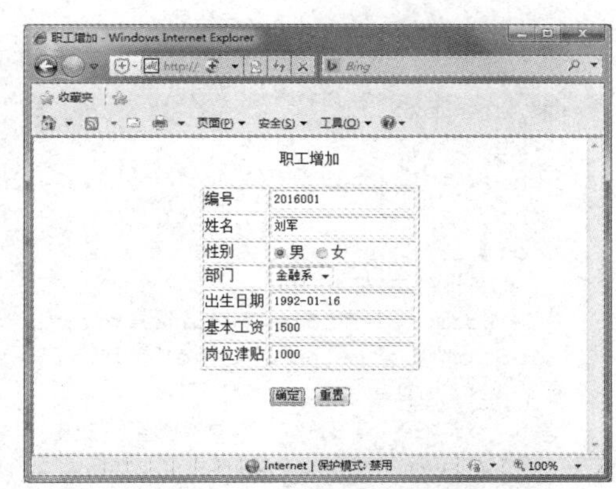

图 3.13 “职工增加”页面

主要步骤(在项目 web_03 中)如下。

(1) 在项目的 WebRoot 文件夹中添加一个新的 JSP 页面 ZgZj.jsp，其代码如下：

```
<%@ page contentType="text/html;charset=gb2312" language="java" %>
<%@ page import="java.sql.*"%>
<%request.setCharacterEncoding("gb2312"); %>
<script language="JavaScript">
function check(theForm)
{
```

```
  if (theForm.bh.value.length != 7)
  {
    alert("职工编号必须为 7 位！");
    theForm.bh.focus();
    return (false);
  }
  if (theForm.xm.value == "")
  {
    alert("请输入姓名！");
    theForm.xm.focus();
    return (false);
  }
  if (theForm.csrq.value == "")
  {
    alert("请输入出生日期！");
    theForm.csrq.focus();
    return (false);
  }
  if (theForm.jbgz.value == "")
  {
    alert("请输入基本工资！");
    theForm.jbgz.focus();
    return (false);
  }
  if (theForm.gwjt.value == "")
  {
    alert("请输入岗位津贴！");
    theForm.gwjt.focus();
    return (false);
  }
  return (true);
}
</script>
<html>
<head><title>职工增加</title></head>
<body>
<div align="center">
  <P>职工增加</P>
  <form id="form1" name="form1" method="post" action="ZgZj0.jsp"
onSubmit="return check(this)">
    <table border="1">
    <tr><td>编号</td><td><input name="bh" type="text" id="bh"
/></td></tr>
    <tr><td>姓名</td><td><input name="xm" type="text" id="xm"
/></td></tr>
    <tr><td>性别</td><td>
    <input type="radio" name="xb" value="男" checked="checked" />男
    <input type="radio" name="xb" value="女" />女
    </td></tr>
<%
        Class.forName("com.microsoft.sqlserver.jdbc.SQLServerDriver");
        String url="jdbc:sqlserver://localhost:1433;DatabaseName=rsgl";
        String user="sa";
        String password="abc123!";
        Connection conn=DriverManager.getConnection(url,user,password);
```

```
        String sql = "select * from bmb order by bmbh";
       Statement stmt=conn.createStatement();
       ResultSet rs = stmt.executeQuery(sql);
%>
    <tr><td>部门</td>
    <td><select name="bm">
<%
       while(rs.next())
       {
        String bmbh=rs.getString("bmbh");
        String bmmc=rs.getString("bmmc");
%>
    <option value="<%=bmbh%>"><%=bmmc%></option>
<%
       }
       rs.close();
       stmt.close();
        conn.close();
%>
    </select>
    </td></tr>
    <tr><td>出生日期</td><td><input name="csrq" type="text" id="csrq"
/></td></tr>
    <tr><td>基本工资</td><td><input name="jbgz" type="text" id="jbgz"
/></td></tr>
    <tr><td>岗位津贴</td><td><input name="gwjt" type="text" id="gwjt"
/></td></tr>
    </table>
    <br>
    <input name="submit" type="submit"  value="确定" />
    <input name="reset" type="reset" value="重置" />
  </form>
</div>
</body>
</html>
```

(2) 在项目的 WebRoot 文件夹中添加一个新的 JSP 页面 ZgZj0.jsp，其代码如下：

```
<%@ page contentType="text/html;charset=gb2312" language="java" %>
<%@ page import="java.sql.*"%>
<%request.setCharacterEncoding("gb2312"); %>
<html>
<head><title></title></head>
<body>
<%
    String bh = request.getParameter("bh");
    String xm = request.getParameter("xm");
    String xb = request.getParameter("xb");
    String bm = request.getParameter("bm");
    String csrq = request.getParameter("csrq");
    String jbgz = request.getParameter("jbgz");
    String gwjt = request.getParameter("gwjt");
    try {
        Class.forName("com.microsoft.sqlserver.jdbc.SQLServerDriver");
        String url="jdbc:sqlserver://localhost:1433;DatabaseName=rsgl";
```

```
        String user="sa";
        String password="abc123!";
        Connection conn=DriverManager.getConnection(url,user,password);
        String sql = "insert into zgb(bh,xm,xb,bm,csrq,jbgz,gwjt)";
       sql=sql+"  values('"+bh+"','"+xm+"','"+xb+"','"+bm+"','"+csrq+"'";
        sql=sql+","+jbgz+","+gwjt+")";
       Statement stmt=conn.createStatement();
       int n = stmt.executeUpdate(sql);
       if (n>0){
           out.print("职工记录增加成功! ");
        }
        else {
           out.print("职工记录增加失败! ");
        }
       stmt.close();
        conn.close();
    }
    catch (Exception e) {
        out.print(e.toString());
    }
%>
</body>
</html>
```

解析：

(1) 在 ZgZj.jsp 页面中，对于在表单中输入的职工信息，提交后先利用 JavaScript 脚本在客户端进行相应的验证。

(2) 在 ZgZj0.jsp 页面中，先获取通过表单提交的各项职工信息，然后再构造 Insert 语句并通过所创建的 Statement 对象执行，从而实现职工记录的增加。其实，职工记录的增加也可通过 PreparedStatement 对象实现，只需将页面中“String sql = "insert into zgb(bh,xm,xb,bm,csrq,jbgz,gwjt)";”与“int n = stmt.executeUpdate(sql);”之间的语句改写为以下语句即可。

```
    String sql="insert into zgb(bh,xm,xb,bm,csrq,jbgz,gwjt)
values(?,?,?,?,?,?,?)";
    PreparedStatement stmt=conn.prepareStatement(sql);
    stmt.setString(1,bh);
    stmt.setString(2,xm);
    stmt.setString(3,xb);
    stmt.setString(4,bm);
    stmt.setDate(5,Date.valueOf(csrq));  //
    stmt.setFloat(6,Float.valueOf(jbgz));
    stmt.setFloat(7,Float.valueOf(gwjt));
    int n = stmt.executeUpdate();
```

3.4.3 数据维护

Web 应用系统的另外一大功能就是数据的维护，包括数据的查询、修改与删除。在此，以职工维护为例进行相应说明，并通过 JSP+JDBC 模式加以实现(基于 SQL Server 2005/2008 数据库 rsgl 中的职工表 zgb 与部门表 bmb)。

图 3.14 所示为“职工查询”页面。在其中的“部门”下拉列表框中选择某个部门后，再单击“确定”按钮，即可打开图 3.15 所示的“职工维护”页面，以分页形式显示相应部门的职工列表，并为每个职工记录提供 3 个操作链接，即“详情”链接、“修改”链接与“删除”链接。若单击某个“详情”链接，则打开图 3.16 所示的“职工信息”页面，以表格形式显示相应职工的详细信息。若单击某个“修改”链接，则打开图 3.17 所示的“职工修改”页面，以表格形式显示相应职工的各项信息，并允许用户进行相应的修改，单击“确定”按钮后再将所做的修改更新到职工表中。若单击某个“删除”链接，则打开图 3.18 所示的“职工删除”页面，以表格形式显示相应职工的各项信息，单击“确定”按钮后再将其从职工表中删除。

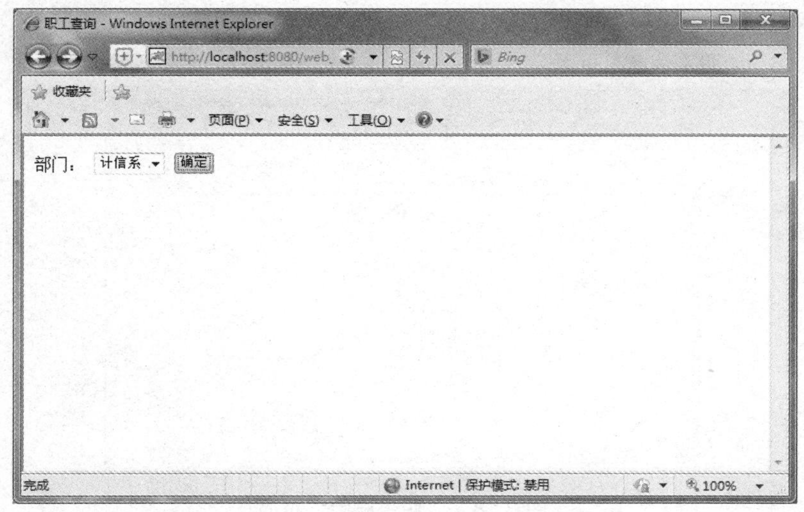

图 3.14　“职工查询”页面

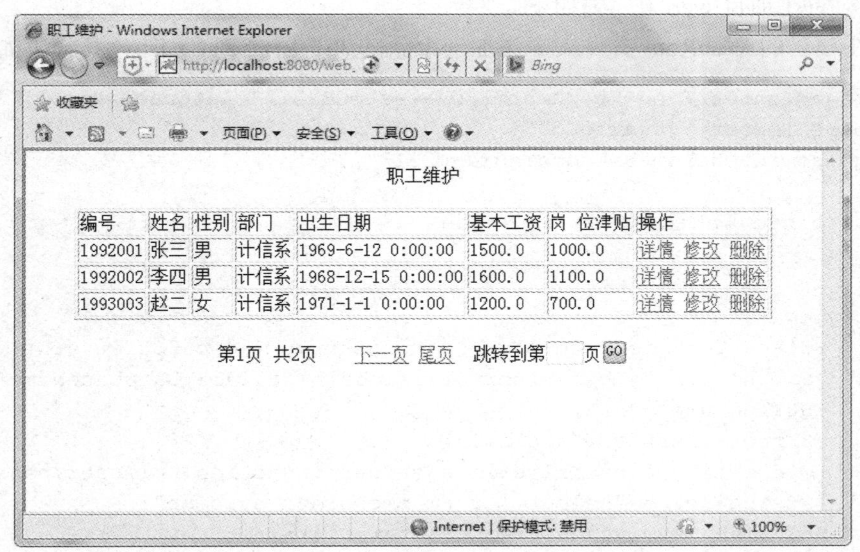

图 3.15　“职工维护”页面

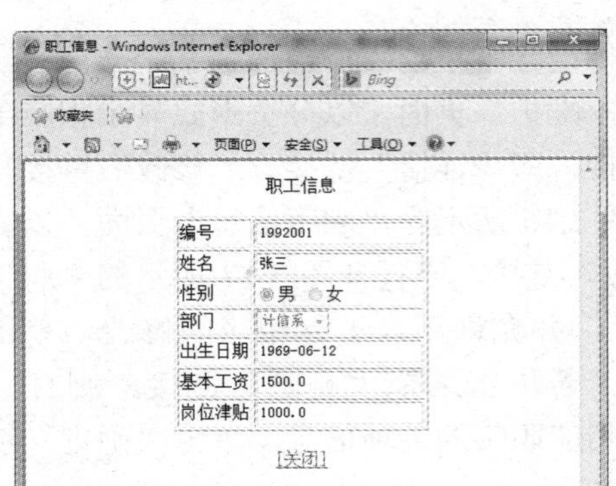

图 3.16 “职工信息”页面

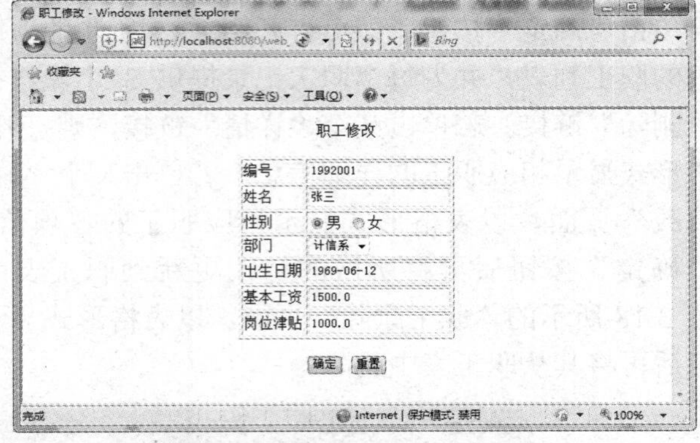

图 3.17 “职工修改”页面

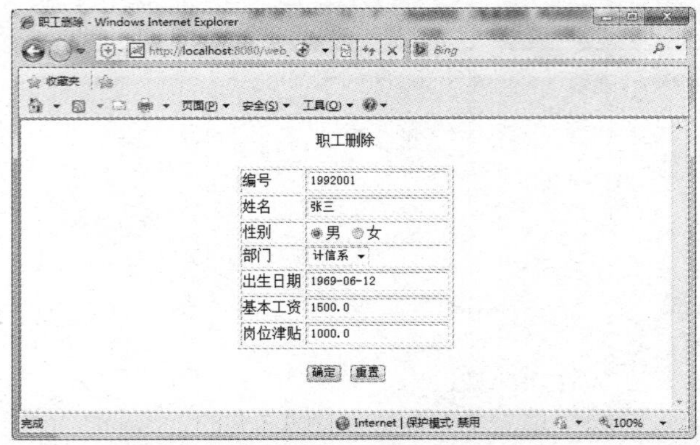

图 3.18 “职工删除”页面

主要步骤(在项目 web_03 中)如下。

(1) 在项目的 WebRoot 文件夹中添加一个新的 JSP 页面 ZgCx.jsp，其代码如下：

```
<%@ page contentType="text/html;charset=GB2312" language="java" %>
<%@ page import="java.sql.*"%>
<%request.setCharacterEncoding("gb2312"); %>
<html>
<head><title>职工查询</title></head>
<body>
<%
    try {
        Class.forName("com.microsoft.sqlserver.jdbc.SQLServerDriver");
        String url="jdbc:sqlserver://localhost:1433;DatabaseName=rsgl";
        String user="sa";
        String password="abc123!";
        Connection conn=DriverManager.getConnection(url,user,password);
        String sql = "select * from bmb order by bmbh";
       Statement stmt=conn.createStatement();
       ResultSet rs = stmt.executeQuery(sql);
%>
```

```
    <form method="post" action="ZgWh.jsp">
    部门：
    <select name="bmbh">
<%
       while(rs.next())
       {
        String bmbh=rs.getString("bmbh");
        String bmmc=rs.getString("bmmc");
%>
    <option value="<%=bmbh%>"><%=bmmc%></option>
<%
       }
%>
    </select>
    <input name="Submit" type="submit" value="确定" />
    </form>
<%
       rs.close();
       stmt.close();
        conn.close();
    }
    catch (Exception e) {
        out.print(e.toString());
    }
%>
</body>
</html>
```

(2)　在项目的 WebRoot 文件夹中添加一个新的 JSP 页面 ZgWh.jsp，其代码如下：

```
<%@ page contentType="text/html;charset=GB2312" language="java" %>
<%@ page import="java.sql.*"%>
<%request.setCharacterEncoding("gb2312"); %>
<html>
<head><title>职工维护</title></head>
<body>
<div align="center">
    <P>职工维护</P>
<%
    String bmbh=request.getParameter("bmbh");
    String pageNo=request.getParameter("pageno");
    int pageSize=3;
    int pageCount;
    int rowCount;
    int pageCurrent;
    int rowCurrent;
    if(pageNo==null||pageNo.trim().length()==0){
        pageCurrent=1;
    }else{
        pageCurrent=Integer.parseInt(pageNo);
    }
```

```
    try {
         Class.forName("com.microsoft.sqlserver.jdbc.SQLServerDriver");
         String url="jdbc:sqlserver://localhost:1433;DatabaseName=rsgl";
         String user="sa";
         String password="abc123!";
         Connection conn=DriverManager.getConnection(url,user,password);
         String sql = "select bh,xm,xb,bmmc,csrq,jbgz,gwjt from zgb,bmb";
         sql=sql+" where zgb.bm=bmb.bmbh and bm='"+bmbh+"'";
         sql=sql+" order by bh";
        Statement
stmt=conn.createStatement(ResultSet.TYPE_SCROLL_INSENSITIVE,ResultSet.CO
NCUR_READ_ONLY);
        ResultSet rs = stmt.executeQuery(sql);
         rs.last();
        rowCount = rs.getRow();
        pageCount = (rowCount + pageSize - 1)/pageSize;
        if(pageCurrent>pageCount)
            pageCurrent=pageCount;
        if(pageCurrent<1)
            pageCurrent=1;
%>
    <table border="1">
    <tr><td>编号</td><td>姓名</td><td>性别</td><td>部门</td>
    <td>出生日期</td><td>基本工资</td><td>岗位津贴</td>
    <td>操作</td></tr>
<%
    rs.beforeFirst();
    rowCurrent=1;
    while(rs.next()){
         if(rowCurrent>(pageCurrent-1)*pageSize &&
rowCurrent<=pageCurrent*pageSize){
         String bh=rs.getString("bh");
         String xm=rs.getString("xm");
         String xb=rs.getString("xb");
         String bmmc=rs.getString("bmmc");
         String csrq=rs.getDate("csrq").toLocaleString();
         String jbgz=String.valueOf(rs.getFloat("jbgz"));
         String gwjt=String.valueOf(rs.getFloat("gwjt"));
%>
    <tr><td><%=bh%></td><td><%=xm%></td><td><%=xb%></td><td><%=bmmc%></t
d>
    <td><%=csrq%></td><td><%=jbgz%></td><td><%=gwjt%></td>
    <td><a href="ZgXq.jsp?bh=<%=bh%>" target="_blank">详情</a> <a
href="ZgXg.jsp?bh=<%=bh%>">修改</a> <a href="ZgSc.jsp?bh=<%=bh%>">删除
</a></td></tr>
<%
         }
        rowCurrent++;
    }
 %>
```

```
    </table>
        <p align="center">
        <form method="POST" action="ZgWh.jsp">
            第<%=pageCurrent %>页 共<%=pageCount %>页 
            <%if(pageCurrent>1){ %>
            <a href="ZgWh.jsp?bmbh=<%=bmbh %>&pageno=1">首页</a>
            <a href="ZgWh.jsp?bmbh=<%=bmbh %>&pageno=<%=pageCurrent-
1 %>">上一页</a>
            <%} %>
             
            <%if(pageCurrent<pageCount){ %>
            <a
href="ZgWh.jsp?bmbh=<%=bmbh %>&pageno=<%=pageCurrent+1 %>">下一页</a>
            <a href="ZgWh.jsp?bmbh=<%=bmbh %>&pageno=<%=pageCount %>">尾
页</a>
            <%} %>
             跳转到第<input type="text" name="pageno" size="3"
maxlength="5">页<input name="submit" type="submit" value="GO">
            <input name="bmbh" type="hidden" value="<%=bmbh %>">
        </form>
<%
      rs.close();
      stmt.close();
       conn.close();
    }
    catch(ClassNotFoundException e)
    {
       out.println(e.getMessage());
    }
    catch(SQLException e)
    {
       out.println(e.getMessage());
    }
    catch (Exception e) {
       out.print(e.toString());
    }
%>
</div>
</body>
</html>
```

(3) 在项目的 WebRoot 文件夹中添加一个新的 JSP 页面 ZgXq.jsp，其代码如下：

```
<%@ page contentType="text/html;charset=gb2312" language="java" %>
<%@ page import="java.sql.*"%>
<%@ page import="java.text.*"%>
<%request.setCharacterEncoding("gb2312"); %>
<html>
<head><title>职工信息</title></head>
<body>
<div align="center">
```

```
 <P>职工信息</P>
<%
       Class.forName("com.microsoft.sqlserver.jdbc.SQLServerDriver");
       String url="jdbc:sqlserver://localhost:1433;DatabaseName=rsgl";
       String user="sa";
       String password="abc123!";
       Connection conn=DriverManager.getConnection(url,user,password);
       String bh0=request.getParameter("bh").trim();
       String sql0 = "select * from zgb where bh='"+bh0+"'";
      Statement stmt0=conn.createStatement();
      ResultSet rs0 = stmt0.executeQuery(sql0);
      rs0.next();
      String xm0=rs0.getString("xm").trim();
      String xb0=rs0.getString("xb").trim();
      String bm0=rs0.getString("bm").trim();
      SimpleDateFormat sdf=new SimpleDateFormat("yyyy-MM-dd");
      String csrq0=sdf.format(rs0.getDate("csrq"));
      String jbgz0=String.valueOf(rs0.getFloat("jbgz"));
      String gwjt0=String.valueOf(rs0.getFloat("gwjt"));
      rs0.close();
      stmt0.close();
%>
<form id="form1" name="form1" method="post" action="">
    <table border="1">
    <tr><td>编号</td><td><input name="bh" type="text" id="bh"
value="<%=bh0%>" readonly="true" /></td></tr>
    <tr><td>姓名</td><td><input name="xm" type="text" id="xm"
value="<%=xm0%>" /></td></tr>
    <tr><td>性别</td><td>
    <input type="radio" name="xb" value="男" <% if (xb0.equals("男")){ %>
checked="checked" <% } %> disabled="disabled"/>男
    <input type="radio" name="xb" value="女" <% if (xb0.equals("女")){ %>
checked="checked" <% } %> disabled="disabled"/>女
    </td></tr>
<%
       String sql = "select * from bmb order by bmbh";
      Statement stmt=conn.createStatement();
      ResultSet rs = stmt.executeQuery(sql);
%>
    <tr><td>部门</td>
    <td><select name="bm" disabled="disabled">
<%
      while(rs.next())
      {
       String bmbh=rs.getString("bmbh").trim();
       String bmmc=rs.getString("bmmc").trim();
%>
    <option value="<%=bmbh%>" <% if (bm0.equals(bmbh)){ %> selected
<% } %>><%=bmmc%></option>
<%
```

```
        }
        rs.close();
        stmt.close();
         conn.close();
%>
    </select>
    </td></tr>
    <tr><td>出生日期</td><td><input name="csrq" type="text" id="csrq"
value="<%=csrq0%>" /></td></tr>
    <tr><td>基本工资</td><td><input name="jbgz" type="text" id="jbgz"
value="<%=jbgz0%>" /></td></tr>
    <tr><td>岗位津贴</td><td><input name="gwjt" type="text" id="gwjt"
value="<%=gwjt0%>" /></td></tr>
    </table>
    <br>
    <a href="javascript:window.close()" >[关闭]</a>
</form>
</div>
</body>
</html>
```

(4) 在项目的 WebRoot 文件夹中添加一个新的 JSP 页面 ZgXg.jsp，其代码如下：

```
<%@ page contentType="text/html;charset=gb2312" language="java" %>
<%@ page import="java.sql.*"%>
<%@ page import="java.text.*"%>
<%request.setCharacterEncoding("gb2312"); %>
<script language="JavaScript">
function check(theForm)
{
  if (theForm.bh.value.length != 7)
  {
   alert("职工编号必须为 7 位！");
   theForm.bh.focus();
   return (false);
  }
  if (theForm.xm.value == "")
  {
   alert("请输入姓名！");
   theForm.xm.focus();
   return (false);
  }
  if (theForm.csrq.value == "")
  {
   alert("请输入出生日期！");
   theForm.csrq.focus();
   return (false);
  }
  if (theForm.jbgz.value == "")
  {
   alert("请输入基本工资！");
```

```
    theForm.jbgz.focus();
    return (false);
  }
  if (theForm.gwjt.value == "")
  {
    alert("请输入岗位津贴！");
    theForm.gwjt.focus();
    return (false);
  }
  return (true);
}
</script>
<html>
<head><title>职工修改</title></head>
<body>
<div align="center">
  <P>职工修改</P>
<%
        Class.forName("com.microsoft.sqlserver.jdbc.SQLServerDriver");
        String url="jdbc:sqlserver://localhost:1433;DatabaseName=rsgl";
        String user="sa";
        String password="abc123!";
        Connection conn=DriverManager.getConnection(url,user,password);
        String bh0=request.getParameter("bh").trim();
        String sql0 = "select * from zgb where bh='"+bh0+"'";
       Statement stmt0=conn.createStatement();
       ResultSet rs0 = stmt0.executeQuery(sql0);
       rs0.next();
       String xm0=rs0.getString("xm").trim();
       String xb0=rs0.getString("xb").trim();
       String bm0=rs0.getString("bm").trim();
       SimpleDateFormat sdf=new SimpleDateFormat("yyyy-MM-dd");
       String csrq0=sdf.format(rs0.getDate("csrq"));
       String jbgz0=String.valueOf(rs0.getFloat("jbgz"));
       String gwjt0=String.valueOf(rs0.getFloat("gwjt"));
       rs0.close();
       stmt0.close();
%>
<form id="form1" name="form1" method="post" action="ZgXg0.jsp"
onSubmit="return check(this)">
    <table border="1">
    <tr><td>编号</td><td><input name="bh" type="text" id="bh"
value="<%=bh0%>" readonly="true" /></td></tr>
    <tr><td>姓名</td><td><input name="xm" type="text" id="xm"
value="<%=xm0%>" /></td></tr>
    <tr><td>性别</td><td>
    <input type="radio" name="xb" value="男" <% if (xb0.equals("男")){ %>
checked="checked" <% } %>/>男
    <input type="radio" name="xb" value="女" <% if (xb0.equals("女")){ %>
checked="checked" <% } %>/>女
```

```
    </td></tr>
<%
        String sql = "select * from bmb order by bmbh";
       Statement stmt=conn.createStatement();
       ResultSet rs = stmt.executeQuery(sql);
%>
    <tr><td>部门</td>
    <td><select name="bm">
<%
       while(rs.next())
       {
        String bmbh=rs.getString("bmbh").trim();
        String bmmc=rs.getString("bmmc").trim();
%>
    <option value="<%=bmbh%>" <% if (bm0.equals(bmbh)){ %> selected
<% } %>><%=bmmc%></option>
<%
       }
       rs.close();
       stmt.close();
        conn.close();
%>
    </select>
    </td></tr>
    <tr><td>出生日期</td><td><input name="csrq" type="text" id="csrq"
value="<%=csrq0%>" /></td></tr>
    <tr><td>基本工资</td><td><input name="jbgz" type="text" id="jbgz"
value="<%=jbgz0%>" /></td></tr>
    <tr><td>岗位津贴</td><td><input name="gwjt" type="text" id="gwjt"
value="<%=gwjt0%>" /></td></tr>
    </table>
    <br>
    <input name="submit" type="submit" value="确定" />
    <input name="reset" type="reset" value="重置" />
</form>
</div>
</body>
</html>
```

(5) 在项目的 WebRoot 文件夹中添加一个新的 JSP 页面 ZgXg0.jsp，其代码如下：

```
<%@ page contentType="text/html;charset=gb2312" language="java" %>
<%@ page import="java.sql.*"%>
<%request.setCharacterEncoding("gb2312"); %>
<html>
<head><title></title></head>
<body>
<%
    String bh = request.getParameter("bh");
    String xm = request.getParameter("xm");
    String xb = request.getParameter("xb");
```

```
    String bm = request.getParameter("bm");
    String csrq = request.getParameter("csrq");
    String jbgz = request.getParameter("jbgz");
    String gwjt = request.getParameter("gwjt");
    try {
        Class.forName("com.microsoft.sqlserver.jdbc.SQLServerDriver");
        String url="jdbc:sqlserver://localhost:1433;DatabaseName=rsgl";
        String user="sa";
        String password="abc123!";
        Connection conn=DriverManager.getConnection(url,user,password);
        String sql = "update zgb set
xm='"+xm+"',xb='"+xb+"',bm='"+bm+"'";
        sql=sql+",csrq='"+csrq+"',jbgz="+jbgz+",gwjt="+gwjt;
        sql=sql+" where bh='"+bh+"'";
       Statement stmt=conn.createStatement();
       int n = stmt.executeUpdate(sql);
       if (n>0){
           out.print("<script Language='JavaScript'>window.alert('职工记
录修改成功! ')</script>");
          out.print("<script Language='JavaScript'>window.localtion
='ZgCx.jsp'</script>");
        }
        else {
           out.print("<script Language='JavaScript'>window.alert('职工记
录修改失败! ')</script>");
          out.print("<script Language='JavaScript'>window.localtion
='ZgCx.jsp'</script>");
        }
       stmt.close();
        conn.close();
    }
    catch (Exception e) {
        out.print(e.toString());
    }
%>
</body>
</html>
```

该页面中“String sql = "update zgb set xm='"+xm+"',xb='"+xb+"',bm='"+bm+"'";”与“int n = stmt.executeUpdate(sql);”之间的语句也可改写为:

```
String sql="update zgb set xm=?,xb=?,bm=?,csrq=?,jbgz=?,gwjt=? where
bh=?";
PreparedStatement stmt=conn.prepareStatement(sql);
stmt.setString(1,xm);
stmt.setString(2,xb);
stmt.setString(3,bm);
stmt.setString(4,csrq);
stmt.setFloat(5,Float.valueOf(jbgz));
stmt.setFloat(6,Float.valueOf(gwjt));
stmt.setString(7,bh);
int n = stmt.executeUpdate();
```

(6) 在项目的 WebRoot 文件夹中添加一个新的 JSP 页面 ZgSc.jsp，其代码如下：

```
<%@ page contentType="text/html;charset=gb2312" language="java" %>
<%@ page import="java.sql.*"%>
<%@ page import="java.text.*"%>
<%request.setCharacterEncoding("gb2312"); %>
<html>
<head><title>职工删除</title></head>
<body>
<div align="center">
  <P>职工删除</P>
<%
        Class.forName("com.microsoft.sqlserver.jdbc.SQLServerDriver");
        String url="jdbc:sqlserver://localhost:1433;DatabaseName=rsgl";
        String user="sa";
        String password="abc123!";
        Connection conn=DriverManager.getConnection(url,user,password);
        String bh0=request.getParameter("bh").trim();
        String sql0 = "select * from zgb where bh='"+bh0+"'";
       Statement stmt0=conn.createStatement();
       ResultSet rs0 = stmt0.executeQuery(sql0);
       rs0.next();
       String xm0=rs0.getString("xm").trim();
       String xb0=rs0.getString("xb").trim();
       String bm0=rs0.getString("bm").trim();
       SimpleDateFormat sdf=new SimpleDateFormat("yyyy-MM-dd");
       String csrq0=sdf.format(rs0.getDate("csrq"));
       String jbgz0=String.valueOf(rs0.getFloat("jbgz"));
       String gwjt0=String.valueOf(rs0.getFloat("gwjt"));
       rs0.close();
       stmt0.close();
%>
<form id="form1" name="form1" method="post" action="ZgSc0.jsp"
onSubmit="return check(this)">
    <table border="1">
    <tr><td>编号</td><td><input name="bh" type="text" id="bh"
value="<%=bh0%>" readonly="true" /></td></tr>
    <tr><td>姓名</td><td><input name="xm" type="text" id="xm"
value="<%=xm0%>" /></td></tr>
    <tr><td>性别</td><td>
    <input type="radio" name="xb" value="男" <% if (xb0.equals("男")){ %>
checked="checked" <% } %>/>男
    <input type="radio" name="xb" value="女" <% if (xb0.equals("女")){ %>
checked="checked" <% } %>/>女
    </td></tr>
<%
        String sql = "select * from bmb order by bmbh";
       Statement stmt=conn.createStatement();
       ResultSet rs = stmt.executeQuery(sql);
%>
```

```
    <tr><td>部门</td>
    <td><select name="bm">
<%
       while(rs.next())
       {
        String bmbh=rs.getString("bmbh").trim();
        String bmmc=rs.getString("bmmc").trim();
%>
    <option value="<%=bmbh%>" <% if (bm0.equals(bmbh)){ %> selected
<% } %>><%=bmmc%></option>
<%
       }
       rs.close();
       stmt.close();
        conn.close();
%>
    </select>
    </td></tr>
    <tr><td>出生日期</td><td><input name="csrq" type="text" id="csrq"
value="<%=csrq0%>" /></td></tr>
    <tr><td>基本工资</td><td><input name="jbgz" type="text" id="jbgz"
value="<%=jbgz0%>" /></td></tr>
    <tr><td>岗位津贴</td><td><input name="gwjt" type="text" id="gwjt"
value="<%=gwjt0%>" /></td></tr>
    </table>
    <br>
    <input name="submit" type="submit" value="确定" />
    <input name="reset" type="reset" value="重置" />
</form>
</div>
</body>
</html>
```

(7) 在项目的 WebRoot 文件夹中添加一个新的 JSP 页面 ZgSc0.jsp，其代码如下：

```
<%@ page contentType="text/html;charset=gb2312" language="java" %>
<%@ page import="java.sql.*"%>
<%request.setCharacterEncoding("gb2312"); %>
<html>
<head><title></title></head>
<body>
<%
    String bh =  request.getParameter("bh");
    try {
        Class.forName("com.microsoft.sqlserver.jdbc.SQLServerDriver");
        String url="jdbc:sqlserver://localhost:1433;DatabaseName=rsgl";
        String user="sa";
        String password="abc123!";
        Connection conn=DriverManager.getConnection(url,user,password);
        String sql = "delete from zgb";
        sql=sql+" where bh='"+bh+"'";
```

```
        Statement stmt=conn.createStatement();
        int n = stmt.executeUpdate(sql);
        if (n>0){
             out.print("<script Language='JavaScript'>window.alert('职工记
录删除成功! ')</script>");
           out.print("<script Language='JavaScript'>window.localtion
='ZgCx.jsp'</script>");
         }
         else {
             out.print("<script Language='JavaScript'>window.alert('职工记
录删除失败! ')</script>");
           out.print("<script Language='JavaScript'>window.localtion
='ZgCx.jsp'</script>");
         }
        stmt.close();
         conn.close();
     }
     catch (Exception e) {
         out.print(e.toString());
     }
%>
</body>
</html>
```

该页面中“String sql = "delete from zgb";”与“int n = stmt.executeUpdate(sql);”之间的语句也可改写为：

```
String sql="delete from zgb where bh=?";
PreparedStatement stmt=conn.prepareStatement(sql);
stmt.setString(1,bh);
int n = stmt.executeUpdate();
```

说明：对于 Java Web 应用系统的开发来说，JSP+JDBC 是最为基本且最为简单的一种模式(见图 3.19)。基于该模式的 Web 应用程序的工作流程如下。

(1) 浏览器发出请求。

(2) JSP 页面接收请求，并根据业务处理的需要通过 JDBC 完成对数据库的操作。

(3) JSP 将业务处理的结果信息动态生成相应的 Web 页面，并发送到浏览器。

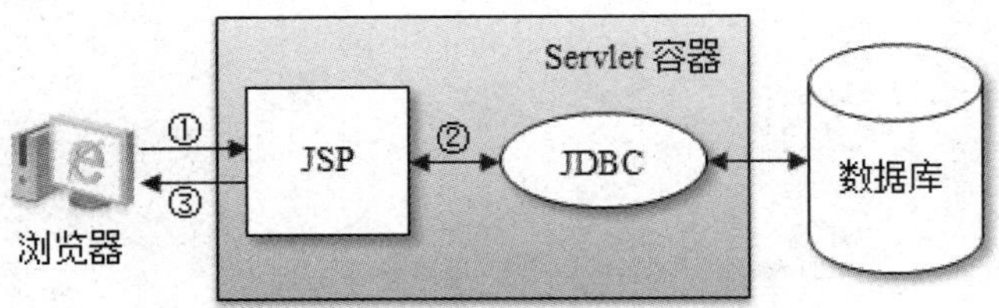

图 3.19　JSP+JDBC 模式

本章小结

本章首先介绍了 JDBC 的概况，然后较为详细地讲解了 JDBC 核心类与接口的主要用法，并通过具体实例与案例深入说明了 JDBC 的数据库编程技术。通过本章的学习，读者应熟练掌握 JDBC 数据库编程的基本步骤与相关技术，并能使用 JSP+JDBC 模式开发相应的 Web 应用系统。

思考题

1. JDBC 的类库包含在哪两个包中？
2. DriverManager 类的主要作用是什么？有哪些常用方法？
3. 在应用程序中，如何加载 JDBC 驱动程序？
4. Connection 接口的主要作用是什么？有哪些常用方法？
5. Statement 接口的主要作用是什么？有哪些常用方法？
6. PreparedStatement 接口的主要作用是什么？有哪些常用方法？
7. CallableStatement 接口的主要作用是什么？有哪些常用方法？
8. ResultSet 接口的主要作用是什么？有哪些常用方法？
9. JDBC 编程的基本步骤是什么？

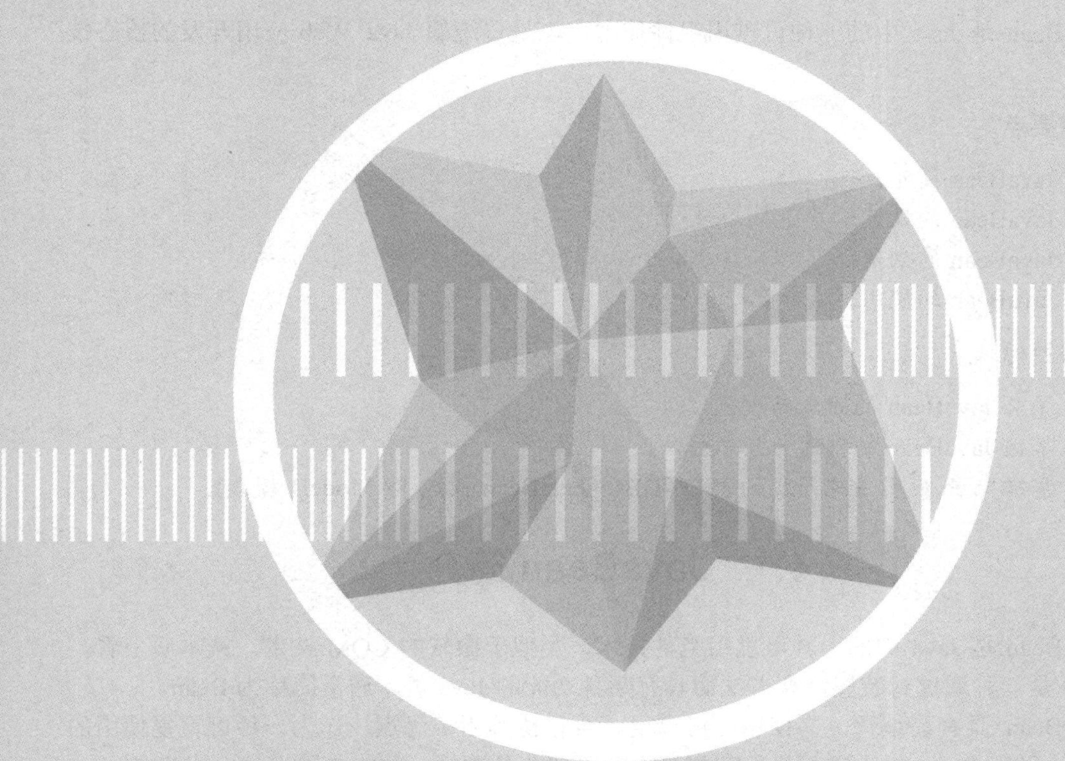

第 4 章

JavaBean 技术

JavaBean 是 Java 中的一种可重用组件技术，也是传统的 Java Web 应用开发的核心技术之一。

本章要点：

- JavaBean 简介；
- JavaBean 的规范；
- JavaBean 的创建；
- JavaBean 的使用。

学习目标：

- 了解 JavaBean 的概念与规范；
- 掌握 JavaBean 的创建与使用方法；
- 掌握 Web 应用系统开发的 JSP+JDBC+JavaBean 模式(即 Model1 模式)。

4.1 JavaBean 简介

JavaBean 是 Java 中的一种可重用组件技术，类似于微软的 COM 技术。从本质上看，JavaBean 是一种通过封装属性和方法而具有某种功能的 Java 类，通常简称为 Bean。

JavaBean 具有易编写、易使用、可重用、可移植等诸多优点。作为一种可重复使用的软件组件，JavaBean 通常用于封装某些特定功能或业务逻辑，如将文件上传、发送 E-mail 以及业务处理或复杂计算分离出来并使之成为可重复利用的独立模块。使用已有的 JavaBean，可有效地减少代码的编写量，缩短应用的开发时间，并提高其可伸缩性。

JSP 对在 Web 应用中集成 JavaBean 组件提供了完善的支持。使用 JSP 所提供的有关动作标记(<jsp:useBean>、<jsp:setProperty>和<jsp:getProperty>)，即可轻松地实现对 JavaBean 组件的调用。因此，在 Java Web 应用开发中，可充分利用 JavaBean 技术，将可重复利用的程序代码封装为相应的 JavaBean，供相关的 JSP 页面直接调用。例如，可将连接数据库、执行 SQL 语句的功能封装为相应的数据库访问 JavaBean，这样，在需要访问数据库的页面中，即可直接调用该 JavaBean 实现对数据库的有关操作。

4.2 JavaBean 的规范

通常，一个标准的 JavaBean 需遵循以下规范。

(1) JavaBean 是一个公共的(public)类。

(2) JavaBean 类必须存在一个不带参数的构造函数。

(3) JavaBean 的属性应声明为 private，方法应声明为 public。

(4) JavaBean 应提供 setXxx()与 getXxx()方法来存取类中的属性，其中，“Xxx”为属性名称(第一个字母应大写)。若属性为布尔类型，则可使用 isXxx()方法代替 getXxx()方法。

4.3　JavaBean 的创建

创建一个 JavaBean 其实就是在遵循 JavaBean 规范的基础上创建一个 Java 类，并将其保存为*.java 文件。

【实例 4-1】 JavaBean 创建示例：创建一个用户 JavaBean——UserBean。

主要步骤如下。

(1) 新建一个 Web 项目 web_04。

(2) 在项目中创建一个包 org.etspace.abc.bean。具体步骤如下。

① 选择 File→New→Package 菜单项，弹出 New Java Package 对话框，如图 4.1 所示。

② 在 Name 文本框中输入包名(在此为 org.etspace.abc.bean)，同时在 Source folder 文本框处指定其存放的文件夹(在此为 web_04/src)。

③ 单击 Finish 按钮，关闭 New Java Package 对话框。

(3) 在 org.etspace.abc.bean 包中创建一个用户 JavaBean，即 UserBean。按以下步骤进行操作。

① 在项目中右击 org.etspace.abc.bean 包，并在其快捷菜单中选择 New→Class 菜单项，弹出 New Java Class 对话框，如图 4.2 所示。

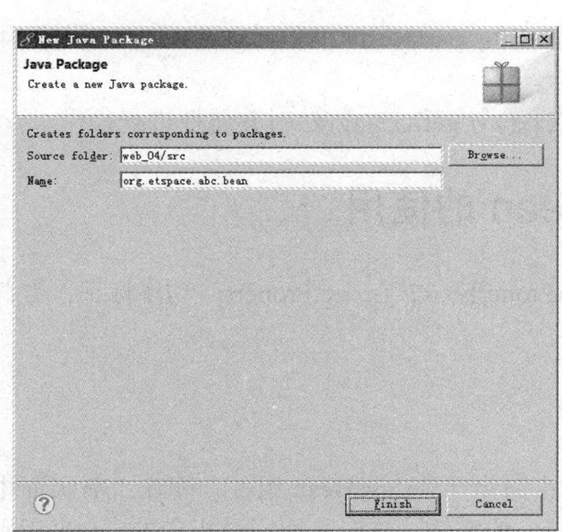

图 4.1　New Java Package 对话框

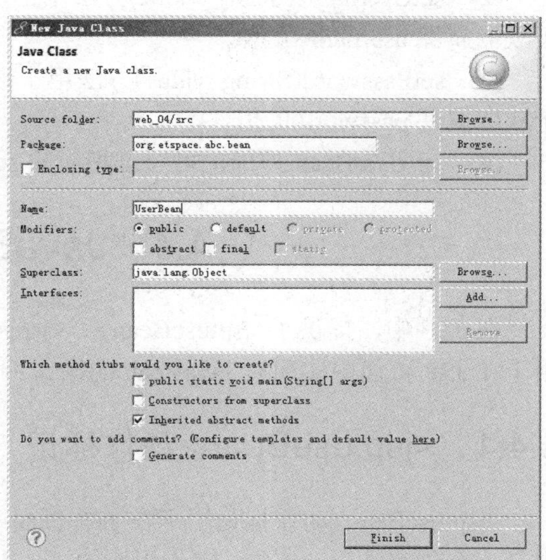

图 4.2　New Java Class 对话框

② 在 Name 文本框中输入类名 UserBean。

③ 单击 Finish 按钮，关闭 New Java Class 对话框。此时，在项目的 org.etspace.abc.bean 包中，将自动创建一个 Java 类文件 UserBean.java。

④ 输入并保存 UserBean 的代码。具体代码为：

```
package org.etspace.abc.bean;
public class UserBean {
```

```
    private String  username = null;
    private String  password = null;
    public UserBean(){
    }
    public void setUsername(String value){
        username = value;
    }
    public String getUsername(){
        return username;
    }
    public void setPassword(String value){
        password = value;
    }
    public String getPassword(){
        return password;
    }
}
```

解析：

(1) UserBean 是一个很典型的 JavaBean，共有两个 String 型的属性，即 username 与 password。

(2) setUsername(String value)方法用于设置属性 username 的值，getUsername()方法用于获取属性 username 的值。

(3) setPassword(String value)方法用于设置属性 password 的值，getPassword()方法用于获取属性 password 的值。

(4) 在 UserBean 外部，可通过相应的 setXxx()与 getXxx()方法对其属性进行操作。

4.4 JavaBean 的使用

在 JSP 中，提供了<jsp:useBean>、<jsp:setProperty>与<jsp:getProperty>动作标记，专门用于在 JSP 页面中实现对 JavaBean 的操作。

4.4.1 <jsp:useBean>动作标记

<jsp:useBean>动作标记用于在 JSP 页面中实例化一个 JavaBean 组件，即在 JSP 页面中定义一个具有唯一 id 与一定作用域的 JavaBean 的实例。JSP 页面通过指定的 id 来识别 JavaBean 或调用其中的方法。

<jsp:useBean>动作标记基本语法格式为：

```
<jsp:useBean id="beanName" scope="page|request|session|application"
class="packageName. className"/>
```

其中，各属性的作用见表 4.1。

表 4.1　<jsp:useBean>动作标记的属性

属性	说　明
id	用于指定 JavaBean 的实例名
scope	用于指定 JavaBean 的作用域，其取值为 page、request、session、application 之一，分别表示当前页面范围、当前用户请求范围、当前用户会话范围、当前 Web 应用范围
class	用于指定 JavaBean 的类名(包括所在包的名称)

4.4.2　<jsp:setProperty>动作标记

<jsp:setProperty>动作标记用于设置 JavaBean 的属性值。JavaBean 属性值的设置也可通过调用其相应方法实现。

<jsp:setProperty>动作标记的语法格式共有 4 种，分别为：

(1)　<jsp:setProperty name= "beanName " property= "*" />。

(2)　<jsp:setProperty name= "beanName " property= "propertyName " />。

(3)　<jsp:setProperty name= "beanName" property= "propertyName" param= "parameterName "/>。

(4)　<jsp:setProperty name= "beanName " property= "propertyName " value= "propertyValue | <%= expression %> " />。

其中，各属性的作用见表 4.2。

表 4.2　<jsp:setProperty>动作标记的属性

属　性	说　明
name	用于指定 JavaBean 的实例名
property	用于指定 JavaBean 的属性名
param	用于指定 HTTP 表单(或请求)的参数名
value	用于指定属性值

使用格式 1，可根据表单参数设置 JavaBean 中所有同名属性的值；使用格式 2，可根据表单参数设置 JavaBean 中指定同名属性的值；使用格式 3，可根据指定的表单参数设置 JavaBean 中指定属性的值；使用格式 4，可用指定的值设置 JavaBean 中指定属性的值。

注意：使用<jsp:setProperty>动作标记之前，必须使用<jsp:useBean>标记得到一个可操作的 JavaBean，而且该 JavaBean 中必须有相应的 setXxx()方法。

4.4.3　<jsp:getProperty>动作标记

<jsp:getProperty>动作标记用于获取 JavaBean 的属性值。JavaBean 属性值的获取也可通过调用其相应方法实现。

<jsp:getProperty>动作标记的语法格式为：

```
<jsp:getProperty name= "beanName" property= "propertyName" />
```

其中，各属性的作用见表 4.3。

表 4.3　<jsp:getProperty>动作标记的属性

属　性	说　明
name	用于指定 JavaBean 的实例名
property	用于指定 JavaBean 的属性名

注意：使用<jsp:getProperty>动作标记之前，必须使用<jsp:useBean>标记得到一个可操作的 JavaBean，而且该 JavaBean 中必须保证有相应的 getXxx()方法。

【实例 4-2】JavaBean 使用示例：在 UserInfo.jsp 页面中，应用 UserBean 实现用户信息的设置与显示。

主要步骤如下。

在项目 web_04 的 WebRoot 文件夹中添加一个新的 JSP 页面 UserInfo.jsp，其代码为：

```
<%@ page language="java" import="java.util.*" pageEncoding="utf-8"%>
<jsp:useBean id="myUserBean" class="org.etspace.abc.bean.UserBean" scope
= "page" />
<html>
  <head>
    <title>用户信息</title>
  </head>
  <body>
    <jsp:setProperty name="myUserBean" property="username" value="admin" />
    <jsp:setProperty name="myUserBean" property="password" value="12345" />
    用户名：<jsp:getProperty name="myUserBean" property="username" /><br>
    密码：<jsp:getProperty name="myUserBean" property="password" /><br>
    <br>
<%
    myUserBean.setUsername("system");
    myUserBean.setPassword("54321");
%>
    用户名：<%=myUserBean.getUsername() %><br>
    密码：<%=myUserBean.getPassword() %><br>
  </body>
</html>
```

运行结果如图 4.3 所示。

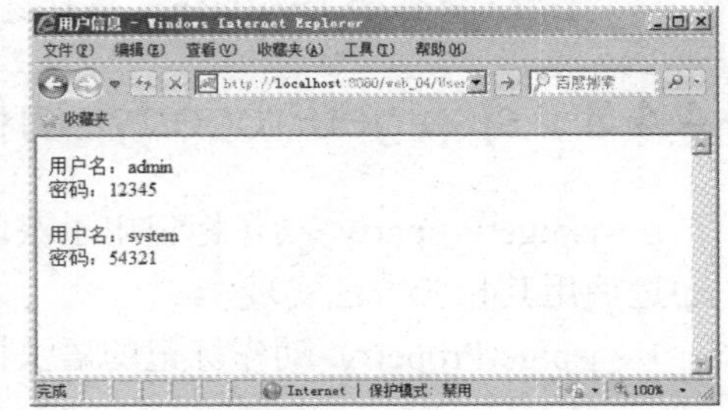

图 4.3　页面 UserInfo.jsp 的运行结果

【实例 4-3】用户登录。图 4.4 所示为“用户登录”页面，在此页面中输入用户名与密码后，再单击“登录”按钮提交表单。若用户名与密码输入正确(在此假定正确的用户名与密码分别为“admin”与“12345”)，则显示图 4.5 所示的登录成功页面，否则显示图 4.6 所示的登录失败页面。

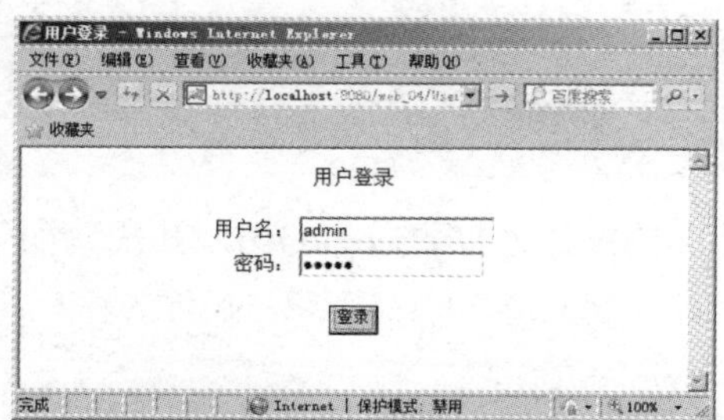

图 4.4　“用户登录”页面

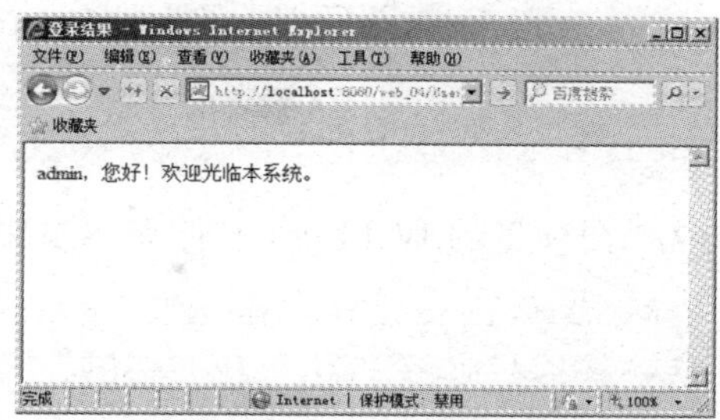

图 4.5　登录成功页面

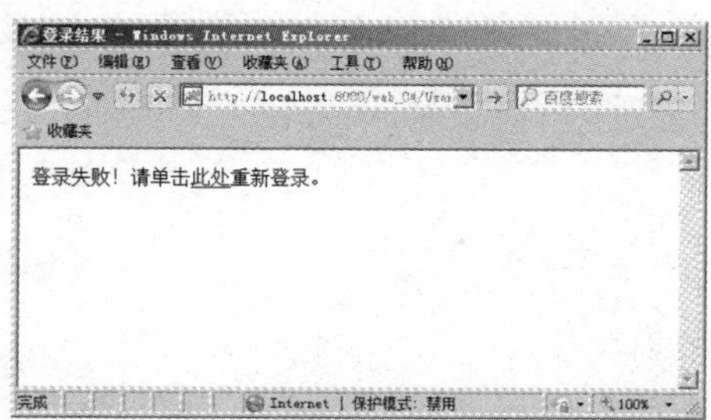

图 4.6　登录失败页面

主要步骤(在项目 web_04 中)如下。

(1) 在项目中的 org.etspace.abc.bean 包中创建一个 JavaBean——UserCheckBean。其文件名为 UserCheckBean.java，代码如下：

```
package org.etspace.abc.bean;
public class UserCheckBean {
    private String  username = null;
    private String  password = null;
    public UserCheckBean(){
    }
    public void setUsername(String value){
        username = value;
    }
    public String getUsername(){
        return username;
    }
    public void setPassword(String value){
        password = value;
    }
    public String getPassword(){
        return password;
    }
    public boolean check(){
        if (username.equals("admin") && password.equals("12345")) {
            return true;
        }
        else {
```

```
            return false;
        }
    }
}
```

(2) 在项目的 WebRoot 文件夹中添加一个新的 JSP 页面 UserLogin.jsp，其代码如下：

```
<%@ page language="java" pageEncoding="utf-8" %>
<script language="JavaScript">
function check(theForm)
{
  if (theForm.username.value == "")
  {
    alert("请输入用户名！");
    theForm.username.focus();
    return (false);
  }
  if (theForm.password.value == "")
  {
    alert("请输入密码！");
    theForm.password.focus();
    return (false);
  }
  return (true);
}
</script>
<html>
    <head>
        <title>用户登录</title>
    </head>
    <body>
        <div align="center">
        <form action="UserLoginResult.jsp" method="post"
onSubmit="return check(this)">
        用户登录<br><br>
        <table>
        <tr><td align="right">用户名：</td><td><input type="text"
name="username"></td></tr>
        <tr><td align="right">密码：</td><td><input type="password"
name="password"></td></tr>
        </table>
        <br>
        <input type="submit" value="登录">
        </form>
        </div>
    </body>
</html>
```

(3) 在项目的 WebRoot 文件夹中添加一个新的 JSP 页面 UserLoginResult.jsp，其代码如下：

```
<%@ page language="java" import="java.util.*" pageEncoding="utf-8"%>
<% request.setCharacterEncoding("utf-8"); %>
<jsp:useBean id="myUserCheckBean"
class="org.etspace.abc.bean.UserCheckBean" scope = "page" />
<jsp:setProperty name="myUserCheckBean" property="*" />
<html>
  <head>
    <title>UserCheckBean</title>
  </head>
  <body>
<%
    if (myUserCheckBean.check()) {
%>
    <font color=blue><%=myUserCheckBean.getUsername()%></font>，您好！欢迎
光临本系统。
<%
    } else {
%>
    登录失败！请单击<a href="javascript:history.back(-1);">此处</a>重新登录。
<%
    }
%>
  </body>
</html>
```

解析：

(1) 在本实例中，UserCheckBean 的属性值是通过 HTTP 表单的同名参数值来自动设置的。HTML 表单与 JavaBean 进行交互的方法由此可见一斑。

(2) 在本实例中，UserCheckBean 中的 check()方法用于检测所输入的用户名与密码是否正确。

4.5　JavaBean 的应用案例

4.5.1　系统登录

下面，以 JSP+JDBC+JavaBean 模式实现 Web 应用系统中的系统登录功能，其运行结果与第 3 章 3.4.1 小节的系统登录案例相同。

实现步骤(在项目 web_04 中)如下。

(1) 将 SQL Server 2005/2008 的 JDBC 驱动程序 sqljdbc4.jar 添加到项目的 WebRoot\WEB-INF\lib 文件夹中。

(2) 在项目的 src 文件夹中新建一个包 org.etspace.abc.jdbc。

(3) 在 org.etspace.abc.jdbc 包中新建一个 JavaBean——DbBean。其文件名为 DbBean.java，代码如下：

```
package org.etspace.abc.jdbc;
import java.sql.*;
```

```
public class DbBean {
    private Statement stmt=null;
    private Connection conn=null;
    ResultSet rs=null;
    //构造方法(函数)
    public DbBean(){}
    //打开链接
    public void openConnection() {
        try {

    Class.forName("com.microsoft.sqlserver.jdbc.SQLServerDriver").newIns
tance();
            String
url="jdbc:sqlserver://localhost:1433;databaseName=rsgl";
            String user="sa";
            String password="abc123!";
            conn=DriverManager.getConnection(url, user, password);
        }
        catch(ClassNotFoundException e){
            System.err.println("openConn:"+e.getMessage());
        }
        catch(SQLException e) {
            System.err.println("openConn:"+e.getMessage());
        }
        catch(Exception e) {
            System.err.println("openConn:"+e.getMessage());
        }
    }
    //执行查询类的 SQL 语句
    public ResultSet executeQuery(String sql) {
        rs=null;
        try {
            stmt=conn.createStatement(ResultSet.TYPE_SCROLL_SENSITIVE,
ResultSet.CONCUR_UPDATABLE);
            rs=stmt.executeQuery(sql);
        }
        catch(SQLException e) {
            System.err.println("executeQuery:"+e.getMessage());
        }
        return rs;
    }
    //执行更新类的 SQL 语句
    public int executeUpdate(String sql) {
        int n = 0;
        try {
            stmt=conn.createStatement();
            n=stmt.executeUpdate(sql);
        }catch(Exception e) {
            System.out.print(e.toString());
        }
```

```
        return n;
    }
    //关闭链接
    public void closeConnection() {
        try {
            if (rs!=null)
                rs.close();
        }
        catch(SQLException e) {
            System.err.println("closeRs:"+e.getMessage());
        }
        try {
            if (stmt!=null)
                stmt.close();
        }
        catch(SQLException e) {
            System.err.println("closeStmt:"+e.getMessage());
        }
        try {
            if (conn!=null)
                conn.close();
        }
        catch(SQLException e) {
            System.err.println("closeConn:"+e.getMessage());
        }
    }
}
```

在此，DbBean 封装了用于访问数据库的有关操作。

(4) 在项目的 WebRoot 文件夹中新建一个子文件夹 syslogin。

(5) 在子文件夹 syslogin 中添加一个新的 JSP 页面 login.jsp。其代码与第 3 章 3.4.1 小节系统登录案例中的相同。

(6) 在子文件夹 syslogin 中添加一个新的 JSP 页面 validate.jsp，其代码如下：

```
<%@ page language="java" pageEncoding="utf-8" import="java.sql.*" %>
<jsp:useBean id="myDbBean" scope="page"
class="org.etspace.abc.jdbc.DbBean"></jsp:useBean>
<html>
    <head>
        <title>验证页面</title>
        <meta http-equiv="Content-Type" content="text/html;charset=utf-8">
    </head>
    <body>
        <%
        String username=request.getParameter("username");  //获取提交的姓名
        String password=request.getParameter("password");  //获取提交的密码
        boolean validated=false;  //验证标识
        //查询用户表中的记录
        String sql="select * from users";
        myDbBean.openConnection();
```

```
        ResultSet rs=myDbBean.executeQuery(sql);  //获取结果集
        while(rs.next())
        {
            if((rs.getString("username").trim().compareTo(username)==0)&&
                (rs.getString("password").trim().compareTo(password)==0))
            {
                validated=true;
            }
        }
        rs.close();
        myDbBean.closeConnection();
        if(validated)
        {
            //验证成功跳转到成功页面
        %>
            <jsp:forward page="welcome.jsp"></jsp:forward>
        <%
        }
        else
        {
            //验证失败跳转到失败页面
        %>
            <jsp:forward page="error.jsp"></jsp:forward>
        <%
        }
        %>
    </body>
</html>
```

(7) 在子文件夹 syslogin 中添加一个新的 JSP 页面 welcome.jsp。其代码与第 3 章 3.4.1 小节系统登录案例中的相同。

(8) 在子文件夹 syslogin 中添加一个新的 JSP 页面 error.jsp。其代码与第 3 章 3.4.1 小节系统登录案例中的相同。

4.5.2 数据添加

下面，以 JSP+JDBC+JavaBean 模式实现“职工增加”功能，其运行结果与第 3 章 3.4.2 小节的职工增加案例相同。

主要步骤(在项目 web_04 中)如下。

(1) 在项目的 WebRoot 文件夹中添加一个新的 JSP 页面 ZgZj.jsp，其代码如下：

```
<%@ page contentType="text/html;charset=gb2312" language="java" %>
<%@ page import="java.sql.*"%>
<jsp:useBean id="myDbBean" scope="page"
class="org.etspace.abc.jdbc.DbBean"></jsp:useBean>
<%request.setCharacterEncoding("gb2312"); %>
<script language="JavaScript">
function check(theForm)
{
```

```
  if (theForm.bh.value.length != 7)
  {
    alert("职工编号必须为 7 位！");
    theForm.bh.focus();
    return (false);
  }
  if (theForm.xm.value == "")
  {
    alert("请输入姓名！");
    theForm.xm.focus();
    return (false);
  }
  if (theForm.csrq.value == "")
  {
    alert("请输入出生日期！");
    theForm.csrq.focus();
    return (false);
  }
  if (theForm.jbgz.value == "")
  {
    alert("请输入基本工资！");
    theForm.jbgz.focus();
    return (false);
  }
  if (theForm.gwjt.value == "")
  {
    alert("请输入岗位津贴！");
    theForm.gwjt.focus();
    return (false);
  }
  return (true);
}
</script>
<html>
<head><title>职工增加</title></head>
<body>
<div align="center">
  <P>职工增加</P>
<form id="form1" name="form1" method="post" action="ZgZj0.jsp"
onSubmit="return check(this)">
    <table border="1">
    <tr><td>编号</td><td><input name="bh" type="text" id="bh"
/></td></tr>
    <tr><td>姓名</td><td><input name="xm" type="text" id="xm"
/></td></tr>
    <tr><td>性别</td><td>
    <input type="radio" name="xb" value="男" checked="checked" />男
    <input type="radio" name="xb" value="女" />女
    </td></tr>
<%
```

```
        String sql = "select * from bmb order by bmbh";
        myDbBean.openConnection();
        ResultSet rs=myDbBean.executeQuery(sql);
%>
    <tr><td>部门</td>
    <td><select name="bm">
<%
       while(rs.next())
       {
        String bmbh=rs.getString("bmbh");
        String bmmc=rs.getString("bmmc");
%>
    <option value="<%=bmbh%>"><%=bmmc%></option>
<%
       }
       rs.close();
        myDbBean.closeConnection();
%>
    </select>
    </td></tr>
    <tr><td>出生日期</td><td><input name="csrq" type="text" id="csrq"
/></td></tr>
    <tr><td>基本工资</td><td><input name="jbgz" type="text" id="jbgz"
/></td></tr>
    <tr><td>岗位津贴</td><td><input name="gwjt" type="text" id="gwjt"
/></td></tr>
    </table>
    <br>
    <input name="submit" type="submit"  value="确定" />
    <input name="reset" type="reset" value="重置" />
</form>
</div>
</body>
</html>
```

(2) 在项目的 WebRoot 文件夹中添加一个新的 JSP 页面 ZgZj0.jsp，其代码如下：

```
<%@ page contentType="text/html;charset=gb2312" language="java" %>
<%@ page import="java.sql.*"%>
<jsp:useBean id="myDbBean" scope="page"
class="org.etspace.abc.jdbc.DbBean"></jsp:useBean>
<%request.setCharacterEncoding("gb2312"); %>
<html>
<head><title></title></head>
<body>
<%
    String bh =  request.getParameter("bh");
    String xm =  request.getParameter("xm");
    String xb =  request.getParameter("xb");
    String bm =  request.getParameter("bm");
    String csrq =  request.getParameter("csrq");
```

```
    String jbgz =  request.getParameter("jbgz");
    String gwjt =  request.getParameter("gwjt");
    try {
        String sql = "insert into zgb(bh,xm,xb,bm,csrq,jbgz,gwjt)";
        sql=sql+"values('"+bh+"','"+xm+"','"+xb+"','"+bm+"','"+csrq+"'";
        sql=sql+","+jbgz+","+gwjt+")";
        myDbBean.openConnection();
      int n = myDbBean.executeUpdate(sql);
      if (n>0){
           out.print("职工记录增加成功！");
        }
        else {
           out.print("职工记录增加失败！");
        }
        myDbBean.closeConnection();
    }
    catch (Exception e) {
        out.print(e.toString());
    }
%>
</body>
</html>
```

4.5.3　数据维护

与“职工增加”功能的实现类似，为避免编写过多的重复代码，“职工增加”功能也可采用 JSP+JDBC+JavaBean 的模式实现，并保证其运行结果与第 3 章 3.4.3 小节的职工维护案例相同。限于篇幅，在此不作详述，请大家自行尝试。

说明： JSP+JDBC+JavaBean(通常又简称为 JSP+JavaBean)作为一种通用的程序结构(服务器端程序的组织结构)，是 Java EE 传统开发中的一种常用模式，即 Model1 模式，如图 4.7 所示。其实，最原始的 Web 应用程序是基于 Java Servlet 的，随着 JSP 与 JavaBean 技术的出现，将 Web 应用程序中的 html/xhtml 文档与 Java 业务逻辑代码有效分离成为可能。通常 JSP 负责动态生成 Web 页面，而业务逻辑则用可重用的 JavaBean 组件来实现。

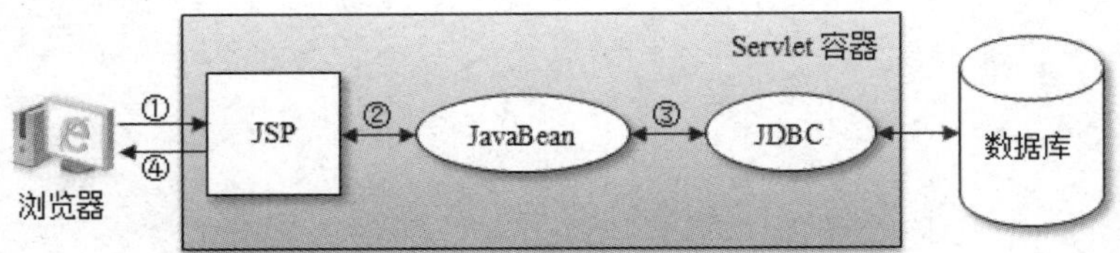

图 4.7　JSP+JDBC+JavaBean 模式(Model1 模式)

基于 Model1 模式的 Web 应用程序的工作流程如下。

(1) 浏览器发出请求。

(2) JSP 页面接收请求，并根据其需要与不用的 JavaBean 进行交互。

(3) JavaBean 执行业务处理，并通过 JDBC 完成对数据库的操作。

(4) JSP 将业务处理的结果信息动态生成相应的 Web 页面，并发送到浏览器。

在 Model1 模式中，JSP 集控制与显示于一体，由此可见，Model1 模式是一种以 JSP 为中心的开发模式。由于 Model1 模式的实现过程较为简单，能够快速开发出很多小型的 Web 项目，因此曾经应用得十分广泛。

本 章 小 结

本章首先介绍了 JavaBean 的概念与规范，然后通过具体实例讲解了 JavaBean 的创建与使用方法，最后再通过具体案例说明了 JavaBean 的综合应用技术。通过本章的学习，读者应熟练掌握 JavaBean 的创建与使用方法，并能使用 JSP+JDBC+JavaBean 模式(即 Model1 模式)开发相应的 Web 应用系统。

思 考 题

1. JavaBean 是什么？
2. 一个标准的 JavaBean 应遵循哪些规范？
3. 如何创建一个 JavaBean？
4. JSP 页面中与 JavaBean 操作相关的动作标记有哪些？
5. <jsp:useBean>动作标记的作用是什么？其基本语法格式是什么？
6. <jsp:setProperty>动作标记的作用是什么？其基本语法格式是什么？
7. <jsp:getProperty>动作标记的作用是什么？其基本语法格式是什么？

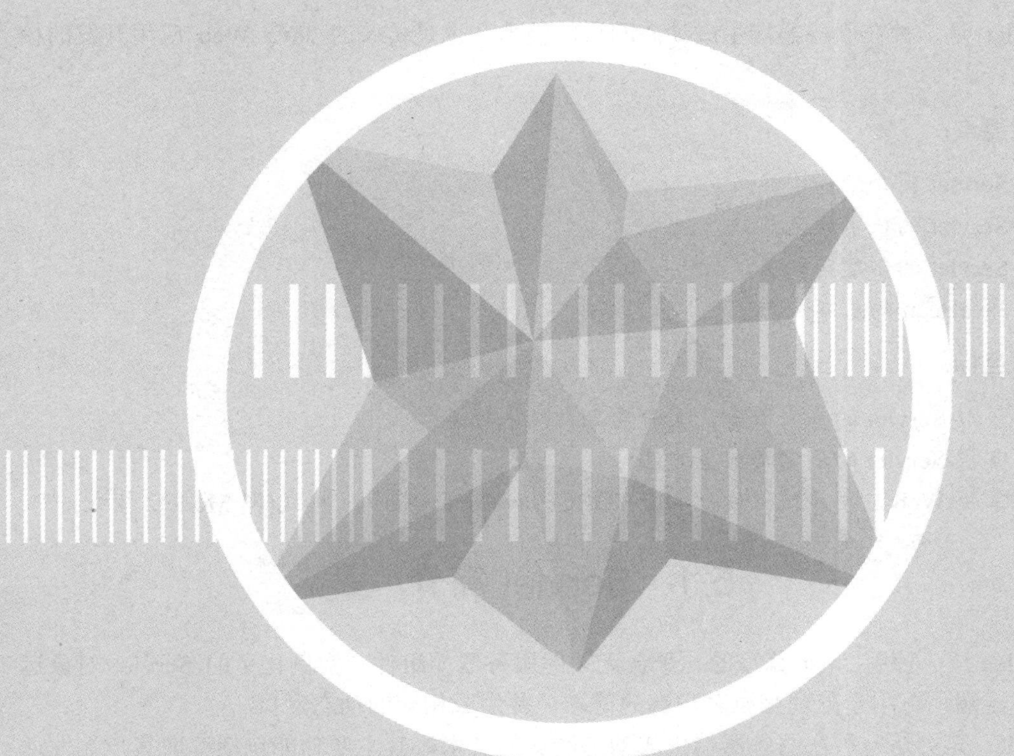

第 5 章

Servlet 技术

Servlet 是一种在服务器端生成动态网页的技术，也是传统的 Java Web 应用开发的核心技术之一。

本章要点：

- Servlet 简介；
- Servlet 的技术规范；
- Servlet 的创建与配置；
- JavaBean 的基本应用。

学习目标：

- 了解 Servlet 的基本概念、生命周期与技术规范；
- 掌握 Servlet 的创建与配置方法；
- 掌握 Web 应用系统开发的 JSP+JDBC+JavaBean+Servlet 模式(即 Model2 模式)。

5.1 Servlet 简介

Servlet 是一种用 Java 编写的与平台无关的服务器端组件。实例化了的 Servlet 对象运行在服务器端，可用于处理来自客户端的请求，并生成相应的动态网页。

Servlet 其实是一种在服务器端生成动态网页的技术，其主要功能包括读取客户端发送到服务器端的显式数据(如表单数据)或隐式数据(如请求报头)、从服务器端发送显式数据(如 HTML)或隐式数据(如状态代码与响应报头)到客户端。Servlet 的主要优点如下。

(1) 易开发。Servlet 支持系统的内置对象(如 request、response 等)，可轻松实现有关功能。

(2) 可移植，稳健性好。Servlet 用 Java 编写，具有 Java 应用程序的优势。

(3) 可节省内存与 CPU 资源。一个 Servlet 进程被客户端发送的第一个相关请求激活，直到 Web 应用程序停止或重启时才会被卸载。在此期间，该 Servlet 进程一直都在等待后续的请求。此外，每当一个相关请求到达时均生成一个新的线程，因此一个进程可同时为多个客户服务。

Servlet 不能独立运行，必须被部署到 Servlet 容器中，由容器进行实例化并调用其方法。Servlet 容器其实是 Web 服务器或 Java EE 应用服务器的一部分，负责在 Servlet 的生命周期内管理 Servlet。

Servlet 的生命周期定义了一个 Servlet 如何被加载和初始化、怎样接收请求、响应请求和提供服务以及如何被卸载等。具体来说，一个 Servlet 的生命周期可分为以下 4 个阶段。

(1) 加载与实例化。Servlet 容器负责加载与实例化 Servlet。在默认情况下，第一次请求访问某个 Servlet 时，容器就会创建一个相应的 Servlet 实例(即进行实例化)。

(2) 初始化。在 Servlet 实例化之后，容器将调用 Servlet 的 init()方法初始化该实例。

(3) 处理请求。Servlet 容器调用 Servlet 实例的 service()方法对请求进行处理。在 service()方法中，Servlet 实例通过 ServletRequest 对象得到客户端的相关信息与请求信息，在对请求进行处理后，再调用 ServletResponse 对象的方法设置响应信息。

(4) 终止服务。当 Web 应用被终止，或 Web 应用重新启动，或 Servlet 容器终止运行，或 Servlet 容器重新装载 Servlet 的新实例时，容器就会调用实例的 destroy()方法，让该实例释放其所占用的资源，完成卸载过程。

Servlet 的运行过程与生命周期均由 Servlet 容器所控制。对于一个来自浏览器的 Servlet 请求，通常按如下顺序进行响应(见图 5.1)。

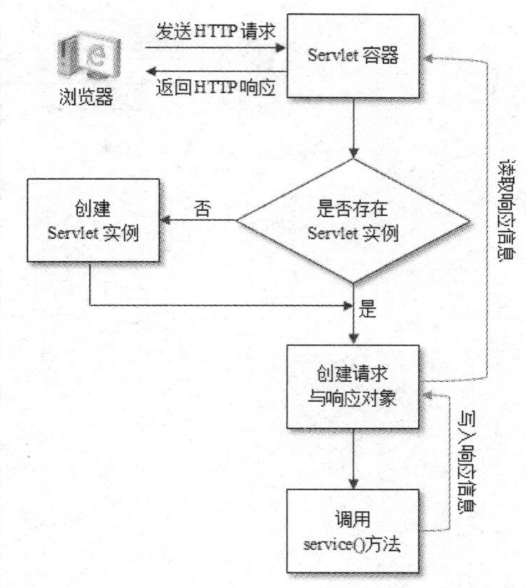

图 5.1　Servlet 请求的响应过程

(1) Servlet 容器检测是否已经装载并创建了该 Servlet 的实例对象。如果是，则直接执行(4)，否则，执行(2)。

(2) 装载并创建该 Servlet 的实例对象。

(3) 调用 Servlet 实例对象的 init()方法。

(4) 创建一个用于封装 HTTP 请求消息的 HttpServletRequest 对象与一个代表响应消息的 HttpServletResponse 对象，然后调用 Servlet 的 service()方法，并将上述请求与响应对象作为参数传递进去。

JSP 的本质就是 Servlet。Web 容器接收到 JSP 页面的访问请求时，会将其交给 JSP 引擎去处理(JSP 引擎就是一个负责解释与执行 JSP 页面的 Servlet 程序)。每个 JSP 页面在第一次被访问时，JSP 引擎会先将其转换成一个 Servlet 源程序(*.java)，接着再将该 Servlet 源程序编译成 Servlet 类文件(*.class)，最后再由 Web 容器加载并解释执行，并将响应结果返回给客户端浏览器。此后再次访问同样的 JSP 页面时，JSP 引擎会检查 JSP 文件是否有更新或被修改。如果是的话，就再次进行转换与编译，然后再执行重新生成的 Servlet 类文件；否则，就直接执行此前所生成的 Servlet 类文件。由此可见，JSP 页面在第一次被访问或修改后首次被访问时，响应速度会稍微慢一些。JSP 页面的执行过程如图 5.2 所示。

直接使用 Servlet 生成页面时，所有的 HTML 页面都要使用页面输出流完成，开发效率较低。另外，对于实现 Servlet 标准的 Java 类，须由 Java 程序员编写或修改，导致不了解 Java 编程技术的美工人员无法参与到页面的设计中。由于 Servlet 擅长流程处理，易于跟踪与排错，而 JSP 则能较为直观地生成动态页面，易于理解与使用，因此在标准的 MVC 模式中，视图(即表示层)通常使用 JSP 技术实现，而 Servlet 则仅作为控制器使用，这样，Servlet 与 JSP 均各得其所，充分发挥了各自的优点。

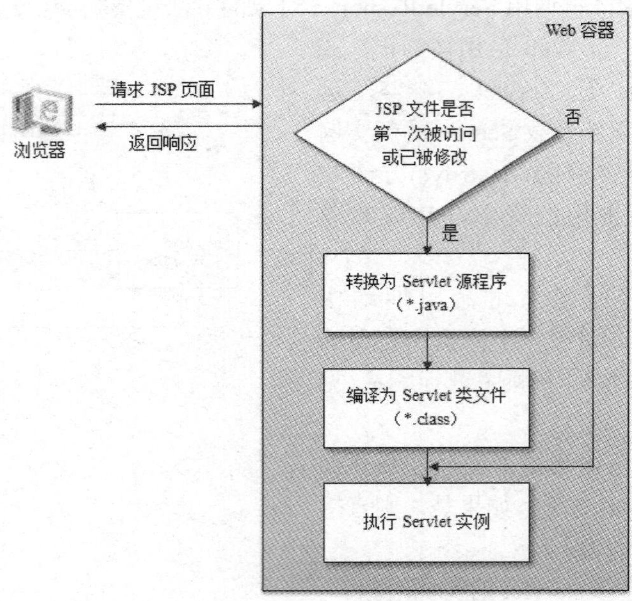

图 5.2　JSP 页面的执行过程

5.2　Servlet 的技术规范

Servlet API(Servlet Application Programming Interface，Servlet 应用编程接口)是 Servlet 规范定义的一套专门用于开发 Servlet 程序的 Java 类和接口，由 javax.servlet 与 javax.servlet.http 两个包组成。

javax.servlet 包所提供的接口、类与异常如下。

(1) 接口。Servlet、ServletRequest、ServletResponse、ServletConfig、ServletContext、RequestDispatcher、SingleThreadModel。

(2) 类。GenericServlet、 ServletInputStream。

(3) 异常。ServletException、UnavailableException。

javax.servlet.http 包所提供的接口与类如下。

(1) 接口。HttpServletRequest、HttpServletResponse、HttpSession、HttpSessionBinding-Listener。

(2) 类。Cookie、HttpServlet。

5.3　Servlet 的创建与配置

5.3.1　Servlet 的创建

创建 Servlet 的常用方法有 3 种，即直接实现 Servlet 接口、继承 GenericServlet 类与继承 HttpServlet 类。

1. 直接实现 Servlet 接口

任何一个 Servlet 类都必须实现 javax.servlet.Servlet 接口。Servlet 接口定义了 5 个方法，各方法及其说明见表 5.1。

表 5.1　Servlet 接口的方法

方　法	说　明
init()	在 Servlet 实例化之后，Servlet 容器会调用 init()方法来初始化该对象
service()	容器调用 service()方法来处理客户端的请求
destroy()	当容器检测到一个 Servlet 对象应该被移除时，会调用该对象的 destroy()方法来释放其所占用的资源，并保存数据到持久存储设备中
getServletConfig()	返回容器调用 init()方法时传递给 Servlet 对象的 ServletConfig 对象。ServletConfig 对象包含了 Servlet 的初始化参数
getServletInfo()	返回一个字符串，其中包括关于 Servlet 的信息，如作者、版本与版权

【实例 5-1】 Servlet 示例——HelloWorld(HelloWorld.java)。

主要步骤如下。

(1) 新建一个 Web 项目 web_05。

(2) 在项目的 src 文件夹中新建一个包 org.etspace.abc.servlet。

(3) 在 org.etspace.abc.servlet 中新建一个 Servlet——HelloWorld。其文件名为 HelloWorld.java，代码如下：

```
package org.etspace.abc.servlet;
import java.io.IOException;
import javax.servlet.Servlet;
import javax.servlet.ServletConfig;
import javax.servlet.ServletException;
import javax.servlet.ServletRequest;
import javax.servlet.ServletResponse;
import java.io.PrintWriter;
public class HelloWorld implements Servlet {
    public void destroy() {
    }
    public ServletConfig getServletConfig() {
        return null;
    }
    public String getServletInfo() {
        return null;
    }
    public void init(ServletConfig con) throws ServletException {
    }
    public void service(ServletRequest req, ServletResponse res)
            throws ServletException, IOException {
        PrintWriter pw=res.getWriter();
        pw.println("<h1><font color='red'>Hello,World!</font></h1>");
    }
}
```

2. 继承 GenericServlet 类

为简化 Servlet 的编写，javax.servlet 包中提供了一个抽象的类 GenericServlet。GenericServlet 类实现了 Servlet 接口与 ServletConfig 接口，给出了 Servlet 接口中除 service()方法外的其他 4 个方法的简单实现。因此，通过继承 GenericServlet 类创建 Servlet，可有效减少代码的编写量。

【实例 5-2】Servlet 示例——HelloWorld1(HelloWorld1.java)。

主要步骤如下。

在项目 web_05 的 org.etspace.abc.servlet 包中新建一个 Servlet——HelloWorld1。其文件名为 HelloWorld1.java，代码如下：

```
package org.etspace.abc.servlet;
import java.io.IOException;
import javax.servlet.GenericServlet;
import javax.servlet.ServletException;
import javax.servlet.ServletRequest;
import javax.servlet.ServletResponse;
import java.io.PrintWriter;
public class HelloWorld1 extends GenericServlet {
    public void service(ServletRequest req, ServletResponse res)
            throws ServletException, IOException {
        PrintWriter pw=res.getWriter();
        pw.println("<h1><font color='green'>Hello,World!</font></h1>");
    }
}
```

3. 继承 HttpServlet 类

javax.servlet.http 包中提供了一个抽象类 HttpServlet(该类继承自 GenericServlet 类)，可用于快速开发应用于 HTTP 协议的 Servlet 类。

在 HttpServlet 类中，重载了 GenericServlet 的 service()方法：

(1) public void service(ServletRequest req, ServletResponse res) throws ServletException, java.io.IOException。

(2) protected void service(HttpServletRequest req, HttpServletResponse res) throws Servlet Exception, java.io.IOException。

根据不同的请求方法，HttpServlet 提供了 7 个处理方法。

(1) protected void doGet(HttpServletRequest req, HttpServletResponse res) throws Servlet Exception, java.io.IOException。

(2) protected void doPost(HttpServletRequest req, HttpServletResponse res) throws Servlet Exception, java.io.IOException。

(3) protected void doHead(HttpServletRequest req, HttpServletResponse res) throws Servlet Exception, java.io.IOException。

(4) protected void doPut(HttpServletRequest req, HttpServletResponse res) throws Servlet

Exception, java.io.IOException。

(5) protected void doDelete(HttpServletRequest req, HttpServletResponse res) throws Servlet Exception, java.io.IOException。

(6) protected void doTrace(HttpServletRequest req, HttpServletResponse res) throws Servlet Exception, java.io.IOException。

(7) protected void doOptions(HttpServletRequest req, HttpServletResponse res) throws Servlet Exception, java.io.IOException。

当容器接收到一个针对 HttpServlet 对象的请求时，该对象就会调用 public 的 service()方法将参数的类型转换为 HttpServletRequest 与 HttpServletResponse，然后调用 protected 的 service()方法将参数传送进去，接着调用 HttpServletRequest 对象的 getMethod()方法获取请求方法名以调用相应的 doXxx()方法。

HttpServletRequest 与 HttpServletResponse 接口包含在 javax.servlet.http 包中，分别继承自 javax.servle.ServletRequest 与 javax.servle. ServletResponse。其中，HttpServletRequest 接口所提供的常用方法如表 5.2 所示。

表 5.2　HttpServletRequest 接口的常用方法

方　法	说　明
void setAttribute(String name,Object obj)	设置指定属性的值
Object getAttribute(String name)	获取指定属性的值。若指定属性并不存在，则返回 null
Enumeration getAttributeNames()	获取所有可用属性名的枚举
Cookie[]getCookies()	获取与请求有关的 Cookie 对象(Cookie 数组)
String getCharacterEncoding()	获取请求的字符编码方式
String getHeader(String name)	获取指定标头的信息
Enumeration getHeaders(String name)	获取指定标头的信息的枚举
Enumeration getHeaderNames()	获取所有标头名的枚举
ServletInputStream getInputStream()	获取请求的输入流
String getMethod()	获取客户端向服务器端传送数据的方法(如 GET、POST 等)
String getParameter(String name)	获取指定参数的值(字符串)
String[] getParameterValues(String name)	获取指定参数的所有值(字符串数组)
Enumeration getParameterNames()	获取所有参数名的枚举
String getRequestURI()	获取请求的 URL(不包括查询字符串)
String getRemoteHost()	获取发送请求的客户端的主机名
String getRemoteAddr()	获取发送请求的客户端的 IP 地址
HttpSession getSession([Boolean create])	获取与当前客户端请求相关联的 HttpSession 对象。若参数 create 为 true，或不指定参数 create，且 session 对象已经存在，则直接返回之，否则就创建一个新的 session 对象并返回之；若参数 create 为 false，且 session 对象已经存在，则直接返回之，否则就返回 null(即不创建新的 session 对象)

续表

方　法	说　明
String getServerName()	获取接受请求的服务器的主机名
int getServerPort()	获取服务器接受请求所用的端口号
String getServletPath()	获取客户端所请求的文件的路径
void removeAttribute(String name)	删除请求中的指定属性

当一个 Servlet 类继承 HttpServlet 时，无须覆盖其 service()方法，只需覆盖相应的 doXxx()方法即可。通常情况下，都是覆盖其 doGet()与 doPost()方法，然后在其中的一个方法中调用另一个方法，这样就可以做到合二为一了。

Servlet(HttpServlet)的生命周期或运行过程如下。

(1) 当 Servlet(HttpServlet)被装载到容器后，生命周期即刻开始。

(2) 首先调用 init()方法进行初始化，然后调用 service()方法，并根据请求的不同调用相应的 doXxx()方法处理客户请求，同时将处理结果封装到 HttpServletResponse 中返回给客户端。

(3) 当 Servlet 实例从容器中移除时调用其 destroy()方法释放资源，生命周期到此结束。

【实例 5-3】 Servlet 示例——HelloWorld2(HelloWorld2.java)。

主要步骤如下。

在项目 web_05 的 org.etspace.abc.servlet 包中新建一个 Servlet——HelloWorld2，其文件名为 HelloWorld2.java，代码如下：

```
package org.etspace.abc.servlet;
import javax.servlet.http.HttpServlet;
import java.io.IOException;
import javax.servlet.ServletException;
import javax.servlet.http.HttpServletRequest;
import javax.servlet.http.HttpServletResponse;
import java.io.PrintWriter;
public class HelloWorld2 extends HttpServlet {
  protected void doGet(HttpServletRequest req, HttpServletResponse res)
          throws ServletException, IOException {
      PrintWriter pw=res.getWriter();
      pw.println("<h1><font color='blue'>Hello,World!</font></h1>");
  }
  protected void doPost(HttpServletRequest req, HttpServletResponse res)
          throws ServletException, IOException {
      doGet(req,res);
  }
}
```

5.3.2 Servlet 的配置

一个 Servlet 只有在 Web 应用程序的配置文件 web.xml 中进行注册并映射其访问路径后，才能被 Servlet 容器加载以及被外界所访问。

在 web.xml 中配置一个 Servlet 需使用<servlet>与<servlet-mapping>元素，其基本格式为：

```
<?xml version="1.0" encoding="UTF-8"?>
<web-app version="2.5"
    xmlns="http://java.sun.com/xml/ns/javaee"
    xmlns:xsi="http://www.w3.org/2001/XMLSchema-instance"
    xsi:schemaLocation="http://java.sun.com/xml/ns/javaee
    http://java.sun.com/xml/ns/JavaEE/web-app_2_5.xsd">
    <servlet>
        <servlet-name>Servlet_Name</servlet-name>
        <servlet-class>Servlet_Class</servlet-class>
    </servlet>
    <servlet-mapping>
        <servlet-name> Servlet_Name </servlet-name>
        <url-pattern>/Servlet_URL </url-pattern>
    </servlet-mapping>
</web-app>
```

其中，<servlet>元素用于注册一个 Servlet，其<servlet-name>子元素用于指定 Servlet 的名称，<servlet-class>子元素用于指定 Servlet 的类名(包括其所在包的包名)；<servlet-mapping>元素用于映射一个已注册的 Servlet 的对外访问路径，其<servlet-name>子元素用于指定 Servlet 的名称，<url-pattern>子元素用于指定相应的访问路径。

【实例 5-4】 Servlet 配置示例：完成 HelloWorld、HelloWorld1 与 HelloWorld2 三个 Servlet 的配置。

主要步骤如下。

打开 Web 项目的配置文件 web.xml，并在其中添加 HelloWorld、HelloWorld1 与 HelloWorld2 三个 Servlet 的配置代码，具体为：

```
    <servlet>
       <servlet-name>HelloWorld</servlet-name>
       <servlet-class>org.etspace.abc.servlet.HelloWorld</servlet-class>
    </servlet>
    <servlet-mapping>
       <servlet-name>HelloWorld</servlet-name>
       <url-pattern>/HelloWorld</url-pattern>
    </servlet-mapping>
    <servlet>
       <servlet-name>HelloWorld1</servlet-name>
       <servlet-class>org.etspace.abc.servlet.HelloWorld1</servlet-class>
    </servlet>
    <servlet-mapping>
       <servlet-name>HelloWorld1</servlet-name>
       <url-pattern>/HelloWorld1</url-pattern>
    </servlet-mapping>
    <servlet>
       <servlet-name>HelloWorld2</servlet-name>
       <servlet-class>org.etspace.abc.servlet.HelloWorld2</servlet-class>
```

```
    </servlet>
    <servlet-mapping>
        <servlet-name>HelloWorld2</servlet-name>
        <url-pattern>/HelloWorld2</url-pattern>
    </servlet-mapping>
```

配置完毕后，再完成项目的部署，并启动 Tomcat 服务器，即可打开浏览器，并在其中实现对有关 Servlet 的访问。在此，HelloWorld、HelloWorld1 与 HelloWorld2 三个 Servlet 的访问结果分别如图 5.3、图 5.4、图 5.5 所示，相应的地址分别为：

http://localhost:8080/web_05/HelloWorld；

http://localhost:8080/web_05/HelloWorld1；

http://localhost:8080/web_05/HelloWorld2。

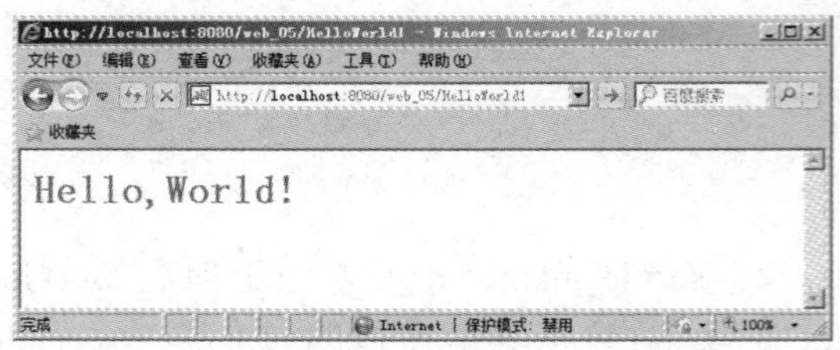

图 5.3　HelloWorld(Servlet)的访问结果

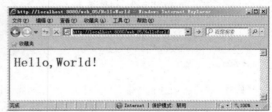

图 5.4　HelloWorld1(Servlet)的访问结果

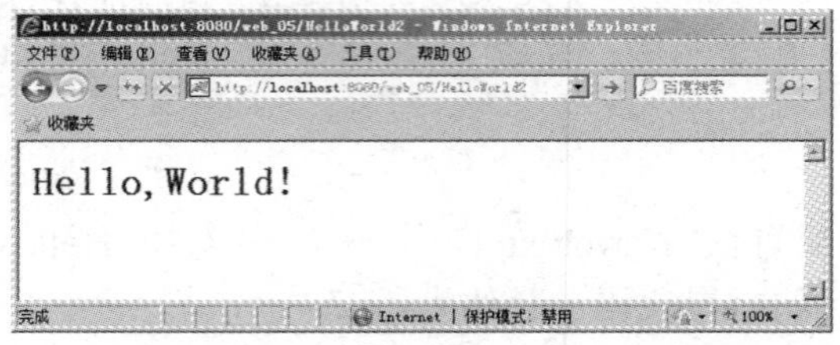

图 5.5　HelloWorld2(Servlet)的访问结果

5.4　Servlet 的基本应用

下面，通过一些具体实例说明 Servlet 的基本应用技术。

【实例 5-5】获取并显示 HTML 表单的数据。图 5.6 所示为“输入内容”页面，在此页面输入要显示的内容后，再单击“提交”按钮提交表单，即可在“显示内容”页面中显示此前所输入的内容，如图 5.7 所示。

图 5.6　“输入内容”页面

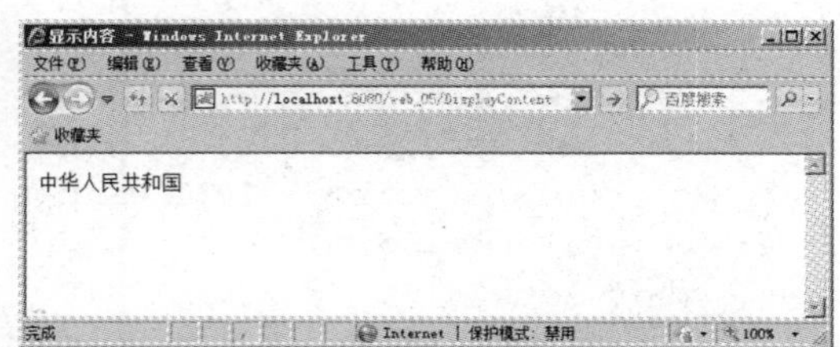

图 5.7　“显示内容”页面

主要步骤(在项目 web_05 中)如下。

(1) 在项目的 WebRoot 文件夹中添加一个新的 JSP 页面 InputContent.jsp，其代码如下：

```
<%@ page language="java" pageEncoding="gbk" %>
<html>
    <head>
        <title>输入内容</title>
    </head>
    <body>
        <div>
        <form action="DisplayContent" method="post">
            请输入您要显示的内容：
            <input type="text" name="content">
            <input type="submit" value="提交">
            <input type="reset" value="重置">
        </form>
        </div>
    </body>
</html>
```

(2) 在项目的 org.etspace.abc.servlet 包中新建一个 Servlet——DisplayContent，其文件名为 DisplayContent.java，代码如下：

```
package org.etspace.abc.servlet;
import javax.servlet.http.HttpServlet;
import java.io.IOException;
import javax.servlet.ServletException;
import javax.servlet.http.HttpServletRequest;
import javax.servlet.http.HttpServletResponse;
import java.io.PrintWriter;
public class DisplayContent extends HttpServlet {
   protected void doGet(HttpServletRequest req, HttpServletResponse res)
            throws ServletException, IOException {
        req.setCharacterEncoding("gbk");
        res.setCharacterEncoding("gbk");
        String content=req.getParameter("content");
        PrintWriter pw=res.getWriter();
        pw.println("<html><head><title>显示内容</title><head><body>");
        pw.println(content);
        pw.println("</body></html>");
    }
    protected void doPost(HttpServletRequest req, HttpServletResponse
res)
            throws ServletException, IOException {
        doGet(req,res);
    }
}
```

(3) 在项目的配置文件 web.xml 中添加 DisplayContent(Servlet)的配置代码，具体为：

```
    <servlet>
        <servlet-name>DisplayContent</servlet-name>
        <servlet-class>org.etspace.abc.servlet.DisplayContent</servlet-
class>
    </servlet>
    <servlet-mapping>
        <servlet-name>DisplayContent</servlet-name>
        <url-pattern>/DisplayContent</url-pattern>
    </servlet-mapping>
```

解析：

(1) 在本实例中，HTML 表单被提交后，将由 DisplayContent(Servlet)进行处理。

(2) 在 DisplayContent(Servlet)中，根据表单元素的名称读取其中所输入的内容，然后按一定的格式输出相应的 HTML 代码。

【实例 5-6】Servlet 访问 Cookie 示例。设计一个 Servlet，其功能为记录并显示用户登录系统的次数，如图 5.8 所示。

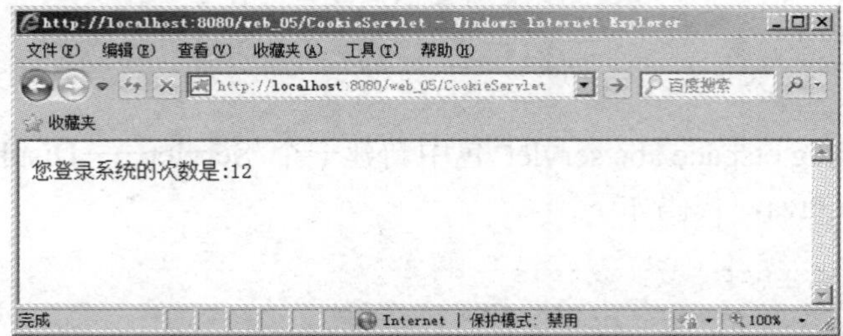

图 5.8　用户登录次数页面

主要步骤(在项目 web_05 中)如下。

(1) 在项目的 org.etspace.abc.servlet 包中新建一个 Servlet——CookieServlet。其文件名为 CookieServlet.java，代码如下：

```
package org.etspace.abc.servlet;
import java.io.*;
import javax.servlet.*;
import javax.servlet.http.*;
public class CookieServlet extends HttpServlet
{
    public void service(HttpServletRequest request,HttpServletResponse
response)throws IOException
    {
        boolean myFound=false;
        Cookie myCookie=null;
        Cookie[] allCookie=request.getCookies();
        response.setContentType("text/html;charset=gbk");
        PrintWriter out=response.getWriter();
        if (allCookie!=null)
        {
            for (int i=0;i<allCookie.length;i++)
```

```
            {
                if (allCookie[i].getName().equals("logincount"))
                {
                    myFound=true;
                    myCookie=allCookie[i];
                }
            }
        }
        out.println("<html>");
        out.println("<body>");
        if (myFound)
        {
            int myLC=Integer.parseInt(myCookie.getValue());
            myLC++;
            out.println("您登录系统的次数是:"+String.valueOf(myLC));
            myCookie.setValue(String.valueOf(myLC));
            myCookie.setMaxAge(30*24*60*60);
            response.addCookie(myCookie);
        }
        else
        {
            out.println("这是您首次登录系统.");
            myCookie=new Cookie("logincount",String.valueOf(1));
            myCookie.setMaxAge(30*24*60*60);
            response.addCookie(myCookie);
        }
        out.println("</body>");
        out.println("</html>");
    }
}
```

(2) 在项目的配置文件 web.xml 中添加 CookieServlet(Servlet)的配置代码，具体为：

```
   <servlet>
      <servlet-name>CookieServlet</servlet-name>
      <servlet-class>org.etspace.abc.servlet.CookieServlet</servlet-
class>
   </servlet>
   <servlet-mapping>
      <servlet-name>CookieServlet</servlet-name>
      <url-pattern>/CookieServlet</url-pattern>
   </servlet-mapping>
```

配置完毕后，再完成项目的部署，并启动 Tomcat 服务器，即可打开浏览器访问该 Servlet，地址为：http://localhost:8080/web_05/CookieServlet。

【实例 5-7】Servlet 访问 Session 示例。设计一个 Servlet，其功能为记录并显示用户访问站点的次数，同时显示有关的请求信息与 Session 信息，如图 5.9 所示。

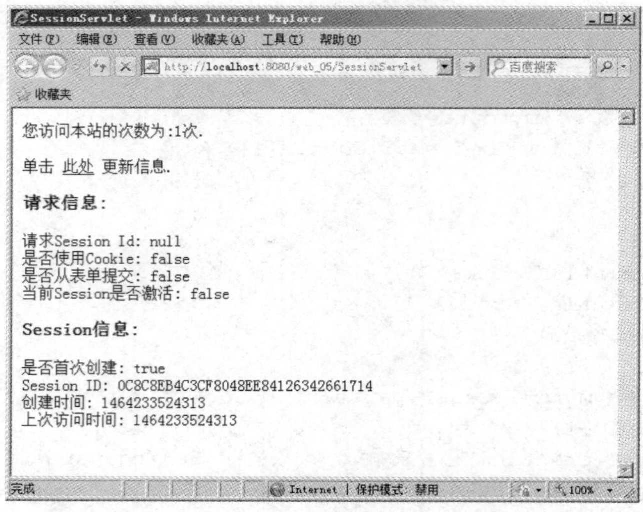

(a)

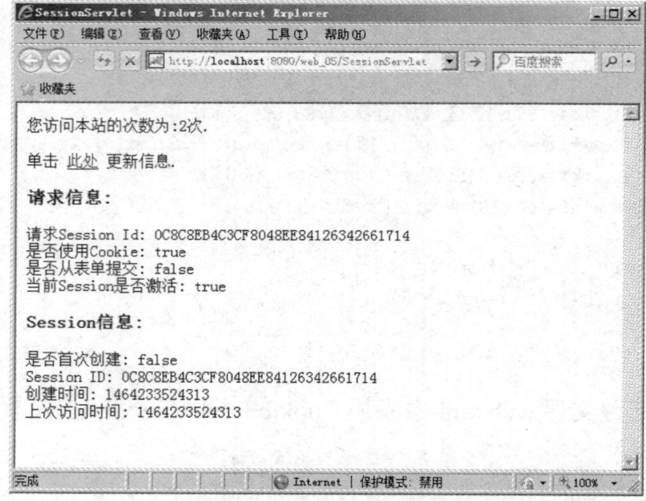

(b)

图 5.9 信息显示页面

主要步骤(在项目 web_05 中)如下。

(1) 在项目的 org.etspace.abc.servlet 包中新建一个 Servlet——SessionServlet，其文件名为 SessionServlet.java，代码如下：

```
package org.etspace.abc.servlet;
import java.io.*;
import java.util.Enumeration;
import javax.servlet.*;
import javax.servlet.http.*;
public class SessionServlet extends HttpServlet
{
  //doGet 方法
```

```
 public void doGet (HttpServletRequest request, HttpServletResponse
response)
      throws ServletException, IOException
  {
    // 获取会话对象(若该对象不存在，则新建之)
    HttpSession session = request.getSession(true);
       // 设置内容类型
    response.setContentType("text/html;charset=gbk");
    // 获取 PrintWriter 对象
    PrintWriter out = response.getWriter();
    out.println("<html>");
       out.println("<head><title>SessionServlet</title></head><body>");
    out.println("<p>");
       // 获取 Session 变量
    Integer myCounter = (Integer) session.getAttribute("counter");
       if (myCounter==null)
        myCounter = new Integer(1);
    else
        myCounter = new Integer(myCounter.intValue() + 1);
    //设置 Session 变量
    session.setAttribute("counter", myCounter);
       out.println("您访问本站的次数为:" + myCounter + "次.<p>");
    out.println("单击 <a href=" + response.encodeURL("SessionServlet")
+">此处</a>");
    out.println(" 更新信息." );
     out.println("<p>");
    out.println("<h3>请求信息:</h3>");
    out.println("请求 Session Id: " +request.getRequestedSessionId());
    out.println("<br>是否使用 Cookie: "
+request.isRequestedSessionIdFromCookie());
    out.println("<br>是否从表单提交: "
+request.isRequestedSessionIdFromURL());
    out.println("<br>当前 Session 是否激活: "
+request.isRequestedSessionIdValid());
    out.println("<h3>Session 信息:</h3>");
    out.println("是否首次创建: " + session.isNew());
    out.println("<br>Session ID: " + session.getId());
    out.println("<br>创建时间: " + session.getCreationTime());
    out.println("<br>上次访问时间: " +session.getLastAccessedTime());
       out.println("</body></html>");
  }
    public void doPost(HttpServletRequest request, HttpServletResponse
response)
            throws ServletException, IOException {
        doGet(request,response);
    }
}
```

(2) 在项目的配置文件 web.xml 中添加 SessionServlet(Servlet)的配置代码，具体为：

```
   <servlet>
```

```
        <servlet-name>SessionServlet</servlet-name>
        <servlet-class>org.etspace.abc.servlet.SessionServlet</servlet-
class>
    </servlet>
    <servlet-mapping>
        <servlet-name>SessionServlet</servlet-name>
        <url-pattern>/SessionServlet</url-pattern>
    </servlet-mapping>
```

配置完毕后，再完成项目的部署，并启动 Tomcat 服务器，即可打开浏览器访问该 Servlet，地址为：http://localhost:8080/web_05/SessionServlet。

5.5 Servlet 的应用案例

5.5.1 系统登录

下面，以 JSP+JDBC+JavaBean+Servlet 模式实现 Web 应用系统中的系统登录功能，其运行结果与第 4 章 4.5.1 小节的系统登录案例相同。

实现步骤(在项目 web_05 中)如下。

(1) 将 SQL Server 2005/2008 的 JDBC 驱动程序 sqljdbc4.jar 添加到项目的 WebRoot\WEB-INF\lib 文件夹中。

(2) 在项目的 src 文件夹中新建一个包 org.etspace.abc.jdbc。

(3) 在 org.etspace.abc.jdbc 包中新建一个 JavaBean——DbBean。其文件名为 DbBean.java，代码与第 4 章 4.5.1 小节系统登录案例中的相同。

(4) 在项目的 WebRoot 文件夹中新建一个子文件夹 syslogin。

(5) 在子文件夹 syslogin 中添加一个新的 JSP 页面 login.jsp。其代码与第 4 章 4.5.1 小节系统登录案例中的类似，只需将其中<form>标记的 action 属性设为 LoginServletByDB 即可，具体为：

```
<form action="LoginServletByDB" method="post">
```

(6) 在项目的 org.etspace.abc.servlet 包中新建一个 Servlet——LoginServletByDB。其文件名为 LoginServletByDB.java，代码如下：

```
package org.etspace.abc.servlet;
import java.io.*;
import java.io.IOException;
import java.io.PrintWriter;
import javax.servlet.RequestDispatcher;
import javax.servlet.ServletException;
import javax.servlet.http.HttpServlet;
import javax.servlet.http.HttpServletRequest;
import javax.servlet.http.HttpServletResponse;
import javax.servlet.http.HttpSession;
import java.sql.*;
import org.etspace.abc.jdbc.DbBean;
```

```
public class LoginServletByDB extends HttpServlet {
    public void doGet(HttpServletRequest request, HttpServletResponse
response)
            throws ServletException, IOException {
        doPost(request,response);
    }
    public void doPost(HttpServletRequest request, HttpServletResponse
response)
            throws ServletException, IOException {
        request.setCharacterEncoding("gbk");
        response.setCharacterEncoding("gbk");
        response.setContentType("text/html;charset=gbk");
        PrintWriter out = response.getWriter();
        try {
            String username = null ;
            String password = null ;
            HttpSession session = null ;
            session = request.getSession(true);
            username = request.getParameter("username");
            password = request.getParameter("password");
            String sql="select * from users where UserName='"+username+"'
and password='"+password+ "'";
            DbBean myDbBean=new DbBean();
            myDbBean.openConnection();
            ResultSet rs= myDbBean.executeQuery(sql);
            if(rs.next()){
                session.setAttribute("userName", username);
                response.sendRedirect("welcome.jsp");
            }else{
                response.sendRedirect("error.jsp");
            }
            rs.close();
            myDbBean.closeConnection();
        }
        catch(SQLException e)
        {
            out.println(e.getMessage());
        }
    }
}
```

(7) 在项目的配置文件 web.xml 中添加 LoginServletByDB(Servlet)的配置代码，具体为：

```
<servlet>
    <servlet-name>LoginServletByDB</servlet-name>
    <servlet-class>org.etspace.abc.servlet.LoginServletByDB</servlet-
class>
</servlet>
<servlet-mapping>
    <servlet-name>LoginServletByDB</servlet-name>
```

```
        <url-pattern>/syslogin/LoginServletByDB</url-pattern>
    </servlet-mapping>
```

(8) 在子文件夹 syslogin 中添加一个新的 JSP 页面 welcome.jsp，其代码如下：

```
<%@ page language="java" pageEncoding="utf-8" %>
<%request.setCharacterEncoding("utf-8"); %>
<html>
    <head>
        <title>登录成功</title>
    </head>
    <body>
        <%=session.getAttribute("userName") %>，您好！欢迎光临本系统。
    </body>
</html>
```

(9) 在子文件夹 syslogin 中添加一个新的 JSP 页面 error.jsp。其代码与第 4 章 4.5.1 小节系统登录案例中的相同。

说明： 在 LoginServletByDB(Servlet)中，可将以下两条语句

```
session.setAttribute("userName", username);
response.sendRedirect("welcome.jsp");
```

修改为：

```
RequestDispatcher
dispatcher=request.getRequestDispatcher("welcome.jsp");
dispatcher.forward(request, response);
```

与此同时，应将 welcome.jsp 页面中的语句

```
<%=session.getAttribute("userName") %>
```

修改为：

```
<%out.print(request.getParameter("username")); %>
```

5.5.2 数据添加

下面，以 JSP+JDBC+JavaBean+Servlet 模式实现“职工增加”功能，其运行结果与第 4 章 4.5.2 小节职工增加案例中的相同。

主要步骤(在项目 web_05 中)如下。

(1) 在项目的 WebRoot 文件夹中添加一个新的 JSP 页面 ZgZj.jsp。其代码与第 4 章 4.5.2 小节职工增加案例中的类似，只需将其中<form>标记的 action 属性设为 zgzjServlet 即可，具体为：

```
<form id="form1" name="form1" method="post" action="zgzjServlet"
onSubmit="return check(this)">
```

(2) 在项目的 org.etspace.abc.servlet 包中新建一个 Servlet——zgzjServlet。其文件名为 zgzjServlet.java，代码如下。

```
package org.etspace.abc.servlet;
import java.io.*;
import java.io.IOException;
import java.io.PrintWriter;
import javax.servlet.RequestDispatcher;
import javax.servlet.ServletException;
import javax.servlet.http.HttpServlet;
import javax.servlet.http.HttpServletRequest;
import javax.servlet.http.HttpServletResponse;
import javax.servlet.http.HttpSession;
import java.sql.*;
import org.etspace.abc.jdbc.DbBean;
public class zgzjServlet extends HttpServlet {
   public void doGet(HttpServletRequest request, HttpServletResponse
response)
         throws ServletException, IOException {
      doPost(request,response);
   }
   public void doPost(HttpServletRequest request, HttpServletResponse
response)
         throws ServletException, IOException {
      request.setCharacterEncoding("gb2312");
      response.setCharacterEncoding("gb2312");
      response.setContentType("text/html;charset=gb2312");
      PrintWriter out = response.getWriter();
      try {
         String bh = request.getParameter("bh");
         String xm = request.getParameter("xm");
         String xb = request.getParameter("xb");
         String bm = request.getParameter("bm");
         String csrq = request.getParameter("csrq");
         String jbgz = request.getParameter("jbgz");
         String gwjt = request.getParameter("gwjt");
         String sql = "insert into zgb(bh,xm,xb,bm,csrq,jbgz,gwjt)";
         sql=sql + " values('" + bh + "','" + xm + "','" + xb + "','"
+ bm + "','" + csrq + "'," + jbgz + "," + gwjt + ")";
         DbBean myDbBean=new DbBean();
         myDbBean.openConnection();
        int n = myDbBean.executeUpdate(sql);
        if (n>0){
           out.print("职工记录增加成功！");
        }
         else {
           out.print("职工记录增加失败！");
         }
         myDbBean.closeConnection();
      }
      catch (Exception e) {
         out.print(e.toString());
      }
```

```
    }
}
```

(3) 在项目的配置文件 web.xml 中添加 zgzjServlet(Servlet)的配置代码，具体为：

```
<servlet>
    <servlet-name>zgzjServlet</servlet-name>
    <servlet-class>org.etspace.abc.servlet.zgzjServlet</servlet-class>
</servlet>
<servlet-mapping>
    <servlet-name>zgzjServlet</servlet-name>
    <url-pattern>/zgzjServlet</url-pattern>
</servlet-mapping>
```

说明： JSP+JDBC+JavaBean+Servlet(通常又简称为 JSP+Servlet+JavaBean)模式即 Model2 模式(见图 5.10)，是对 Model1 模式进行改造后发展出来的一种新模式。Model2 模式从根本上克服了 Model1 模式的缺陷，其工作流程则可分为以下 5 个步骤。

(1) Servlet 接收浏览器发出的请求。

(2) Servlet 根据不同的请求调用相应的 JavaBean 进行处理。

(3) JavaBean 按自己的业务逻辑通过 JDBC 操作数据库。

(4) Servlet 将处理结果传递给 JSP 视图。

(5) JSP 动态生成相应的 Web 页面，并发送到浏览器加以呈现。

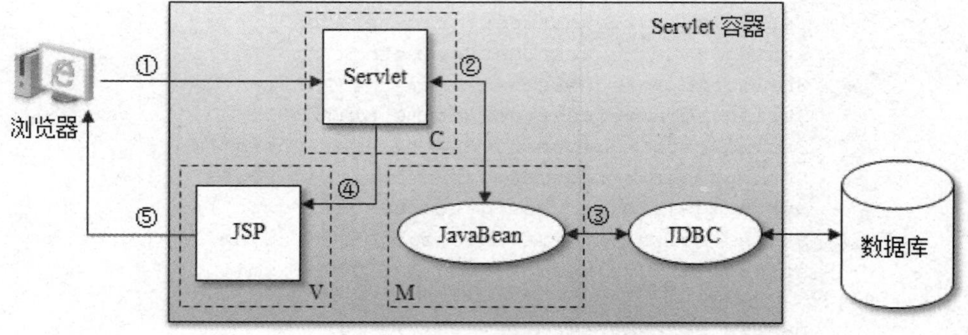

图 5.10　JSP+JDBC+JavaBean+Servlet 模式(Model2 模式)

与 Model1 模式相比较，Model2 引入了 Servlet 组件，并将控制功能交由 Servlet 实现，而 JSP 只负责显示功能，从而实现控制逻辑与显示逻辑的分离，提高了程序的可维护性。

Model2 其实就是 JSP 中的 MVC 模式。其中，模型(M)、视图(V)与控制器(C)的具体实现如下。

(1) 模型(Model)。一个或多个 JavaBean 对象，用于存储数据，提供简单的 setXxx()方法与 getXxx()方法。

(2) 视图(View)。一个或多个 JSP 页面，为模型提供数据显示。JSP 页面主要使用 HTML/XHTML 标记和 JavaBean 的相关标记来显示数据。

(3) 控制器(Controller)。一个或多个 Servlet 对象根据视图提交的请求进行数据处理操作，并将有关的结果存储到 JavaBean 中，然后使用重定向方式请求视图中的某个 JSP 页面更新显示。

不过，Model2 模式虽然成功地克服了 Model1 的缺陷，但却是以重新引入原始的 Servlet 编程为代价的，而暴露 Servlet API 必然会导致编程难度的增加。在此背景下，为屏蔽 Servlet API 的复杂性，减少使用 Model2 模式开发系统的工作量，Struts 等框架便应运而生了。

本 章 小 结

本章首先介绍了 Servlet 的基本概念、生命周期与技术规范，然后通过具体实例讲解了 Servlet 的创建、配置与应用方法，最后再通过具体案例说明了 Servlet 的综合应用技术。通过本章的学习，读者应熟练掌握 Servlet 的创建、配置与应用方法，并能使用 JSP+JDBC+JavaBean+Servlet 模式(即 Model2 模式)开发相应的 Web 应用系统。

思　考　题

1. Servlet 是什么？
2. 一个 Servlet 的生命周期可分为哪几个阶段？
3. 请简述 Servlet 请求的响应过程。
4. Servlet 的常用创建方法有哪些？
5. 如何配置一个 Servlet？

第 6 章

Struts 2 框架

Struts 2 是一个基于 MVC 设计模式的 Web 应用框架，在基于框架的 Java EE 应用开发中使用得相当普遍。

本章要点：

- Struts 2 概述、Struts 2 基本应用、Struts 2 拦截器；
- Struts 2 OGNL、Struts 2 标签库、Struts 2 数据验证、Struts 2 文件上传与下载。

学习目标：

- 了解 Struts 2 的基本执行流程；
- 掌握 Struts 2 的基本应用技术；
- 掌握 Struts 2 拦截器的实现与配置方法；
- 掌握 Struts 2 OGNL 与常用标签的基本用法；
- 掌握 Struts 2 的数据验证技术；
- 掌握 Struts 2 的文件上传与下载方法；
- 掌握 Web 应用系统开发的 JSP+JDBC+JavaBean+Struts 2 模式。

6.1 Struts 2 概述

Struts 2 是一个基于 MVC 设计模式的 Web 应用框架，本质上相当于一个 Servlet。在 MVC 设计模式中，Struts 2 作为控制器(Controller)使用，并可用于建立模型(Mode)与视图(View)之间的数据交互。对于 Struts 2 框架来说，其应用重点在于“控制”，而最简单的流程则为“页面→控制器→页面”，其中最重要的问题就是控制器的数据获取与数据传送。

Struts 是 Apache 软件基金会赞助的一个开源项目(最初是 Jakarta 项目中的一个子项目)。作为 Struts 1 的换代产品，Struts 2 是 Struts 1 与另外一个优秀的 Web 框架 WebWork 相互融合的结果。虽然 Struts 2 是在 Struts1 的基础上发展起来的，但二者的体系结构差别巨大。实际上，Struts 2 是以 Webwork 为核心的，因此可理解为 WebWork 的更新产品。从结果来看，Struts 2 为传统的 Struts 1 注入了 WebWork 的先进的设计理念，并将这两个框架有效地统一起来。在 Struts 2 中，采用拦截器机制来处理用户的请求，可使业务逻辑控制器能够与 Servlet API 完全脱离。相对于 WebWork 来说，Struts 2 的变化则较小。

Struts 2 框架内部的运作基于一种称为“过滤器”的机制，其工作原理如图 6.1 所示。由图可见，Struts 2 的基本执行流程如下。

(1) 客户端提交一个请求(HttpServletRequest)。

(2) 请求被提交到一系列的过滤器(Filter)中。过滤器主要有 3 层，依次为 ActionContextCleanUp、其他过滤器(SiteMesh 等)、FilterDispatcher。

(3) FilterDispatcher 接收到请求后，询问 ActionMapper 是否需要调用某个 Action 来处理这个请求。如果 ActionMapper 决定需要调用某个 Action，那么 FilterDispatcher 就将请求的处理交给 ActionProxy。

(4) ActionProxy 通过 Configuration Manager 询问框架的配置文件 struts.xml，找到需要调用的 Action 实现类(该 Action 实现类通常为程序员自定义的用于处理请求的类)，并创建一个 ActionInvocation 实例。

(5) ActionInvocation 通过代理模式调用 Action，但在调用之前，ActionInvocation 会根据配置加载与 Action 相关的所有拦截器(Interceptor)。

(6) 执行 Action 的有关方法对请求进行处理，并返回相应的结果代码。

(7) ActionInvocation 根据结果代码以及配置文件 struts.xml 中的相关配置，找到对应的结果(Result)，确定结果类型与处理方式，并完成后续的处理过程(如跳转到相应的 JSP 页面等)。

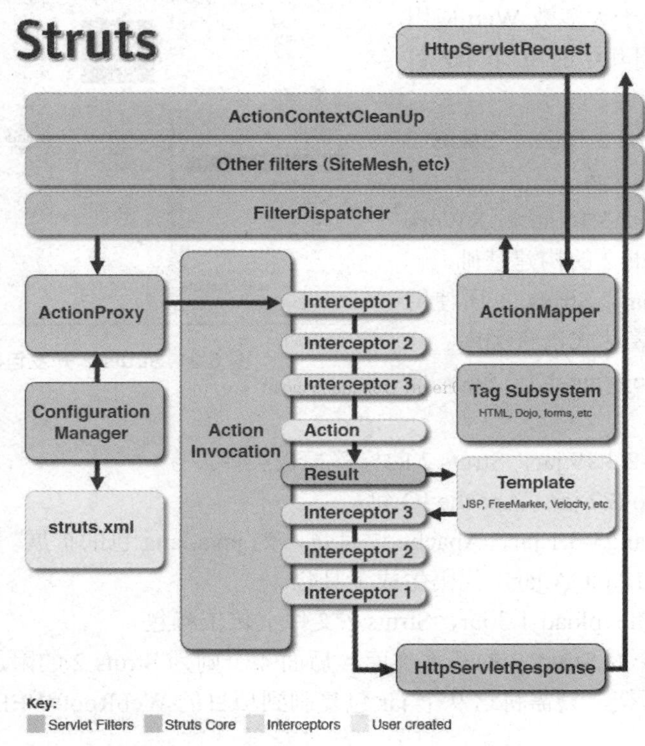

图 6.1　Struts 2 的工作原理

6.2　Struts 2 基本应用

6.2.1　Struts 2 开发包

MyEclipse 10 内置了对 Struts 2 的支持，但只限于某些特定的版本。若要使用更高的版本，则需要自行下载相应的 Struts 2 开发包。Struts 2 开发包可从其官网(http://struts.apache.org/)或其他有关网站下载，在此下载的开发包为 struts-2.3.16-all.zip(完整版)。

解压缩 Struts 2 开发包后，即可查看到其目录结构，如图 6.2 所示。Struts 2 开发包具有典型的 Web 结构，包含有 4 个子目录。

(1) apps。内含基于 Struts 2 框架的应用示例，可作为 Struts 2 的学习资料。

(2) docs。内含 Struts 2 框架的有关文档，如 Struts 2 快速入门、Struts 2 文档、API 文档等。

(3) lib。内含 Struts 2 框架的各种类库以及 Struts 2 的第三方插件类库。

(4) src。内含 Struts 2 框架的全部源代码。

通常，开发基于 Struts 2 的 Web 应用并不需要用到 Struts 2 的全部特性，因此无须加载 lib 目录下的所有类库(即 jar 包)。实际上，对于大多数 Web 应用开发来说，只需使用其中的 9 个 jar 包即可，具体如下。

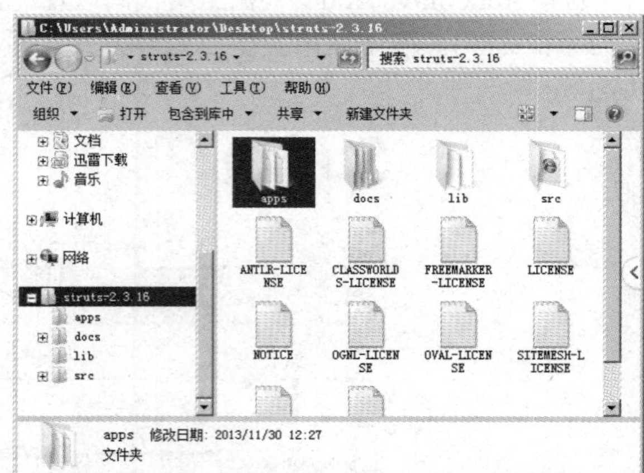

图 6.2　Struts 2 开发包的目录结构

- struts2-core-2.3.16.jar，Struts 2 框架的核心类库。
- xwork-core-2.3.16.jar，XWork 类库，Struts 2 的构建基础。
- ognl-3.0.6.jar，Struts 2 所使用的 OGNL 表达式语言类库。
- commons-logging-1.1.3.jar，日志类库。
- freemarker-2.3.19.jar，Struts 2 的标签模板类库。
- commons-io-2.2.jar，Apache IO 包。
- commons-lang3-3.1.jar，Apache 语言包，为 java.lang 包的扩展。
- javassist-3.11.0.GA.jar，代码生成工具包。
- commons-fileupload-1.3.jar，Struts 2 文件上传依赖包。

其中，前面 5 个为 Struts 2 的基本类库，后面 4 个则为 Struts 2 的附加类库。为使 Web 项目支持 Struts 2 框架，只需将这 9 个 jar 包复制到项目的 WebRoot\WEB-INF\lib 文件夹之中即可。

6.2.2　Struts 2 基本用法

下面，先通过一个简单的实例说明 Struts 2 框架的基本用法。

【实例 6-1】Struts 2 应用示例：用户注册。图 6.3 所示为“用户注册”页面，用户输入相应的用户名与密码后，再单击“确定”按钮后，即可显示相应的结果。若用户名不是“admin”，则允许注册，结果如图 6.4 所示；反之，若用户名为“admin”，则不允许注册，结果如图 6.5 所示。

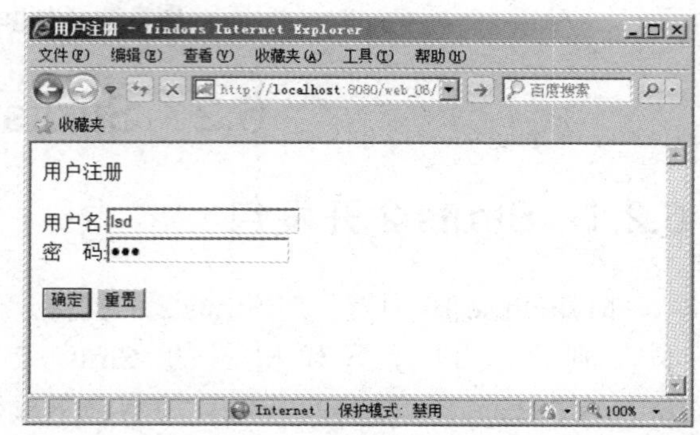

图 6.3　“用户注册”页面

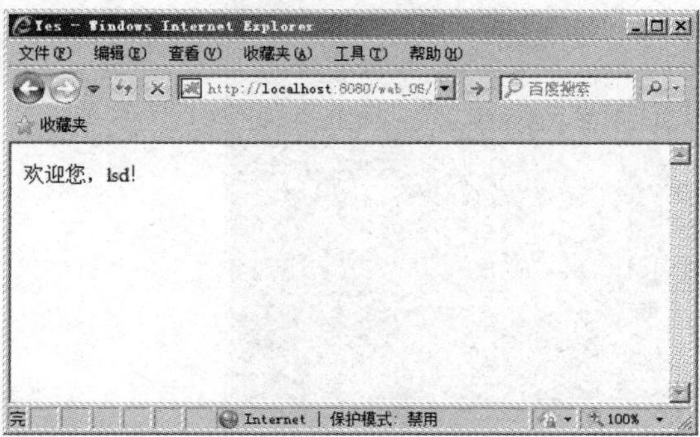

图 6.4　注册成功页面

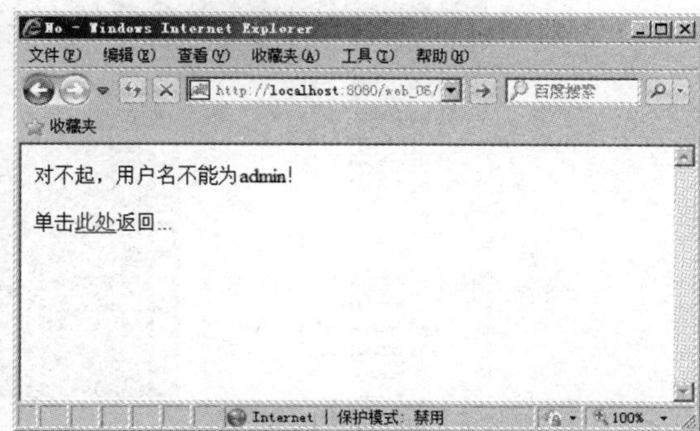

图 6.5　注册失败页面

主要步骤如下。

(1) 新建一个 Web 项目 web_06。

(2) 加载 Struts 2 类库。为此，只需将 Struts 2 的 5 个基本类库与 4 个附加类库(共 9 个 jar 包)复制到项目的 WebRoot\WEB-INF\lib 文件夹即可，如图 6.6 所示。

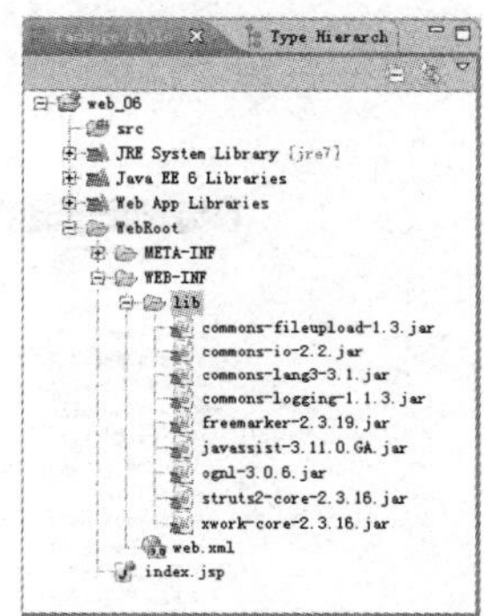

图 6.6　Struts 2 的相关类库

提示： 在 Web 项目 web_06 中加载 Struts 2 类库也可按以下方法进行。

(1) 右击项目名 web_06，并在其快捷菜单中选择 Build Path→Configure Build Path 菜单，打开 Properties for web_06 对话框，并在其中切换至 Libraries 选项卡，如图 6.7 所示。

(2) 单击 Add External JARs 按钮，并在随之弹出的 JAR Selection 对话框中选中 Struts 2 的 5 个基本类库与 4 个附加类库，如图 6.8 所示，然后单击“打开”按钮，返回 Properties for web_06 对话框，如图 6.9 所示。

(3) 单击 OK 按钮，关闭 Properties for web_06 对话框。

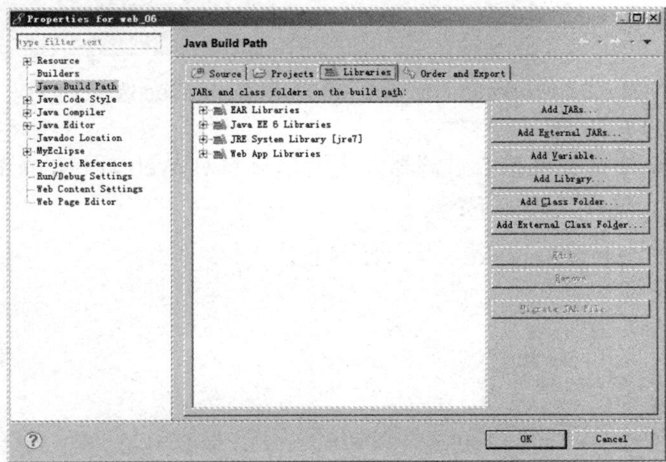

图 6.7　Properties for web_06 对话框(Libraries 选项卡)(1)

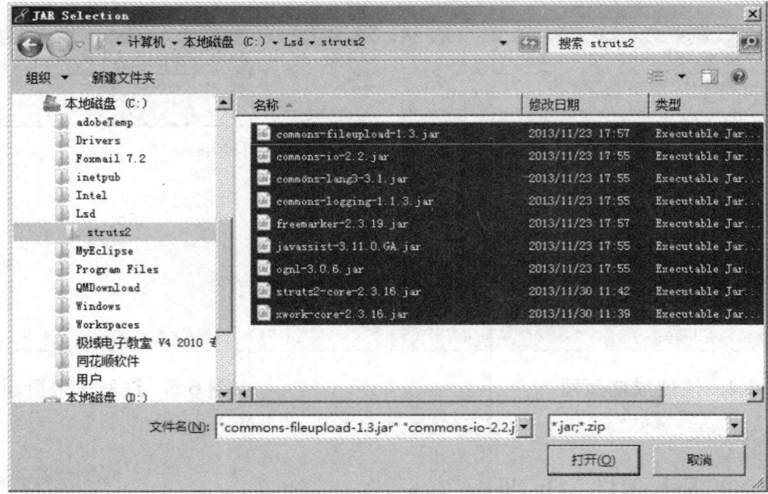

图 6.8　JAR Selection 对话框

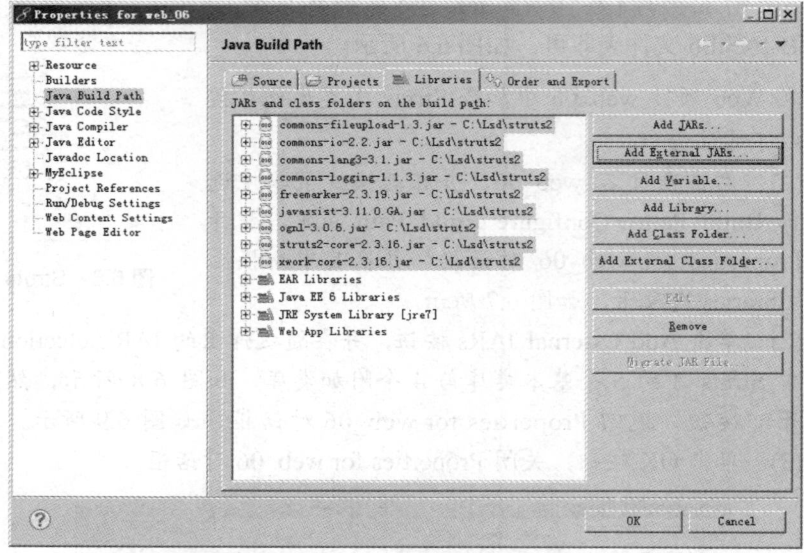

图 6.9　Properties for web_06 对话框(Libraries 选项卡)(2)

(3) 配置 Struts 2 框架的核心过滤器。为此，只需对 Web 项目的配置文件 web.xml 进行相应的修改即可。具体代码如下：

```
<?xml version="1.0" encoding="UTF-8"?>
<web-app …>
    …
    <filter>
        <filter-name>struts2</filter-name>
        <filter-
class>org.apache.struts2.dispatcher.FilterDispatcher</filter-class>
    </filter>
    <filter-mapping>
```

```
        <filter-name>struts2</filter-name>
        <url-pattern>/*</url-pattern>
    </filter-mapping>
…
</web-app>
```

(4) 在项目的 WebRoot 文件夹中添加一个用户注册页面 register.jsp，其代码如下：

```
<%@ page language="java" pageEncoding="utf-8" %>
<html>
    <head>
        <title>用户注册</title>
    </head>
    <body>
        用户注册<br>
        <form action="register.action" method="post">
            用户名:<input type="text" name="username"><br>
            密  码:<input type="password"
name="password"><br><br>
            <input type="submit" value="确定">
            <input type="reset" value="重置">
        </form>
    </body>
</html>
```

(5) 创建 Action 实现类 RegisterAction。为此，可按以下步骤进行操作。

① 在项目的 src 文件夹中新建一个包 org.etspace.abc.action。

② 在 org.etspace.abc.action 包中新建一个 Java 类(即 Action 实现类)RegisterAction，如图 6.10 所示。其文件名为 RegisterAction.java，代码如下：

```
package org.etspace.abc.action;
import com.opensymphony.xwork2.ActionSupport;
import java.util.Map;
import com.opensymphony.xwork2.ActionContext;
public class RegisterAction extends ActionSupport {
    private String username;
    private String password;
    public String getUsername(){
        return username;
    }
    public void setUsername(String username){
        this.username=username;
    }
    public String getPassword(){
        return password;
    }
    public void setPassword(String password){
        this.password=password;
    }
    public String execute() throws Exception {
        if(!username.equals("admin")){
```

```
                //获取 request 对象的 MAP
                Map request=(Map)ActionContext.getContext().get("request");
                //将用户名保存到 request 对象的 MAP 中
                request.put("username",username);
                return "success";
            }else{
                return "error";
            }
        }
}
```

说明： 在 Web 项目中使用 Struts 2 框架，其主要的编码工作就是编写有关的 Action。Action 可以是一个普通的 Java 类，也可以继承自 ActionSupport 类，其属性定义必须符合 JavaBean 规范。在一个 Action 中，通常要包含一个 execute()方法，该方法即为 Action 被调用后会默认执行的方法。

(6) 创建配置文件 struts.xml。struts.xml 是 Struts 2 框架的核心配置文件，主要用于配置开发人员所编写的各个 Action。为创建 struts.xml 文件，只需在项目中右击 src 文件夹，并在其快捷菜单中选择 New→XML(Basic Templates)菜单项，打开 New XML File 对话框(见图 6.11)，然后在 File name 文本框中输入文件名 struts.xml，最后再单击 Finish 按钮即可。在此，struts.xml 文件的代码为：

```
<?xml version="1.0" encoding="UTF-8"?>
<!DOCTYPE struts PUBLIC
    "-//Apache Software Foundation//DTD Struts Configuration 2.0//EN"
    "http://struts.apache.org/dtds/struts-2.0.dtd">
<struts>
    <package name="struts" extends="struts-default">
        <action name="register"
class="org.etspace.abc.action.RegisterAction">
            <result name="success">/register_success.jsp</result>
            <result name="error">/register_error.jsp</result>
        </action>
    </package>
</struts>
```

(7) 在项目的 WebRoot 文件夹中添加一个注册成功页面 register_success.jsp，其代码如下：

```
<%@ page language="java" import="java.util.*" pageEncoding="UTF-8"%>
<%@ taglib uri="/struts-tags" prefix="s" %>
<html>
  <head>
    <title>Yes</title>
  </head>
  <body>
     欢迎您，<s:property value="#request.username"/>!
  </body>
</html>
```

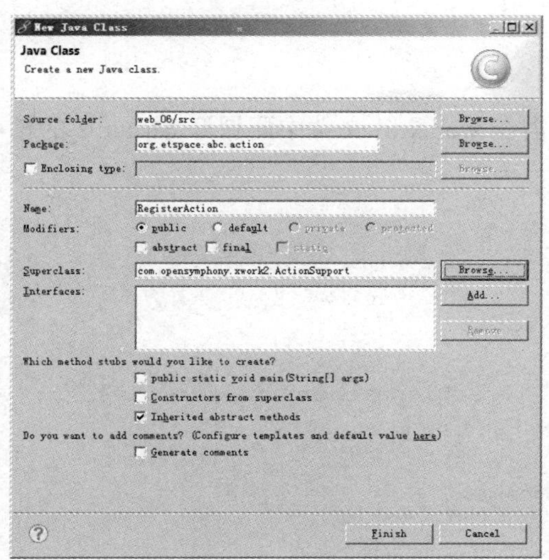

图 6.10　Action 实现类的创建

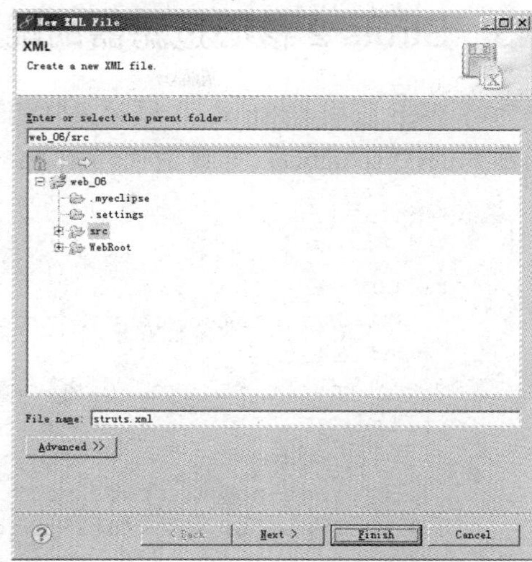

图 6.11　New XML File 对话框

(8) 在项目的 WebRoot 文件夹中添加一个注册失败页面 register_error.jsp，其代码如下：

```
<%@ page language="java" import="java.util.*" pageEncoding="UTF-8"%>
<%@ taglib uri="/struts-tags" prefix="s" %>
<html>
  <head>
    <title>No</title>
  </head>
<body>
    对不起，用户名不能为<s:property value="username"/>!
    <br><br>
    单击<a href="register.jsp">此处</a>返回…
  </body>
</html>
```

至此，项目创建完毕。完成项目的部署后，启动 Tomcat 服务器，然后在浏览器中输入地址 http://localhost:8080/web_06/register.jsp 并按 Enter 键，即可打开图 6.3 所示的“用户注册”页面，开始其运行流程。

在具体应用中，Struts 2 框架的配置文件 struts.xml 主要起到两个方面的映射作用。

(1) 根据请求“*. action”的“*”来决定调用哪个 Action。例如，在【实例 6-1】中，请求为“register.action”，其中“*”部分为“register”。Struts 2 框架接收到该请求后，即可在 struts.xml 文件中查找到名称(name)为“register”的 Action，并获知其实现类(class)为 org.etspace.abc.action.RegisterAction，然后再调用之。

(2) 根据 Action 返回的结果代码确定结果类型与处理方式，并完成相应的处理过程。例如，在【实例 6-1】中，若返回值为“success”，则跳转至页面 register_success.jsp；若返回值为“error”，则跳转至页面 register_error.jsp。

6.2.3 Struts 2 核心过滤器的配置

为将 Web 应用与 Struts 2 框架关联起来，需在 web.xml 文件中配置好 Struts 2 的核心过滤器 FilterDispatcher。其典型代码如下：

```
<web-app …>
    …
    <filter>
        <filter-name>struts2</filter-name>
        <filter-
class>org.apache.struts2.dispatcher.FilterDispatcher</filter-class>
    </filter>
    <filter-mapping>
        <filter-name>struts2</filter-name>
        <url-pattern>/*</url-pattern>
    </filter-mapping>
    …
</web-app>
```

其中，<filter>元素用于指定过滤器的名称及其实现类，<filter-mapping>用于指定相应过滤器的关联方式。在此，过滤器的名称为 struts2(通过<filter-name>子元素指定)，其实现类为 struts2 中的 org.apache.struts2.dispatcher.FilterDispatcher(通过<filter-class>子元素指定)，关联方式为/*(通过<url-pattern>子元素指定，“/*”表示该 URL 目录下的所有请求都交由 struts2 处理)。

提示： 从 Struts 2.1.3 以上版本开始，建议将核心过滤器由 FilterDispatcher 替换为更先进的 StrutsPrepareAndExecuteFilter。StrutsPrepareAndExecuteFilter 的全称类名为 org.apache.struts2.dispatcher.ng.filter.StrutsPrepareAndExecuteFilter，因此，配置 Struts 2 核心过滤器的<filter>元素也可改写为：

```
 <filter>
 <filter-name>struts2</filter-name>
 <filter-class>

 org.apache.struts2.dispatcher.ng.filter.StrutsPrepareAndExecuteF
ilter
 </filter-class>
 </filter>
```

说明： Filter 过滤器是 Java 项目开发中的一种常用技术。作为用户请求与处理程序之间的一层处理逻辑，过滤器可对用户请求与处理程序的响应内容进行处理，通常用于权限控制、编码转换等场合。

Servlet 过滤器是在 Java Servlet 规范中定义的，可对过滤器所关联的 URL 请求与响应进行检查与修改。Servlet 被调用后，Servlet 过滤器能够检查 response 对象，修改 response Header 与 response 内容。Servlet 过滤器所过滤的 URL 资源可以是 Servlet、JSP/HTML 文件，或者是整个路径下的任何资源。必要时，可将多个过滤器构成一个过滤器链，这样，当请求过滤器关联的 URL 时，过滤器链中的各个过滤器就会逐一被作用。

所有的过滤器都必须实现 java.Serlvet.Filter 接口。该接口包含 3 个过滤器实现类必须实现的方法。

(1) init(FilterConfig)。该方法为 Servlet 过滤器的初始化方法，在 Servlet 容器创建 Servlet 过滤器实例后被调用。

(2) doFilter(ServletRequest,ServletResponse,FilterChain)。该方法用于完成实际的过滤操作。当用户请求与过滤器关联的 URL 时，Servlet 容器将调用过滤器的 doFilter()方法。此外，在返回响应之前也会调用此方法。其中，FilterChain 参数用于访问过滤器链上的下一个过滤器。

(3) destroy()。该方法为 Servlet 过滤器的销毁方法(主要用于释放 Servlet 过滤器所占用的资源)，在 Servlet 容器销毁过滤器实例前被调用。

编写好过滤器的实现类后，还要在 web.xml 中进行相应的配置。其基本格式如下：

```
<filter>
    <!--自定义的过滤器名称-->
    <filter-name>过滤器名</filter-name>
    <!--自定义的过滤器实现类(若类在包中，则要加完整的包名)-->
    <filter-class>过滤器实现类</filter-class>
    <init-param>
        <!--类中的参数名-->
        <param-name>参数名</param-name>
        <!--对应的参数值-->
        <param-value>参数值</param-value>
    </init-param>
</filter>
```

过滤器必须与特定的 URL 关联后才能发挥作用。具体来说，关联方式有以下 3 种。

(1) 与一个 URL 资源关联。其配置的基本格式如下：

```
<filter-mapping>
    <filter-name>过滤器名</filter-name>
    <url-pattern>URL 地址</url-pattern>
</filter-mapping>
```

(2) 与一个URL目录下的所有资源关联。其配置的基本格式如下：

```
<filter-mapping>
    <filter-name>过滤器名</filter-name>
    <url-pattern>/*</url-pattern>
</filter-mapping>
```

(3) 与一个Servlet关联。其配置的基本格式如下：

```
<filter-mapping>
    <filter-name>过滤器名</filter-name>
```

```
    <url-pattern>Servlet 名称</url-pattern>
</filter-mapping>
```

6.2.4 Struts 2 Action 的实现

Struts 2 的核心功能为 Action，因此，在 Web 项目中使用 Struts 2 框架时，主要的编码工作就是编写有关的 Action 实现类。Action 实现类可以是一个普通的 Java 类，也可以继承自 ActionSupport 类。正如【实例 6-1】中的 RegisterAction 类所示，在大多数情况下，Action 实现类均会继承 ActionSupport 类。

Action 的属性定义必须符合 JavaBean 规范。对于每个属性，均应生成相应的 getter 与 setter 方法，这样，当 Action 被调用时，即可通过其属性自动获取表单中相应元素的值。

在一个 Action 实现类中，通常要包含一个 execute()方法，该方法即为 Action 被调用后会默认执行的方法。除此之外，还可根据需要包含其他完成特定功能的方法。实际上，为有效减少应用中 Action 实现类的数量，往往会将相关功能置于同一个 Action 中，并通过不同的方法加以实现。

Action 实现类中的各个方法在执行完毕后，均应使用 return 语句返回一个相应的结果代码(实际上是一个可自行指定的字符串)，该结果代码相当于逻辑视图，Struts 2 框架将据此确定对应的物理视图，并完成后续的处理过程。

说明： Action 的实现类可继承自 ActionSupport 类。ActionSupport 类实现了 Action、Validateable、ValidationAware、TextProvider、LocaleProvider 与 Serializable 接口，为 Action 提供了一些默认实现，主要包括预定义常量、从资源文件中读取文本资源、接收验证错误信息以及验证的默认实现等。在开发过程中，若要用到 ActionSupport 类所提供的某些功能，只需重写相应的方法即可。ActionSupport 类的基本定义如下：

```
public class ActionSupport implements Action, Validateable,
ValidationAware,
TextProvider, LocaleProvider,Serializable {
}
```

1. Action 接口

Action 接口包含在 com.opensymphony.xwork2 包中，其定义如下：

```
public interface Action {
    public static final String SUCCESS="success";
    public static final String NONE="none";
    public static final String ERROR="error";
    public static final String INPUT="input";
    public static final String LOGIN="login";
    public String execute() throws Exception;
}
```

可见，在 Action 接口中定义了 5 个常量(SUCCESS、NONE、ERROR、INPUT、LOGIN)与一个 execute()方法。若 Action 实现类继承了 ActionSupport 类，则可直接使用

Action 接口所定义的几个常量。

在【实例 6-1】中，由于 RegisterAction 继承了 ActionSupport 类，因此其 execute()方法的代码可修改为：

```
public String execute() throws Exception {
    if(!username.equals("admin")){
        return SUCCESS;
    }else{
        return ERROR;
    }
}
```

其中，常量 SUCCESS 的值为字符串“success”，常量 ERROR 的值为字符串“error”。

2. Validateable 接口

Validateable 接口提供了一个 validate()方法，该方法主要用于校验表单数据，是在执行 execute()方法之前被执行的。

3. ValidationAware 接口

ValidationAware 接口定义了一些用于对 Action 执行过程中产生的信息进行处理的方法。例如，该接口提供了 addFieldError(String fieldname,String errorMessage)方法，用于在验证出错时保存错误信息。

4. TextProvider 接口

TextProvider 接口提供了一系列的 getText()方法，用于获取相应的国际化信息资源。在 Struts 2 中，国际化信息资源都是以 key-value 对的形式出现的，通过使用 TextProvider 接口中的 getText()方法，可根据 key 来获取相应的 value 值。

5. LocaleProvider 接口

LocaleProvider 接口提供了一个 getLocale()方法。该方法用于在开发国际化应用时获取语言/地区信息。

注意： 在将表单元素的值传递给 Action 实现类的相应属性时，Struts 2 框架其实是根据表单元素的名称在 Action 实现类中查找并执行相应的 setter 方法来进行赋值的。例如，假定页面的表单中存在一个名称为“username”的文本框，那么在将该表单提交给某个 Action 时，Struts 2 框架就会在该 Action 的实现类中查找并执行 setUsername()方法以便将在文本框中输入的值赋给相应的属性。因此，在 Action 实现类中，属性名不一定要与相应表单元素的名称相同，但必须要存在与相应表单元素名称相对应的 setter 方法。不过，为方便起见，通常会将 Action 实现类的属性名与相应的表单元素名称统一起来。

Action 的传值其实有两种方式，即字段传值方式与模型传值方式。其中，字段传值方式是根据表单元素的名称直接在 Action 实现类中编写相应的 setter 方法来对属性赋值，主要用于表单元素不多的场合。若表单元素较多，则最好使用模型传值方式。模型传值方式的具体示例请参见 6.9.2 小节的“职工增加”案例。

提示： 为降低程序的耦合性，Struts 2 并未将 Action 与任何的 Servlet API 相关联。但在 Action 实现类中，有时需用到某些 Servlet API，如 request，response，session，application 等，因此，可先调用 ActionContext 类的静态方法 getContext()来获取一个 ActionContext 对象，然后再通过该 ActionContext 对象来获取所需要的 Servlet API 对象。例如：

```
ActionContext ac=ActionContext.getContext();  //获取 ActionContext
对象
Map application=ac.getApplication();  //获取 application 对象
Map session=ac.getSession();  //获取 session 对象
Map request=ac.get();  //获取 request 对象
```

在此，ActionContext 对象所获取的对象均为 Map 类型，而并非 Application、HttpSession、HttpServletRequest 等。其实，正因为 Struts 2 采取将 Map 对象模拟成 Application 等对象的方法，才成功地将 Servlet 从 Action 中分离出来。

其实，Struts 2 提供了专门的 ServletActionContext 类，可用来获取 HttpServletRequest、HttpServletResponse 或 HttpSession 对象。如以下代码：

```
//获取 HttpServletRequest 对象
HttpServletRequest request=ServletActionContext.getRequest();
//获取 HttpServletResponse 对象
HttpServletResponse  response =ServletActionContext.getResponse();
//获取 HttpSession 对象
HttpSession session=request.getSession();
```

此外，为获取 HttpServletRequest、HttpServletResponse 或 HttpSession 对象，也可使用以下方法：

```
ActionContext ac = ActionContext.getContext();
HttpServletRequest request =
(HttpServletRequest)ac.get(ServletActionContext.HTTP_REQUEST);
HttpServletResponse response =
(HttpServletResponse)ac.get(ServletActionContext.HTTP_RESPONSE);
HttpSession session = request.getSession();
```

6.2.5 Struts 2 Action 的配置

对于编写好的每一个 Action，均应根据具体情况进行相应的配置。通常，Action 的配置是在 struts.xml 文件中进行的。该文件的基本格式为：

```
<?xml version="1.0" encoding="UTF-8"?>
<!DOCTYPE struts PUBLIC
    "-//Apache Software Foundation//DTD Struts Configuration 2.0//EN"
```

```
    "http://struts.apache.org/dtds/struts-2.0.dtd">
<struts>
    <package …>
        <action …>
            <result …>…</result>
            …
        </action>
        …
    </package>
    …
</struts>
```

可见，Struts 2 中的 Action 是使用 package 元素的子元素 action 进行定义的，而且 action 元素中通常会包含相应的 result 子元素。其中，package 元素用于定义 Struts 2 中的包，而在包中可进一步定义 Action 与拦截器等。具体示例可参见【实例 6-1】，其 struts.xml 文件的代码为：

```
<?xml version="1.0" encoding="UTF-8"?>
<!DOCTYPE struts PUBLIC
    "-//Apache Software Foundation//DTD Struts Configuration 2.0//EN"
    "http://struts.apache.org/dtds/struts-2.0.dtd">
<struts>
    <package name="struts" extends="struts-default">
        <action name="register"
class="org.etspace.abc.action.RegisterAction">
            <result name="success">/register_success.jsp</result>
            <result name="error">/register_error.jsp</result>
        </action>
    </package>
</struts>
```

1. package 元素

在一个 Web 应用中，可包含不同的 Struts 2 包。Struts 2 中的包是通过 package 元素进行定义的，该元素的属性见表 6.1。

表 6.1　package 元素的属性

属　性	说　明
name	必选属性，用于指定包的名称。包与包之间的引用是通过其名称进行的
extends	可选属性，用于指定所继承的父包。通过继承父包，即可继承父包中所包含的 Action 与拦截器等。由于包信息的获取是按其在配置文件中的先后顺序进行的，因此父包必须在子包之前被定义
namespace	可选属性，用于指定包的命名空间。未指定时，将使用默认的命名空间“""”
abstract	可选属性，用于指定是否为抽象包。若将其值设为 true，则为抽象包。抽象包不能包含 action 配置信息，但可以被继承

Struts 2 中的包与 Java 中的包不同，可以通过扩展另外的包而继承其中的所有定义，

并可另行添加某些特有配置，以及修改原有包的部分配置等。由此看来，Struts 2 中的包更像 Java 中的类。通常，在定义包时都会继承一个名为“struts-default”的 Struts 2 内置的包。该包定义在 Struts 2 的 struts-default.xml 文件中，而包中则定义了一些结果类型、拦截器与拦截器栏。

必要时，可为包指定命名空间。命名空间可以是根命名空间“/”，也可以是自定义的命名空间“/*”(在此“*”表示自定义的字符串)。未指定任何命名空间时，则使用默认的命名空间“""”。如以下示例：

```
<package name="mypackage1">
    <action name="register"
class="org.etspace.abc.action.RegisterAction">
        <result name="success">/register_success.jsp</result>
    </action>
    <action name="login" class="org.etspace.abc.action.LoginAction">
        <result name="success">/login_success.jsp</result>
    </action>
</package>
<package name="mypackage2" namespace="/">
    <action name="register"
class="org.etspace.abc.action.RegisterAction">
        <result name="success">/register_success.jsp</result>
    </action>
</package>
<package name="mypackage3" namespace="/mynamespace">
    <action name="register"
class="org.etspace.abc.action.RegisterAction">
        <result name="success">/register_success.jsp</result>
    </action>
</package>
```

在此示例中，mypackage1 包的命名空间为默认命名空间，mypackage2 包的命名空间为根命名空间“/”，mypackage3 包的命名空间为自定义的命名空间“/mynamespace”。

为包指定了命名空间后，对于包中 Action 的请求也要指明相应的命名空间。实际上，当 Struts 2 接收到请求后，会将请求信息解析为命名空间名与 Action 名两个部分，然后根据命名空间名在 struts.xml 中查找指定命名空间的包，并在该包中查找与 Action 名相同的配置；若未找到，则继续在默认命名空间中查找；若依然未找到，则给出错误信息。如果在请求中没有指定命名空间，那么会先在根命名空间“/”中查找相应的 Action；未找到时，再到默认命名空间中查找。

对于以上示例，若页面中的请求为“mynamespace/register.action”，则“/mynamespace”命名空间中的名称为 register 的 Action 会被执行；若请求为“mynamespace/login.action”，则默认命名空间中的名称为 login 的 Action 会被执行；若请求为“register.action”，则根命名空间“/”中的名称为 register 的 Action 会被执行；若请求为“login.action”，则默认命名空间中的名称为 login 的 Action 会被执行。

2. action 元素

一个 Struts 2 包可根据需要包含一个或多个 Action 与拦截器等。其中，Action 的定义是通过 package 元素的子元素 action 进行的，该元素的属性见表 6.2。

表 6.2　action 元素的属性

属　性	说　明
name	必选属性，用于指定 Action 的名称
class	可选属性，用于指定 Action 的实现类(应使用带包名的全称类名)。未指定时，将自动引用自定义默认类或系统默认类(ActionSupport)
method	可选属性，用于指定请求所要调用的 Action 中的方法。未指定时，将默认调用 Action 中的 execute()方法
converter	可选属性，用于指定 Action 所使用的类型转换器

默认情况下，Action 在处理请求时，会自动执行其实现类中的 execute()方法。实际上，在 Action 实现类中，除了 execute()方法以外，还可以包含其他方法。如果一个请求要调用 Action 实现类中的其他方法，那么只需在相应的 action 元素中使用 method 属性加以指定即可。例如，假定在 RegisterAction 类中有 execute()与 regist()两个方法，则以下配置可让请求调用其中的 regist()方法：

```
<action name="register" class="org.etspace.abc.action.RegisterAction"
method="regist">
    …
</action>
```

如果不想在 action 元素中使用 method 属性指定需调用的方法，那么也可以在请求中直接加以指定。例如，以下表单所提交的请求将自动调用 RegisterAction 类中的 regist()方法：

```
<form action="register!regist.action" method="post">
    …
</form>
```

该表单所提交的请求为“register!regist.action”。其中，“!”前面的“register”为相应 Action 的名称，“!”后面的“regist”则为相应 Action 实现类中的方法名。

此外，在使用 Struts 2 标签实现提交按钮时，也可以直接指定需调用的 Action 实现类中的方法。例如，单击以下表单所提供的“注册”按钮，即可直接调用 RegisterAction 类中的 regist()方法：

```
<s:form action="register.action" method="post">
    …
    <s:submit value="注册" method="regist"/>
</s:form>
```

说明： 在配置 Action 时，Struts 2 允许使用通配符，让系统自动识别需调用的 Action 实现类中的方法。例如，假定在 RegisterAction 类中有 execute()与 regist()两个

方法，并有以下配置:

```
<action name="*" class="org.etspace.abc.action.RegisterAction"
method="{1}">
…
</action>
```

其中，“{1}”的值就是其前通配符“*”所匹配的值。在这种情况下，以下表单所提交的请求将自动调用 RegisterAction 类中的 regist()方法:

```
<form action="regist.action" method="post">
    …
</form>
```

显然，使用通配符可以减少 action 元素的数量，从而有效减少配置文件的内容。不过，对于 Action 的编写来说，其方法的命名必须与请求名称相对应。因此，在实际开发中，应根据具体情况来决定是否使用通配符。

其实，Action 的方法与返回值均可使用通配符进行匹配。例如，假定 regist()方法返回“error”时就跳转到 regist.jsp 页面，则相应的配置可修改为:

```
<action name="*" class="org.etspace.abc.action.RegisterAction"
method="{1}">
   …
   <result name="error">/{1}.jsp</result>
   …
</action>
```

说明: 在配置 Action 时，如果 action 元素中并未使用 class 属性指定其实现类，那么将自动引用相应的默认类。Struts 2 系统的默认类为 ActionSupport。必要时，也可使用 package 元素的子元素 default-class-ref 来自定义默认类。如以下示例:

```
<package name="default" extends="struts-default">
     <default-class-ref
class="org.etspace.abc.action.RegisterAction"></default-class-
ref>
     <action name="regist">
         …
     </action>
     …
</package>
```

在此，将 RegisterAction 类自定义为默认类。对于名称为“regist”的 Action 来说，由于没有指定其实现类，则在请求该 Action 时将自动引用自定义的默认类 RegisterAction。当然，对于指定了实现类的 Action 来说，默认类是不起作用的。

3. result 元素

当 Action 实现类中的方法执行完毕时，将返回一个相应的结果代码(实际上是一个可自行指定的字符串)。对于每一个结果代码，均应使用 action 元素的子元素 result 进行定义，以明确其结果类型与后续的跳转目标或处理方式。在此，Action 所返回的结果代码相当于逻辑视图。根据 result 元素的定义，即可确定各个逻辑视图所对应的物理视图，并通过物理视图完成后续的处理过程。result 元素的属性见表 6.3。

表 6.3　result 元素的属性

属　性	说　明
name	可选属性，用于指定结果代码。未指定时，将取默认值“success”。该属性的常用取值见表 6.4
type	可选属性，用于指定结果类型。未指定时，将取默认值“dispatcher”。该属性的有关取值见表 6.5

表 6.4　result 元素的常用结果代码

代　码	说　明
success	默认代码，表示请求处理成功
error	表示请求处理失败
none	表示请求处理完成后不跳转到任何页面
input	表示输入验证失败
login	表示登录失败

表 6.5　result 元素的结果类型

类　型	说　明
chain	用于处理 Action 链
dispatcher	默认类型，用于转发至(跳转到)一个页面，如 JSP 页面等
freemarker	用于处理 FreeMarker 模板
httpheader	用于控制特殊的 HTTP 行为
redirect	用于重定向到一个 URL
redirectAction	用于重定向到一个 Action
stream	用于向浏览器发送 InputStream 对象，通常用来处理文件下载、返回 AJAX 数据
velocity	用于处理 Velocity 模板
xslt	用于处理 XML/XSLT 模板
plainText	用于显示原始文件内容，如文件源代码等
postback	用于将请求参数作为表单提交到指定的目的地

result 元素的基本格式为：

```
<result name ="结果代码" type ="结果类型">
        <param name ="参数名">参数值</param>
    …
</result>
```

其中，param 子元素用于指定具体的跳转目标(相当于物理视图)及相关设置，其 name 属性的取值为相应的参数名。不过，在实际应用中，为简洁起见，对于 result 元素，通常无须明确添加 param 子元素，而是直接在<result>与</result>之间指定相应的物理视图位置。

说明： Struts 2 所支持的结果类型可在其开发包中的 struts-default.xml 文件中进行查看。例如，对于 struts-2.3.16-all.zip，在其 src\core\src\main\resources 路径下，即可找到 struts-default.xml 文件，其中与结果类型相关的代码为：

```
<result-types>
   <result-type name="chain"
class="com.opensymphony.xwork2.ActionChainResult"/>
   <result-type name="dispatcher"
class="org.apache.struts2.dispatcher.ServletDispatcherResult"
default="true"/>
   <result-type name="freemarker"
class="org.apache.struts2.views.freemarker.FreemarkerResult"/>
   <result-type name="httpheader"
class="org.apache.struts2.dispatcher.HttpHeaderResult"/>
   <result-type name="redirect"
class="org.apache.struts2.dispatcher.ServletRedirectResult"/>
   <result-type name="redirectAction"
class="org.apache.struts2.dispatcher.ServletActionRedirectResult
"/>
   <result-type name="stream"
class="org.apache.struts2.dispatcher.StreamResult"/>
   <result-type name="velocity"
class="org.apache.struts2.dispatcher.VelocityResult"/>
   <result-type name="xslt"
class="org.apache.struts2.views.xslt.XSLTResult"/>
   <result-type name="plainText"
class="org.apache.struts2.dispatcher.PlainTextResult" />
   <result-type name="postback"
class="org.apache.struts2.dispatcher.PostbackResult" />
</result-types>
```

其中，结果类型 dispatcher 所在的元素具有属性“default="true"”，说明该类型为默认类型。

提示： 一般来说，结果类型不同，可供设置的参数也会有所不同。例如，当结果类型为 dispatcher 或 redirect 时，可供设置的参数有两个，即 location 与 parse(location 用于指定请求处理完成后的跳转地址，如“/welcome.jsp”；parse 用于指定是否允许在相应跳转地址中使用 OGNL 表达式，其默认值为 true，即允许使用表达式，如“/welcome.jsp?name=${name}”)。又如，当结果类型为 chain 或 redirectAction 时，可供设置的参数有两个，即 actionName 与 namespace(actionName 用于指定请求处理完成后需跳转到的 Action 的名称，namespace 用于指定相应 Action 所在的命名空间)。

在实际应用中，最为常用的结果类型是 dispatcher(该类型是 Struts 2 默认的结果类型，可省略不写)。dispatcher 类型主要用于实现页面的转发(或跳转)，其物理视图通常为 JSP 页面。使用 dispatcher 类型所实现的跳转类似于 JSP 中的“forward”，属于同一请求，即请求转发时浏览器的地址栏会保持不变，请求参数与请求属性等数据也不会丢失。

redirect 类型主要用于重定向到另外一个页面或 Action。与 dispatcher 类型不同的是，使用 redirect 类型时，当 Action 结束对用户请求的处理后，将重新生成一个请求，然后再进行跳转，因此地址栏会发生改变。

redirectAction 类型与 redirect 类型相似，均为重定向而非转发。不过，redirectAction 类型通常用于重定向到一个新的 Action，而不是 JSP 页面。如以下代码：

```
…
<package name="regist" extends="struts-default">
    <action name="regist" class="org.abc.RegistAction">
        <result name="success" type="redirectAction">
            <param name="actionName">login</param>
            <param name="namespace">/login</param>
        </result>
    </action>
</package>
<package name="login" extends="struts-default" namespace="/login">
    <action name="login" class="org.abc.LoginAction">
        <result name="success">/index.jsp</result>
    </action>
</package>
…
```

在此，对 regist 包中的名称为 regist 的 Action 的请求处理完毕后，将直接重定向到 login 包中名称为 login 的 Action(login 包的命名空间为“/login”)。

虽然 redirect 与 redirectAction 类型都可以重定向到另外的 Action，但是在重定向过程中，请求参数与请求属性等数据均会丢失，因此无法实现数据的传递。与这两种类型不同，chain 类型专门用于实现 Action 之间的跳转(而非重定向)，并在跳转过程中保证数据不丢失。显然，使用 chain 类型可根据需要组成一条 Action 链，并可实现数据的共享，这对于编程来说，是极为有利的。其实，Action 的跳转之所以能够共享数据，是因为处于同一个 Action 链的各个 Action 都共享同一个值栈，由于每个 Action 在执行完毕后都会把数据压入值栈，因此在需要时即可直接到值栈中去获取相应的数据。

提示：使用 action 元素的子元素 result 配置的结果只对当前的 Action 有效，均属于局部结果。对于不同的 Action，若存在着完全相同的返回结果，并继续使用局部结果的方式进行配置，肯定会出现较多的冗余。为此，Struts 2 提供了全局结果的配置方式。全局结果是使用 package 元素的子元素 global-results 进行配置的。在 global-results 元素中，可包含一个或多个 result 子元素。例如，假定各 Action 返回“error”时均跳转到错误页面 error.jsp，则相关的配置代码如下：

```
<package name="default" extends="struts-default">
  <global-results>
```

```
    <result name="error">/error.jsp</result>
  </global-results>
  <action …>
    …
  </action>
  <action …>
    …
  </action>
  …
</package>
```

说明： 在实际开发中，对于小型项目，通常会将所有的配置信息都放在 struts.xml 文件中。但对于大型项目，由于配置信息较多，为方便管理与维护，最好使用分而治之的策略，按照一定的划分原则将有关配置信息分别放在不同的配置文件中，然后在 struts.xml 文件中使用 include 元素将各个配置文件导入其中。例如，对于【实例 6-1】，可将 struts.xml 复制为 register.xml(同样置于项目的 src 文件夹)，然后再将 struts.xml 文件的内容修改为：

```
<?xml version="1.0" encoding="UTF-8"?>
<!DOCTYPE struts PUBLIC
    "-//Apache Software Foundation//DTD Struts Configuration
2.0//EN"
    "http://struts.apache.org/dtds/struts-2.0.dtd">
<struts>
    <include file="register.xml"></include>
</struts>
```

如果要导入的配置文件被置于某个包中，那么在导入时还要指明相应包所对应的路径。例如，若 register.xml 置于 org.etspace.abc.action 包中，则相应的 include 元素应修改为：

```
<include file="org/etspace/abc/action/register.xml"></include>
```

6.3 Struts 2 拦截器

拦截器是 Struts 2 的核心所在，Struts 2 框架中的绝大部分功能都是通过拦截器来实现的。如果要扩展 Struts 2 的功能，那么只需提供相应的拦截器，并将其配置在 Struts 2 容器之中即可。反之，如果不再需要某个功能了，那么只需取消相应的拦截器即可。

当 Struts 2 的核心过滤器拦截到用户请求后，大量的拦截器就会对其进行处理，然后再调用用户自定义的 Action 实现类中的方法。必要时，可通过自定义拦截器的方式来完成相应的功能，如身份验证、权限控制等。

6.3.1 拦截器的实现

为便于自定义拦截器的实现，Struts 2 提供了一些相关的接口或类，如 Interceptor 接口、AbstractInterceptor 类等。基于这些接口或类，可轻松实现所需要的拦截器。

1. Interceptor 接口

Interceptor 接口(com.opensymphony. xwork2.interceptor.Interceptor)的代码如下：

```
    import java.io.Serializable;
    import com.opensymphony.xwork2.ActionInvocation;
    public interface Interceptor extends Serializable{
     void init();
     String intercept(ActionInvocation invocation) throws Exception;
     void destroy();
}
```

可见，Interceptor 接口定义有 3 个方法，分别如下。

(1) init()。该方法在拦截器实例化之后、执行之前被调用，主要用于初始化资源。

(2) intercept(ActionInvocation invocation)。该方法用于实现拦截的具体动作。通过调用其参数 invocation 的 invoke()方法，可将控制权交给下一个拦截器，或者交给 Action 实现类的方法。

(3) destroy()。该方法与 init()方法相对应，在拦截器实例销毁之前被调用，用于销毁在 init()方法中打开的资源。

通过实现 Interceptor 接口，即可完成有关拦截器的具体设计。

2. AbstractInterceptor 类

AbstractInterceptor 类提供了 Interceptor 接口中 init()与 destroy()方法的空实现。因此，拦截器也可通过继承 AbstractInterceptor 类来实现。与实现 Interceptor 接口的方式相比，这种方法可以简化代码的编写，对于无须打开资源的拦截器，应优先使用这种方法来实现。

6.3.2　拦截器的配置

对于用户自定义的拦截器，应在 struts.xml 文件中进行具体的配置。其实，Struts 2 内建有大量的拦截器，并在 struts-default.xml 文件中进行了相应的配置。在 struts.xml 中进行配置时，若继承 struts-default 包，则可自动应用其中所定义的拦截器。

为定义拦截器，应使用<interceptor .../>元素。其基本格式为：

```
<interceptor name="拦截器名" class="拦截器实现类">
    <param name="参数名">参数值</param>
    …
</interceptor>
```

必要时，可将多个拦截器组成一个拦截器栈。为定义拦截器栈，应使用<interceptor-stack .../>元素，并在其下使用<interceptor-ref .../>子元素来引用构成该拦截器栈的各个事先已配置好的拦截器。其基本格式如下：

```
<interceptor-stack name="拦截器栈名">
    <interceptor-ref name="拦截器 1"></interceptor-ref>
    <interceptor-ref name="拦截器 2"></interceptor-ref>
    <interceptor-ref name="拦截器 3"></interceptor-ref>
    …
```

```
</interceptor-stack>
```

其实，一个拦截器栈也可以引用其他拦截器栈，其实质就是将被引用的拦截器栈中的各个拦截器包含到当前拦截器栈中。

在 struts.xml 中配置一个包时，可为其指定默认的拦截器。由于每个包只能指定一个默认拦截器，因此若要指定多个拦截器共同作为默认拦截器，则应先将其定义为拦截器栈，然后再将该拦截器栈配置为默认拦截器。为定义默认拦截器，应使用<default-interceptor-ref …/>元素。其基本格式如下：

```
 <package name="包名">
    <interceptors>
        <interceptor name="拦截器 1" class="拦截器实现类"></interceptor>
        <interceptor name="拦截器 2" class="拦截器实现类"></interceptor>
        …
        <interceptor-stack name="拦截器栈名">
            <interceptor-ref name="拦截器 1"></interceptor-ref>
            <interceptor-ref name="拦截器 2"></interceptor-ref>
            …
        </interceptor-stack>
    </interceptors>
    <default-interceptor-ref name="拦截器名或拦截器栈名"></default-
interceptor-ref>
    …
</package>
```

注意： 如果为包指定了默认的拦截器，那么该拦截器会对包中的每个 Action 起作用。不过，如果显式地为某个 Action 配置了拦截器，那么默认的拦截器将不会起作用。

【实例 6-2】 Struts 2 自定义拦截器示例。在【实例 6-1】的基础上，再添加两项限制，即用户名不能为 system，且密码不能与用户名相同。图 6.12 所示为“用户注册”页面，要求用户输入相应的用户名与密码，单击“确定”按钮后，即可显示相应的结果。若用户名为 system，或密码与用户名相同，则不允许注册，结果如图 6.13 所示。

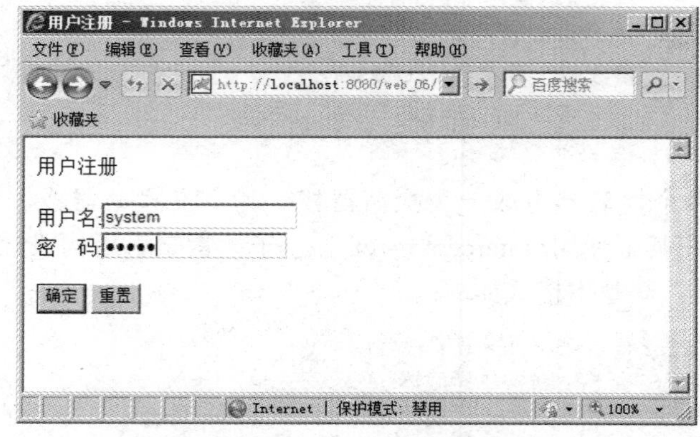

图 6.12 “用户注册”页面

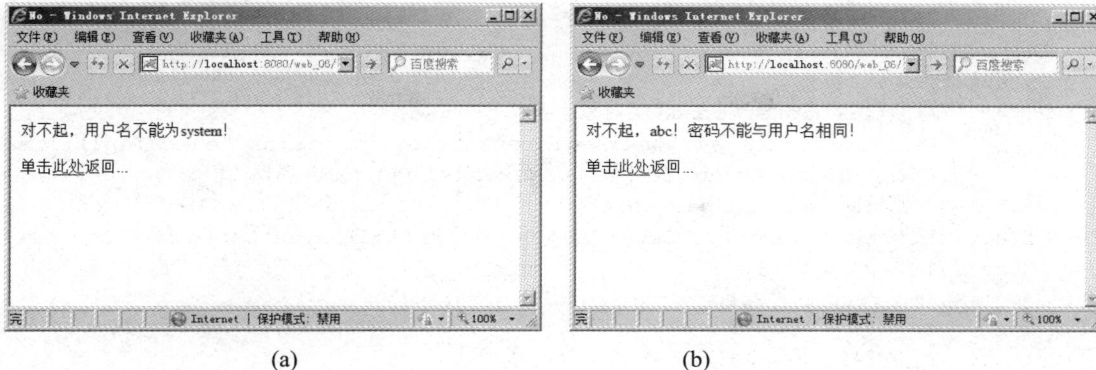

(a)　　　　　　(b)

图 6.13　拦截结果页面

分析：在【实例 6-1】的基础上，只需自定义相应的拦截器并完成其配置，即可实现本实例所要求的功能。

主要步骤(在项目 web_06 中)如下。

(1)　在项目的 src 文件夹中新建一个包 org.etspace.abc.interceptor。

(2)　在 org.etspace.abc.interceptor 包中新建一个拦截器实现类 RegisterInterceptor。其文件名为 RegisterInterceptor.java，代码如下：

```
package org.etspace.abc.interceptor;
import com.opensymphony.xwork2.ActionInvocation;
import com.opensymphony.xwork2.interceptor.AbstractInterceptor;
import com.opensymphony.xwork2.Action;
import org.etspace.abc.action.RegisterAction;
public class RegisterInterceptor extends AbstractInterceptor {
    public String intercept(ActionInvocation invocation) throws
Exception {
        //获取 RegisterAction 类对象
        RegisterAction action=(RegisterAction)invocation.getAction();
        //若 RegisterAction 类对象的 username 属性的值为“system”,
        //则返回“error”。
        if(action.getUsername().toLowerCase().equals("system")){
            return Action.ERROR;
        }
        //继续执行其他拦截器或 Action 类中的方法
        return invocation.invoke();
    }
}
```

(3)　在 org.etspace.abc.interceptor 包中新建一个拦截器实现类 RegisterInterceptorOfPassword。其文件名为 RegisterInterceptorOfPassword.java，代码如下：

```
package org.etspace.abc.interceptor;
import com.opensymphony.xwork2.ActionInvocation;
import com.opensymphony.xwork2.interceptor.AbstractInterceptor;
import com.opensymphony.xwork2.Action;
import org.etspace.abc.action.RegisterAction;
```

```
public class RegisterInterceptorOfPassword extends AbstractInterceptor {
    public String intercept(ActionInvocation invocation) throws
Exception {
        //获取 RegisterAction 类对象
        RegisterAction action=(RegisterAction)invocation.getAction();
        //若 RegisterAction 类对象的 password 与 username 属性的值相同，
        //则返回“errorofpassword”。
if(action.getPassword().toLowerCase().equals(action.getUsername().toLowe
rCase())){
            return "errorofpassword";
        }
        //继续执行其他拦截器或 Action 类中的方法
        return invocation.invoke();
    }
}
```

(4) 在项目的配置文件 struts.xml 中配置拦截器。有关代码如下：

```
…
<struts>
    <package name="struts" extends="struts-default">
        <!—定义拦截器-->
        <interceptors>
            <interceptor name="registerInterceptor"
class="org.etspace.abc.interceptor.RegisterInterceptor"></interceptor>
            <interceptor name="registerInterceptorOfPassword" class=
"org.etspace.abc.interceptor.RegisterInterceptorOfPassword"></interceptor>
        </interceptors>
        <action name="register"
class="org.etspace.abc.action.RegisterAction">
            <result name="success">/register_success.jsp</result>
            <result name="error">/register_error.jsp</result>
            <result
name="errorofpassword">/register_errorofpassword.jsp</result>
            <!--使用系统默认的拦截器栈 -->
            <interceptor-ref name="defaultStack"></interceptor-ref>
            <!--使用自定义的拦截器 -->
            <interceptor-ref name="registerInterceptor"></interceptor-ref>
            <interceptor-ref
name="registerInterceptorOfPassword"></interceptor-ref>
        </action>
    </package>
</struts>
```

(5) 在项目的 WebRoot 文件夹中添加一个注册失败页面 register_errorofpassword.jsp，其代码如下：

```
<%@ page language="java" import="java.util.*" pageEncoding="UTF-8"%>
<%@ taglib uri="/struts-tags" prefix="s" %>
<html>
  <head>
```

```
    <title>No</title>
  </head>
  <body>
    对不起，<s:property value="username"/>！密码不能与用户名相同！
    <br><br>
    单击<a href="register.jsp">此处</a>返回…
  </body>
</html>
```

至此，项目创建完毕。完成项目的部署后，再启动 Tomcat 服务器，然后在浏览器中输入地址 http://localhost:8080/web_06/register.jsp 并按 Enter 键，即可打开图 6.12 所示的“用户注册”页面，开始其运行流程。

6.4　Struts 2 OGNL

OGNL(Object Graphic Navigation Language，对象图导航语言)是一个开源项目。作为一种功能强大的 EL(Expression Language，表达式语言)，OGNL 可通过极为简单的方式访问 Java 对象中的有关属性。OGNL 首先应用在 Webwork 项目中，后来则成为 Struts 2 框架视图默认的表达式语言。

6.4.1　OGNL 表达式

标准的 OGNL 要求设定一个根对象(root 对象)。现假定使用标准 OGNL(而不是 Struts 2 OGNL)表达式来求值，若 OGNL 上下文中有两个对象 A 与 B，同时 A 对象被设置为根对象(root)，则可通过以下 OGNL 表达式访问有关对象的 xxx 属性。

(1)　A.xxx。该表达式返回对象 A 的 xxx 属性值，即返回 A.getXxx()。

(2)　#B.xxx。该表达式返回对象 B 的 xxx 属性值，即返回 B.getXxx()。

(3)　xxx。因为对象 A 为根对象，因此该表达式返回对象 A 的 xxx 属性值，即返回 A.getXxx()。

可见，OGNL 的使用非常简单，对于根对象，可直接访问(如示例中的对象 A)。而对于非根对象，则需要使用命名空间“#”才能进行访问(如示例中的#B)。

OGNL 表达式的使用是 Struts 2 框架的特点之一。在 Struts 2 框架中，值栈(Value Stack)即为 OGNL 的根对象。因此，Struts 2 OGNL 是从值栈的顶部元素开始进行搜索的。

现假设值栈中存在两个对象 teacher 与 student，且 teacher 在值栈的顶部，而 student 则置于 teacher 之后。若 teacher 的属性为 name 与 salary，student 的属性为 name 与 age，则相应的值栈如图 6.14 所示。此时，可通过以下 OGNL 表达式访问有关对象的属性。

(1)　salary。该表达式返回对象 teacher 的 salary 属性值，即返回 teacher.getSalary()。

(2)　age。该表达式返回对象 student 的 age 属性值，即返回 student.getAge()。

(3)　name。因为对象 teacher 位于值栈的顶部，而 Struts 2 OGNL 从值栈的顶部元素开始进行搜索，因此该表达式返回对象 teacher 的 name 属性值，即返回 teacher.getName()。

(4)　student.name。该表达式返回对象 student 的 name 属性值，即返回 student.getName()。

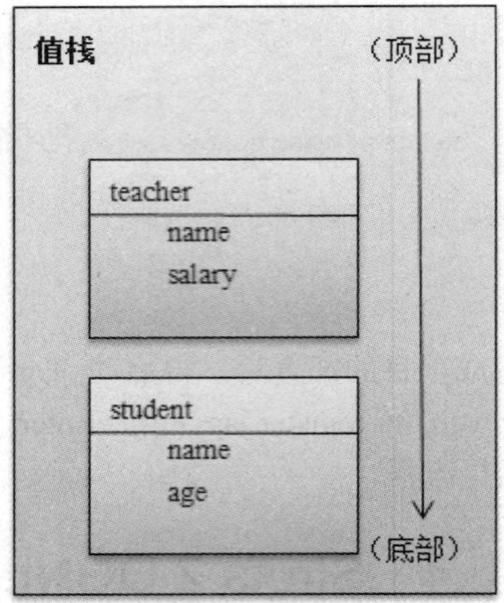

图 6.14　包含 teacher 与 student 对象的值栈

此外，Struts 2 允许使用索引访问值栈中的对象。例如：

(1) [0].name。相当于调用 teacher.getName()。

(2) [1].name。相当于调用 student.getName()。

在 Struts 2 中，OGNL Context 就是 ActionContext，如图 6.15 所示。其中，值栈是 Struts 2 中 OGNL 的根对象。除此之外，其他对象均为非根对象。

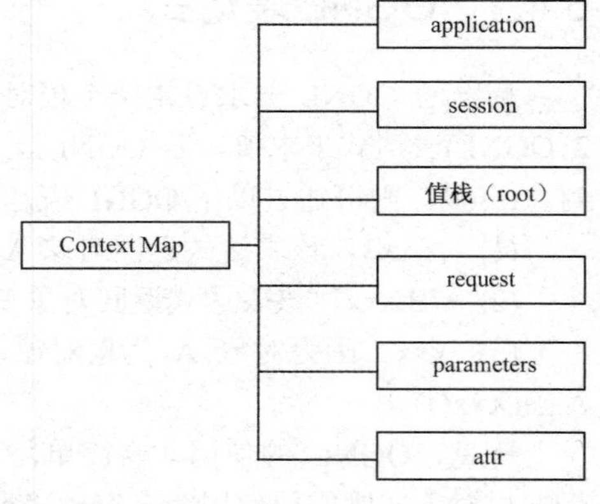

图 6.15　Struts 2 的 OGNL Context 结构

若要访问 Context 中的非根对象，则应加上前缀“#”。各个对象的作用与基本用法如下。

(1) application 对象。用于访问 ServletContext。如 #application.userName 或者 #application ["userName"]，相当于调用 Servlet 的 getAttribute("userName")。

(2) session 对象。用于访问 HttpSession。如 #session.userName 或者#session["userName"]，相当于调用 session.getAttribute("userName")。

(3) request 对象。用于访问 HttpServletRequest 属性的 Map。如#request.userName 或者#request["userName"]，相当于调用 request.getAttribute("userName")。

(4) parameters 对象。用于访问 HTTP 请求的参数的 Map。如#parameters.userName 或者#parameters["userName"]，相当于调用 request.getParameter("userName")。

(5) attr 对象。用于按 request、session、application 的顺序访问各种作用域内的属性。如#attr.id、#attr.userName 等。

例如，在【实例 6-1】的 register_success.jsp 页面中，有以下代码：

```
欢迎您，<s:property value="#request.username"/>!
```

其中，#request.username 用于访问 request 对象的属性 username，相当于调用了

request.getAttribute("username")。在此，request 对象 username 属性的值其实是在 RegisterAction 类中通过以下代码存入其中的：

```
Map request=(Map)ActionContext.getContext().get("request");
request.put("username",username);
```

6.4.2　OGNL 集合

在 OGNL 中可以使用集合，包括 List 对象、Map 对象等。

要生成一个 List 对象，可使用具有以下格式的 OGNL 表达式：

```
{e1, e2, e3, …}
```

其中，e1、e2、e3 等为该 List 对象所包含的各个元素，各元素间用逗号隔开。例如：

```
{'apple','orange','pear','banana'}
```

要生成一个 Map 对象，可使用具有以下格式的 OGNL 表达式：

```
#{key1:value1, key2:value2, key3:value3,…}
```

其中，key1:value1、key2:value2、key3:value3 等为该 Map 对象所包含的各个元素，各元素间用逗号隔开。与 List 对象不同，Map 对象的元素为 key-value 对(即键-值对)，各元素的 key、value 要用冒号隔开。例如：

```
#{0:'apple',1:'orange',2:'pear',3:'banana'}
```

对于集合类型，在 OGNL 表达式中可以使用 in 与 not in 两个操作符。其中，in 用于判断某个元素是否在指定的集合对象中，而 not in 则用于判断某个元素是否不在指定的集合对象中。例如：

```
<s: if test="e1 in {'e1', 'e2', 'e3'}">
  …
</s: if>
或
<s: if test=" e1 not in {'e1', 'e2', 'e3'}">
  …
</s: if>
```

此外，在 OGNL 中还允许使用某个规则来获取指定集合对象的子集。其中，常用的相关操作符有以下 3 个。

(1)　?，获取所有符合逻辑的元素。

(2)　^，获取符合逻辑的第一个元素。

(3)　$，获取符合逻辑的最后一个元素。

例如：

```
Person.relatives.{?# this.gender=='male'}
```

该代码的作用为获取 Person 的所有 gender(性别)为 male(男性)的 relatives(亲属)集合。

6.5　Struts 2 标签库

Struts 2 的标签主要定义在 URI 为/struts-tags 的命名空间下。借助于 Struts 2 标签库，可避免在 JSP 页面中使用 Java 脚本代码，从而简化 JSP 页面逻辑的实现。

Struts 2 标签可分为 3 类，即 UI 标签、非 UI 标签与 Ajax 标签。其中，UI 标签主要用于生成 HTML 元素，又可分为表单标签与非表单标签两类；非 UI 标签主要用于数据访问与逻辑控制等，又可分为控制标签与数据标签两类；Ajax 标签则主要用于 Ajax 功能或特性的支持。

6.5.1　数据标签

Struts 2 的数据标签属于非 UI 标签，主要用于提供与数据访问相关的功能。Struts 2 所提供的主要数据标签见表 6.6。

表 6.6　Struts 2 的常用数据标签

标　签	作　用
property	用于输出某个值
param	用于设置参数，通常作为 bean 标签与 action 标签的子标签
bean	用于创建一个 JavaBean 实例
action	用于在 JSP 页面直接调用一个 Action
set	用于设置一个新的变量
date	用于格式化输出一个日期
include	用于在 JSP 页面中包含其他的 JSP 或 Servlet 资源
push	用于将某个值放入值栈的栈顶
url	用于生成一个 URL 地址
debug	用于在页面上生成一个调试链接，当单击该链接时，可以看到当前值栈与 Stack Context 中的内容
il8n	用于指定国际化资源文件的 baseName
text	用于输出国际化

下面，仅对几个常用的数据标签的基本用法进行简要说明。

1. <s:property>标签

property 标签用于输出指定的值，其有关属性见表 6.7。

表 6.7　property 标签的属性

属　性	说　明
default	可选属性。若要输出的属性值为 null，则输出该属性指定的值
escape	可选属性，用于指定是否进行 HTML 转义。其默认值为 true

续表

属　性	说　明
value	可选属性，用于指定要输出的属性值。若未指定该属性，则默认输出值栈栈顶的值
id	可选属性，用于指定该元素的引用 id(标识)

通常，property 标签所要输出的值由其 value 属性指定(未指定 value 属性时，则以值栈栈顶的值代替之)。例如：

```
<s:property value="#request.username"/>
```

2. <s:param>标签

param 标签通常作为 bean、action 等其他标签的子标签使用，主要用于为这些标签提供参数，其有关属性见表 6.8。

表 6.8　param 标签的属性

属　性	说　明
name	可选属性，用于指定参数名
value	可选属性，用于指定参数值
id	可选属性，用于指定该元素的引用 id(标识)

例如，以下代码用于将参数 fruit 的值设置为对象 apple 的值(若 apple 对象不存在，则 fruit 参数的值为 null)：

```
<s:param name="fruit" value="apple" />
```

若要将参数 fruit 的值设为字符串 apple，则应使用以下代码(即 apple 还要用单引号括起来)：

```
<s:param name="fruit" value="'apple'" />
```

该代码与以下代码是等价的：

```
<s:param name= "fruit">apple</s:param>
```

3. <s:bean>标签

bean 标签用于创建一个 JavaBean 实例，其有关属性见表 6.9。

表 6.9　bean 标签的属性

属　性	说　明
name	必选属性，用于指定要实例化的 JavaBean 的实现类
id	可选属性，用于指定相应 JavaBean 实例的引用 id(标识)

使用 bean 标签创建 JavaBean 实例时，可在其内使用 param 标签为 JavaBean 实例的属性赋值。为此，需在 JavaBean 中为有关属性提供相应的 setter 方法。若要对有关属性进行访问，则还应提供相应的 getter 方法。

使用 bean 标签时，通常应指定其 id 属性值，这样，相应的 JavaBean 实例会被放入

Stack Context 中，从而允许直接通过指定的 id 来访问。

【实例 6-3】 bean 标签使用示例：设置并显示学生的姓名。

主要步骤(在项目 web_06 中)如下。

(1) 在项目的 src 文件夹中新建一个包 org.etspace.abc.bean。

(2) 在 org.etspace.abc.bean 包中创建一个 Student 类，其文件名为 Student.java，代码如下：

```
package org.etspace.abc.bean;
public class Student {
    private String name;
    public String getName() {
        return name;
    }
    public void setName(String name) {
        this.name=name;
    }
}
```

说明： Student 类其实就是一个 JavaBean，内含一个属性 name，并具有相应的 getter 与 setter 方法。

(3) 在项目的 WebRoot 文件夹中添加一个 Student.jsp 页面，其代码如下：

```
<%@ page language="java" import="java.util.*" pageEncoding="UTF-8"%>
<%@ taglib uri="/struts-tags" prefix="s" %>
<html>
  <head>
    <title>Student</title>
  </head>
  <body>
    <s:bean name="org.etspace.abc.bean.Student" id="student">
        <s:param name="name" value="'张三'"></s:param>
        姓名：<s:property value="name"/><br>
        <s:param name="name">李四</s:param>
        姓名：<s:property value="name"/><br>
        <s:param name="name" value="'王五'"/>
    </s:bean>
    姓名：<s:property value="#student.name"/><br>
  </body>
</html>
```

运行结果如图 6.16 所示。

说明： JavaBean 实例的属性既可在 bean 标签内部输出，也可在 bean 标签外部输入。

4. <s:action>标签

action 标签用于在 JSP 页面中直接调用 Action，其有关属性见表 6.10。

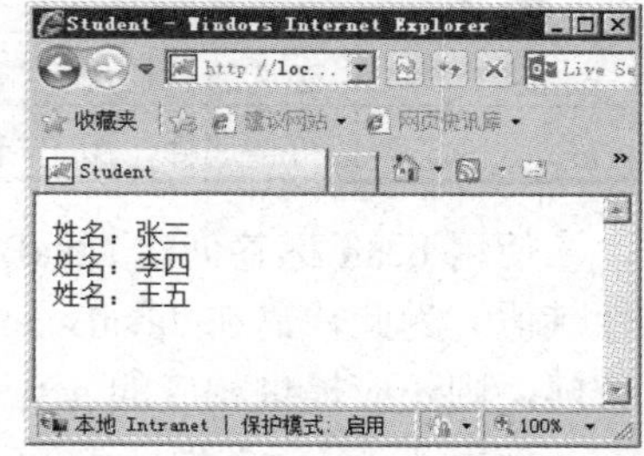

图 6.16 Student 页面

表 6.10　action 标签的属性

属　性	说　明
name	必选属性，用于指定所要调用的 Action
namespace	可选属性，用于指定相应 Action 所在的命名空间
executeResult	可选属性，用于指定是否要将相应 Action 的处理结果页面包含到本页面中。若值为 true，则要包含；反之，若值为 false(默认值)，则不用包含
ignoreContextParam	可选属性，用于指定是否忽略页面中的请求参数(即不将请求参数传入相应的 Action 中)。若值为 true，则不用传入；反之，若值为 false(默认值)，则要传入
id	可选属性，用于指定相应 Action 的引用 id(标识)

使用 action 标签调用 Action 时，可在其内使用 param 标签为调用 Action 的属性直接赋值。

【实例 6-4】action 标签使用示例：在页面中直接调用【实例 6-1】中所创建的 Action——RegisterAction。

主要步骤(在项目 web_06 中)如下。

在项目的 WebRoot 文件夹中添加一个 register_action.jsp 页面，其代码如下：

```
<%@ page language="java" pageEncoding="utf-8" %>
<%@ taglib prefix="s" uri="/struts-tags" %>
<html>
    <head>
        <title>用户注册</title>
    </head>
    <body>
        <s:action name="register" executeResult="true">
            <s:param name="username">Lsd</s:param>
            <s:param name="password">123</s:param>
        </s:action>
        <s:action name="register">
            <s:param name="username">Abc</s:param>
            <s:param name="password">123</s:param>
        </s:action>
    </body>
</html>
```

运行结果如图 6.17 所示。

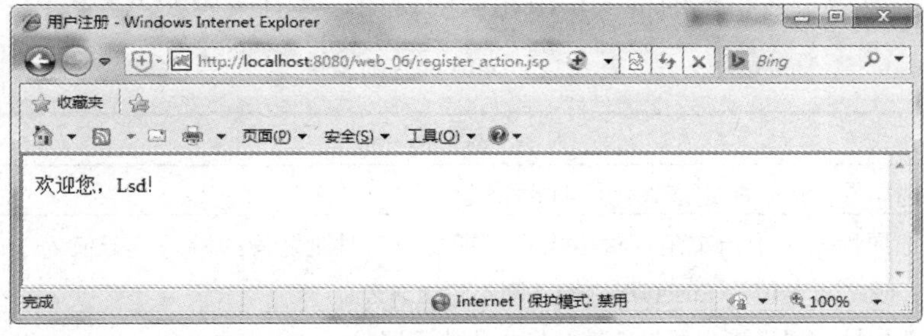

图 6.17　“用户注册”页面

解析：

在 register_action.jsp 页面，两次调用名称为“register”的 Action(即 RegisterAction)，第一次调用时，设置了“executeResult="true"”，因此可显示相应的处理结果。而第二次调用时，并没有设置“executeResult="true"”，因此无法显示其处理结果。

5. <s:set>标签

set 标签用于将指定值栈表达式的值赋给特定作用域中的某个变量，其有关属性见表 6.11。

表 6.11　set 标签的属性

属　性	说　明
name	必选属性，用于指定要为其赋值的变量的名称
scope	可选属性，用于指定相应变量的作用域(action、page、request、session 或 application)。若未指定该属性，则默认放置在值栈中
value	可选属性，用于指定要赋给相应变量的值。若未指定该属性，则将值栈栈顶的值赋给相应变量
id	可选属性，用于指定该元素的引用 id(标识)

例如，以下代码用于访问 session 中的 user 对象的 username、password 与 usertype 等属性：

```
<s:property value="#session['user'].username"/>
<s:property value="#session['user'].password"/>
<s:property value="#session['user'].usertype"/>
```

可将以上代码改写为：

```
<s:set name="user" value="#session['user']" />
<s:property value="#user.username"/>
<s:property value="#user.password " />
<s:property value="#user.usertype " />
```

可见，适当使用 set 标签可让代码更加简洁、更易阅读。

6. <s:date>标签

date 标签主要用于按指定的格式输出相应的日期，其有关属性见表 6.12。

表 6.12　date 标签的属性

属　性	说　明
format	可选属性，用于指定格式化日期的格式
nice	可选属性，用于指定是否输出指定日期与当前时刻之间的时差。若值为 true，则输出时差；反之，若值为 false(默认值)，则不输出时差
name	必选属性，用于指定要对其进行格式化的日期
id	可选属性，用于指定该元素的引用 id(标识)

对于 date 标签，当其 nice 属性为 true 时，会输出指定日期与当前时刻的时差，而不会输出指定日期。因此，使用 date 标签时，若将 nice 属性设置为 true，则一般不指定 format 属性。若未指定 format 属性，也未指定 nice 属性值为 true，则系统会在国际化资源文件中查找 key 为 struts.date.format 的消息，并将其作为格式化文本来格式化指定日期。若未找到该消息，则系统默认采用 DateFormat.MEDIUM 的格式输出指定日期。

7. <s:include>标签

include 标签用于将指定的 JSP 页面或 Servlet 包含到本页面中，其有关属性见表 6.13。

表 6.13　include 标签的属性

属　性	说　明
value	必选属性，用于指定要被包含的 JSP 页面或 Servlet
id	可选属性，用于指定该元素的引用 id(标识)

6.5.2　控制标签

Struts 2 的控制标签属于非 UI 标签，主要用于完成对执行流程或值栈的控制。Struts 2 所提供的主要控制标签见表 6.14。

表 6.14　Struts 2 的常用控制标签

标　签	作　用
if	用于控制选择输出
elseif	用于控制选择输出，必须与 if 标签结合使用
else	用户控制选择输出，必须与 if 标签结合使用
iterator	用于将集合迭代输出
append	用于将多个集合拼接成一个新的集合
merge	用于将多个集合拼接成一个新的集合(但与 append 的拼接方式不同)
generator	用于将一个字符串按指定的分隔符拆分为多个字符串，所生成的多个子字符串可使用 iterator 标签来迭代输出
sort	用于对集合进行排序
subset	用于截取集合的部分元素，以生成新的子集合

下面，仅对几个常用的控制标签的基本用法进行简要说明。

1. <s:if>/<s:elseif>/<s:else>标签

<s:if>/<s:elseif>/<s:else>标签用于实现分支控制，即根据指定表达式的值(boolean 型)来决定是否计算或输出相应标签体的内容。其中，if 标签可以单独使用，而 elseif 与 else 标签必须与 if 标签组合使用。此外，if 标签只能与一个 else 标签组合使用，但可以与多个 elseif 标签组合使用。其完整格式如下：

```
<s:if test="测试表达式">
```

```
    标签体
</s:if>
<s:elseif test="测试表达式">
    标签体
</s:elseif>
… <!-- 其他 elseif 标签 -->
<s:else>
    标签体
</s:else>
```

其中，if 标签与 elseif 标签中的 test 属性用于指定相应的测试表达式。在执行过程中，首先计算 if 标签中的测试表达式，若其值为 true，则 if 标签的标签体有效；否则，再按顺序计算各个 elseif 标签中的测试表达式。若其中的某个测试表达式的值为 true，则相应的 elseif 标签的标签体有效。反之，若所有的测试表达式的值均为 false，则 else 标签的标签体有效。

2. <s:iterator>标签

iterator 标签主要用于对包括 List、Set 等在内的各种集合或 Map 类型的对象进行迭代输出，其有关属性见表 6.15。

表 6.15 iterator 标签的属性

属 性	说 明
value	可选属性，用于指定被迭代的集合(通常由 OGNL 表达式指定)。若未指定该属性，则默认使用值栈栈顶的集合
id	可选属性，用于指定集合元素的引用 id
status	可选属性，用于指定迭代时的 IteratorStatus 实例

对于 iterator 标签，可由其 status 属性指定迭代时的 IteratorStatus 实例，这样，通过相应的 IteratorStatus 实例，即可获取当前迭代元素的有关属性。IteratorStatus 实例所包含的方法见表 6.16。

表 6.16 IteratorStatus 实例的方法

方 法	说 明
int getCount()	返回当前迭代了多少个元素
int getIndex()	返回当前被迭代元素的索引
boolean isEven	返回当前被迭代元素的索引是否为偶数
boolean isOdd	返回当前被迭代元素的索引是否为奇数
boolean isFirst	返回当前被迭代元素是否为第一个元素
boolean isLast	返回当前被迭代元素是否为最后一个元素

【实例 6-5】显示水果列表。

主要步骤(在项目 web_06 中)如下。

在项目的 WebRoot 文件夹中添加一个 fruit jsp 页面，其代码如下：

```
<%@ page language="java" import="java.util.*" pageEncoding="UTF-8"%>
<%@ taglib uri="/struts-tags" prefix="s" %>
<html>
  <head>
    <title>Fruit</title>
  </head>
  <body>
     <table border="1" width="200">
     <s:iterator value="{'apple','orange','pear','banana'}" id="fruit"
status="fruit_status">
     <tr <s:if test="#fruit_status.even">style="background-
color:silver"</s:if>>
         <td><s:property value="#fruit_status.count"/></td>
         <td><s:property value="fruit"/></td>
     </tr>
    </s:iterator>
     </table>
  </body>
</html>
```

运行结果如图 6.18 所示。

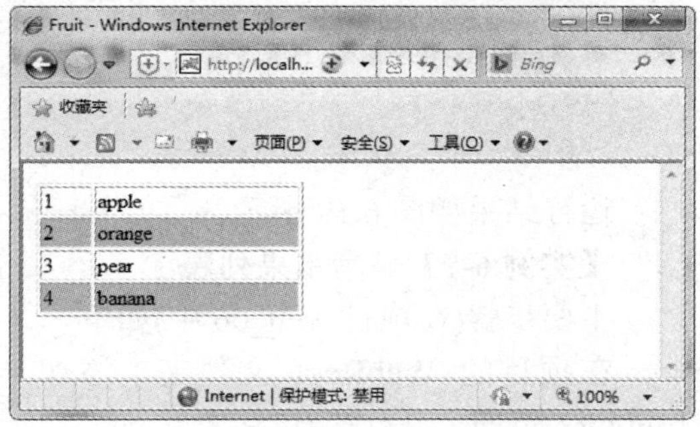

图 6.18　Fruit 页面

3. <s:append>与<s:merge>标签

append 与 merge 标签均用于将多个集合拼接成一个新的集合，但产生新集合的方式有所不同。其中，前者是将各个集合作为整体按顺序拼接在一起，而后者则是将各个集合中对应的元素按顺序拼接在一起。

现假定有 2 个集合，第一个集合与第二个集合分别包含有 3 个元素与 2 个元素。若以 append 方式进行拼接，则新集合中的元素依次为：

- 第 1 个集合的第 1 个元素
- 第 1 个集合的第 2 个元素
- 第 1 个集合的第 3 个元素
- 第 2 个集合的第 1 个元素
- 第 2 个集合的第 2 个元素

若以 merge 方式进行拼接，则新集合中的元素依次为：

- 第 1 个集合的第 1 个元素
- 第 2 个集合的第 1 个元素
- 第 1 个集合的第 2 个元素
- 第 2 个集合的第 2 个元素
- 第 1 个集合的第 3 个元素

【实例 6-6】显示水果列表。

主要步骤(在项目 web_06 中)如下。

在项目的 WebRoot 文件夹中添加一个 fruit1 jsp 页面，其代码如下：

```
<%@ page language="java" import="java.util.*" pageEncoding="UTF-8"%>
<%@ taglib uri="/struts-tags" prefix="s" %>
<html>
  <head>
    <title>Fruit</title>
  </head>
  <body>
        <s:append id="fruitList">
        <s:param value="{'apple','orange','pear','banana'}"/>
        <s:param value="{'葡萄','草莓','柿子'}"/>
    </s:append>
    <table border="1" width="200">
        <s:iterator value="#fruitList" id="fruit" status="fruit_status">
        <tr <s:if test="#fruit_status.even">style="background-
color:silver"</s:if>>
            <td><s:property value="#fruit_status.count"/></td>
            <td><s:property value="fruit"/></td>
        </tr>
    </s:iterator>
    </table>
  </body>
</html>
```

运行结果如图 6.19 所示。

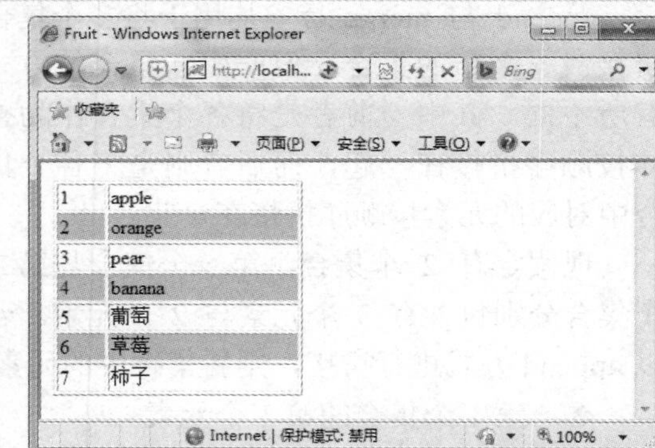

图 6.19　Fruit 页面

【实例 6-7】显示水果列表。

主要步骤(在项目 web_06 中)如下。

在项目的 WebRoot 文件夹中添加一个 fruit2 jsp 页面，其代码如下：

```
<%@ page language="java"
import="java.util.*"
pageEncoding="UTF-8"%>
<%@ taglib uri="/struts-tags"
prefix="s" %>
<html>
  <head>
    <title>Fruit</title>
  </head>
  <body>
        <s:merge id="fruitList">
        <s:param value="{'apple','orange','pear','banana'}"/>
        <s:param value="{'葡萄','草莓','柿子'}"/>
    </s:merge>
    <table border="1" width="200">
        <s:iterator value="#fruitList" id="fruit" status="fruit_status">
        <tr <s:if test="#fruit_status.even">style="background-
color:silver"</s:if>>
            <td><s:property value="#fruit_status.count"/></td>
```

```
            <td><s:property value="fruit"/></td>
        </tr>
    </s:iterator>
    </table>
  </body>
</html>
```

运行结果如图 6.20 所示。

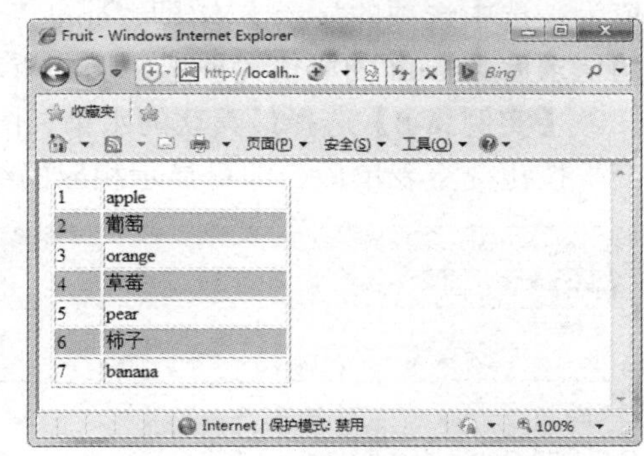

图 6.20 Fruit 页面

6.5.3 表单标签

Struts 2 的表单标签属于 UI 标签，用于生成页面中的表单元素。实际上，大部分的 Struts 2 表单标签与 HTML 的表单标记是一一对应的，包括<s:form>、<s:textfield>、<s:password>标签等。如以下代码：

```
<s:form action="login.action"
method="post">
   <s:textfield name="username" label="用户名" />
   <s:password name="password" label="密码" />
</form>
```

该段代码等价于：

```
<form action="login.action" method="post">
   用户名：<input type="text" name="username">
   密码：<input type="password" name="password">
</s:form>
```

此外，在 Struts 2 的表单标签中，也有一些与 HTML 表单标记并不是一一对应的，主要包括<s:radio>、<s:select>、<s:combobox>、<s:checkboxlist>、<s:head>与<s:datetimepicker>等标签。

1. <s:radio>标签

radio 标签用于生成一组单选按钮，所包含的单选按钮由指定的集合决定。例如：

```
<s:radio list="{'男','女'}" name="sex" label="性别" value="'女
'"></s:radio>
<s:radio list="#{0:'男',1:'女'}" name="sex" label="性别"
value="1"></s:radio>
```

以上代码分别展示了 radio 标签的两种用法。其中，前者所产生的请求参数 sex 的取值为被选中单选按钮上的显示文本(即“男”或“女”)，而后者所产生的请求参数 sex 的取值为被选中单选按钮所对应的 key(即“0”或“1”)。

2. <s:select>标签

select 标签用于生成一个下拉列表框，所包含的选项由指定的集合决定。例如：

```
<s:select list="{'apple','orange','pear','banana'}" name="fruit"
value="'pear'" label="最喜欢的水果"></s:select>
<s:select list="#{0:'apple',1:'orange',2:'pear',3:'banana'}"
name="fruit" value="2" label="最喜欢的水果"></s:select>
```

以上代码分别展示了 select 标签的两种用法。其中，前者所产生的请求参数 fruit 的取值为被选中选项的显示文本(即“apple”“orange”“pear”或“banana”)，而后者所产生的请求参数 fruit 的取值为被选中选项所对应的 key(即“0”“1”“2”或“3”)。

【实例 6-8】选择最喜欢的水果。图 6.21 所示为选择最喜欢的水果的表单，单击“提交”按钮提交表单后，即可显示相应的结果，如图 6.22 所示。

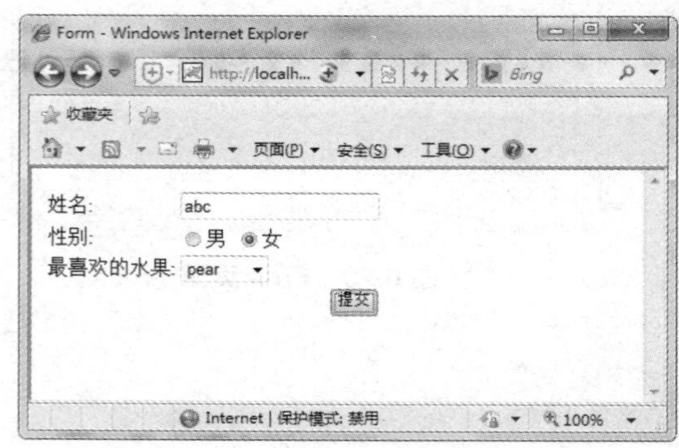

图 6.21 表单页面

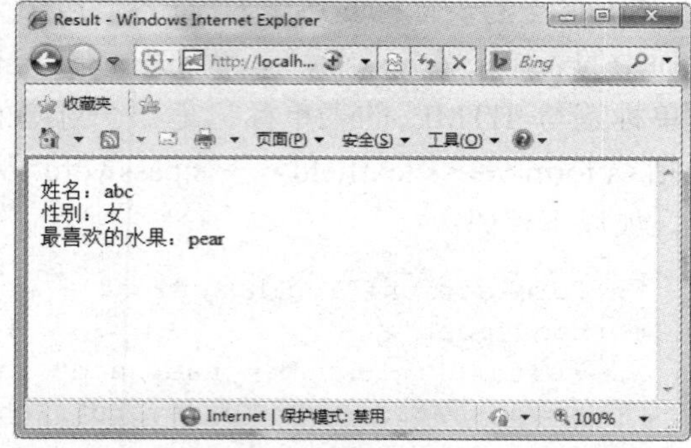

图 6.22 结果页面

主要步骤(在项目 web_06 中)如下。

(1) 在项目的 WebRoot 文件夹中添加一个 form_s1.jsp 页面，其代码如下：

```
<%@ page language="java" import="java.util.*" pageEncoding="UTF-8"%>
<%@ taglib uri="/struts-tags" prefix="s" %>
<html>
  <head>
    <title>Form</title>
  </head>
  <body>
    <s:form action="form_s_result.jsp" method="post">
        <s:textfield name="name" label="姓名"></s:textfield>
        <s:radio list="{'男','女'}" name="sex" label="性别" value="'女
'"></s:radio>
        <s:select list="{'apple','orange','pear','banana'}" name="fruit"
value="'pear'" label="最喜欢的水果"></s:select>
        <s:submit value="提交"></s:submit>
    </s:form>
  </body>
</html>
```

(2) 在项目的 WebRoot 文件夹中添加一个 form_s_result.jsp 页面，其代码如下：

```
<%@ page language="java" import="java.util.*" pageEncoding="UTF-8"%>
<%@ taglib uri="/struts-tags" prefix="s" %>
<html>
```

```
  <head>
    <title>Result</title>
  </head>
  <body>
<%
    String name=request.getParameter("name");
    String sex=request.getParameter("sex");
    String fruit=request.getParameter("fruit");
%>
姓名：<%=name %><br>
性别：<%=sex %><br>
最喜欢的水果：<%=fruit %><br>
  </body>
</html>
```

【实例 6-9】选择最喜欢的水果。图 6.23 所示为选择最喜欢的水果的表单。单击“提交”按钮提交表单后，即可显示相应的结果，如图 6.24 所示。

图 6.23　表单页面

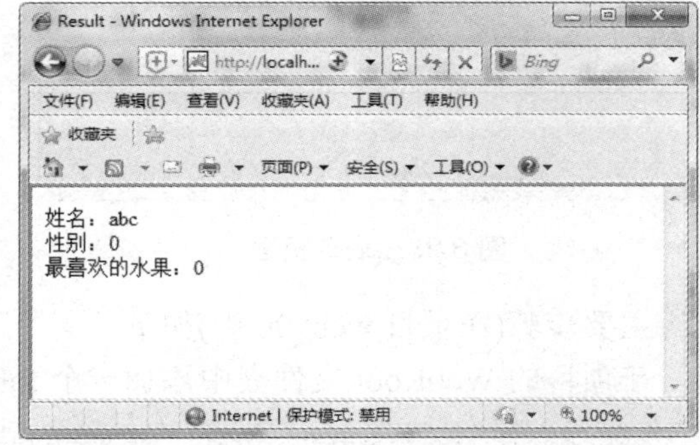

图 6.24　结果页面

主要步骤(在项目 web_06 中)如下。

在项目的 WebRoot 文件夹中添加一个 form_s2.jsp 页面，其代码如下：

```
<%@ page language="java" import="java.util.*" pageEncoding="UTF-8"%>
<%@ taglib uri="/struts-tags" prefix="s" %>
<html>
  <head>
    <title>Form</title>
  </head>
  <body>
    <s:form action="form_s_result.jsp" method="post">
        <s:textfield name="name" label="姓名"></s:textfield>
        <s:radio list="#{0:'男',1:'女'}" name="sex" label="性别"
value="1"></s:radio>
        <s:select list="#{0:'apple',1:'orange',2:'pear',3:'banana'}"
name="fruit" value="2" label="最喜欢的水果"></s:select>
        <s:submit value="提交"></s:submit>
    </s:form>
  </body>
```

```
</html>
```

3. <s:combobox>标签

combobox 标签用于生成一个组合框，即一个单行文本框与一个下拉列表框的组合。其中，单行文本框用于产生请求参数，而下拉列表框则只用于辅助输入。例如：

```
<s:combobox list="{'apple','orange','pear','banana'}" label="最喜欢的水果"
value="pear" name="fruit"></s:combobox>
```

【实例 6-10】选择最喜欢的水果。如图 6.25 所示，为选择最喜欢的水果的表单。单击“提交”按钮提交表单后，即可显示相应的结果，如图 6.26 所示。

图 6.25 表单页面

图 6.26 结果页面

主要步骤(在项目 web_06 中)如下。

在项目的 WebRoot 文件夹中添加一个 form_s3.jsp 页面，其代码如下：

```
<%@ page language="java" import="java.util.*" pageEncoding="UTF-8"%>
<%@ taglib uri="/struts-tags" prefix="s" %>
<html>
  <head>
    <title>Form</title>
  </head>
  <body>
    <s:form action="form_s_result.jsp" method="post">
        <s:textfield name="name" label="姓名"></s:textfield>
        <s:radio list="{'男','女'}" name="sex" label="性别" value="'女
'"></s:radio>
        <s:combobox list="{'apple','orange','pear','banana'}" label="最喜
欢的水果" value="pear" name="fruit"></s:combobox>
        <s:submit value="提交"></s:submit>
    </s:form>
  </body>
</html>
```

4. <s:checkboxlist>标签

checkboxlist 标签用于生成一组复选框，所包含的复选框由指定的集合决定。例如：

```
<s:checkboxlist list="{'apple','orange','pear','banana'}" label="喜欢的水
果" name="fruit" value="{'apple','pear'}"></s:checkboxlist>
<s:checkboxlist list="#{0:'apple',1:'orange',2:'pear',3:'banana'}"
label="喜欢的水果" name="fruit" value="{0,2}"></s:checkboxlist>
```

以上代码，分别展示了 checkboxlist 标签的两种用法。其中，前者所产生的请求参数 fruit 的取值为被选中复选框上的显示文本(即“apple”“orange”“pear”或“banana”)，而后者所产生的请求参数 fruit 的取值为被选中复选框所对应的 key(即“0”“1”“2”或“3”)。

【实例 6-11】选择喜欢的水果。图 6.27 所示为选择喜欢的水果的表单，单击“提交”按钮提交表单后，即可显示相应的结果，如图 6.28 所示。

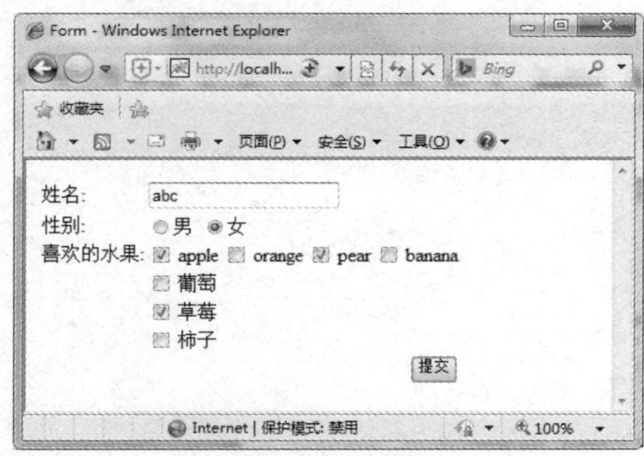

图 6.27　表单页面

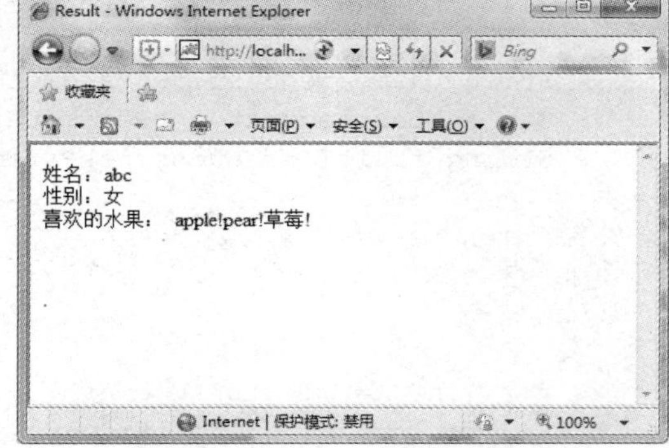

图 6.28　结果页面

主要步骤(在项目 web_06 中)如下。

(1)　在项目的 WebRoot 文件夹中添加一个 form_s_checkbox.jsp 页面，其代码如下：

```
<%@ page language="java" import="java.util.*" pageEncoding="UTF-8"%>
<%@ taglib uri="/struts-tags" prefix="s" %>
<html>
  <head>
    <title>Form</title>
  </head>
  <body>
     <s:form action="form_s_checkbox_result.jsp" method="post">
         <s:textfield name="name" label="姓名"></s:textfield>
         <s:radio list="{'男','女'}" name="sex" label="性别" value="'女
'"></s:radio>
         <s:checkboxlist list="{'apple','orange','pear','banana'}"
label="喜欢的水果" name="fruit" value="{'apple','pear'}"></s:checkboxlist>
         <s:checkbox label="葡萄" name="fruit" value="false" fieldValue="
葡萄"></s:checkbox>
         <s:checkbox label="草莓" name="fruit" value="true" fieldValue="草
莓"></s:checkbox>
         <s:checkbox label="柿子" name="fruit" value="false" fieldValue="
柿子"></s:checkbox>
         <s:submit value="提交"></s:submit>
```

```
    </s:form>
  </body>
</html>
```

(2) 在项目的 WebRoot 文件夹中添加一个 form_s_checkbox _result.jsp 页面，其代码如下：

```
<%@ page language="java" import="java.util.*" pageEncoding="UTF-8"%>
<%@ taglib uri="/struts-tags" prefix="s" %>
<html>
  <head>
    <title>Result</title>
  </head>
  <body>
<%
    String name=request.getParameter("name");
    String sex=request.getParameter("sex");
    String fruit[]=request.getParameterValues("fruit");
%>
姓名：<%=name %><br>
性别：<%=sex %><br>
喜欢的水果：
<%
    for (int i=0;i<fruit.length;i++){
        out.print(fruit[i]);
        out.print("!");
    }
%>
  </body>
</html>
```

5. <s:head>标签

head 标签主要用于生成页面的 head 部分。特别地，如果需要在页面中使用 Ajax 组件，那么就需要在 head 标签中添加 theme 属性，并将其设置为“ajax”(即 theme="ajax")，这样，即可将标准 Ajax 的头信息包含到页面中。

6. <s:datetimepicker>标签

datetimepicker 标签用于生成一个日历小控件。例如：

```
<s:datetimepicker name="date" label="请选择日期" displayFormat="yyyy-MM-
dd"></s:datetimepicker>
```

由于 datetimepicker 标签所提供的日历小控件包含了 JavaScript 代码，因此在使用该标签时，应在页面中添加“<s:head />”标签。

注意： 在 Struts 2 的 2.1.6 版本以前，只需在 head 标签中设置“<s:head theme="ajax"/>”，即可直接使用 datetimepicker 标签。但在 Struts 2 的 2.1.6 及后续版本中，要使用 datetimepicker 标签必须加载一个相应的 Jar 包(或类库)，

并进行相应的设置。例如，若使用的 Struts 2 开发包为 struts-2.3.16-all.zip，则所要加载的 Jar 包为 struts2-dojo-plugin-2.3.16.jar。此外，要先在页面中使用指令“<%@ taglib uri="/struts-dojo-tags" prefix="sd"%>”引用相应的标签库并指定其标签前缀(在此为“sd”)，然后再添加“<sd:head/>”标签，这样，才能通过“<sd:datetimepicker ...></sd:datetimepicker>”的方式使用 datetimepicker 标签。

6.5.4　非表单标签

Struts 2 的非表单标签属于 UI 标签，主要用于在页面中生成一些非表单的可视化元素。Struts 2 所提供的主要非表单标签见表 6.17。

表 6.17　Struts 2 的常用非表单标签

标　签	作　用
a	用于生成超链接
div	用于生成一个 div
fielderror	用于输出表单域的类型转换错误或数据校验错误信息
actionerror	用于输出 Action 实例的 getActionMessage()方法返回的消息
component	用于生成一个自定义组件
tablePanel	用于生成 HTML 页面的 Tab 页
tree	用于生成一个树形结构
treenode	用于生成树形结构的节点

6.6　Struts 2 数据验证

所谓数据验证，是指在将用户所输入的数据提交给处理程序前，先检查其合法性。若数据未能通过合法性检查，则要求用户重新输入或进行修改。显然，数据验证有利于提高应用的可靠性及其运行效率。

在 Struts 2 中，既可进行服务器端验证，也可进行客户端验证。在此，仅介绍服务器端验证的常用方法。

6.6.1　数据校验

ActionSupport 类实现了 Action、Validateable、ValidationAware、TextProvider、LocaleProvider 与 Serializable 接口，其中，Validateable 接口定义了一个 validate()方法，若 Action 实现类继承自 ActionSupport 类，则只要在 Action 实现类中重写 validate()方法，即可实现验证功能。

在 Action 实现类中定义了校验方法 validate()后，该方法会在执行 execute()方法之前首先被执行。若执行该方法后，Action 实现类的 fieldError 中已包含了数据校验错误信息，则会将请求转发到 input 逻辑视图处。

Struts 2 框架中的表单标签<s:form.../>已提供了输出校验错误信息的能力，因此，为显示数据校验错误信息，应在 input 逻辑视图所对应的 JSP 页面中使用相应的 Struts 2 表单标签。

【实例 6-12】Struts 2 数据校验示例。在【实例 6-2】的基础上，再添加相应的数据校验功能，要求用户名与密码不能为空。图 6.29 所示为“用户注册”页面，要求用户输入相应的用户名与密码，单击“确定”按钮后，即可显示相应的结果。若用户名或密码为空，则不允许注册，结果如图 6.30 所示。

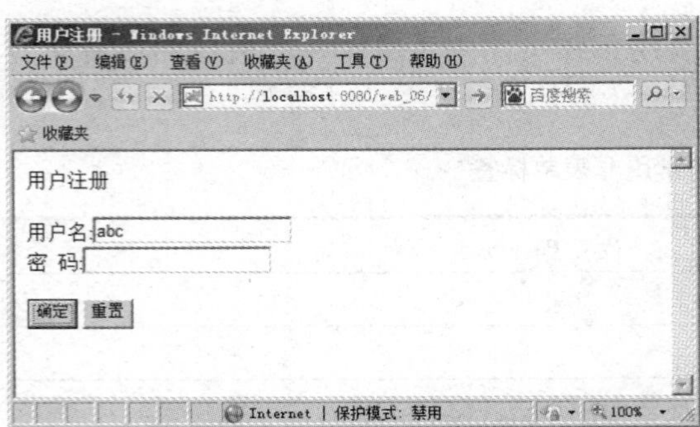

图 6.29 “用户注册”页面

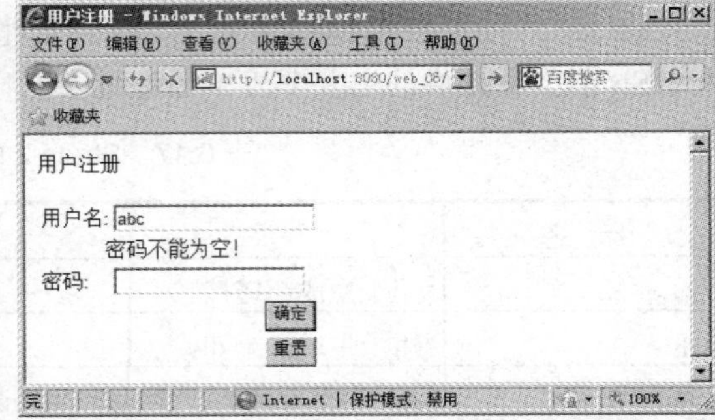

图 6.30 校验信息

主要步骤(在项目 web_06 中)如下。

(1) 修改 Action 实现类——RegisterAction(其文件名为 RegisterAction.java)，在其中添加一个校验方法 validate()。有关代码如下：

```
...
public class RegisterAction extends ActionSupport {
    ...
    //数据校验
    public void validate() {
        //若用户名为空，则把错误信息添加到 Action 类的 fieldErrors

    if(this.getUsername()==null||this.getUsername().trim().equals("")){
            addFieldError("username","用户名不能为空！"); //将错误信息保存起来
        }
        //若密码为空，则把错误信息添加到 Action 类的 fieldErrors

    if(this.getPassword()==null||this.getPassword().trim().equals("")){
            addFieldError("password","密码不能为空！"); //将错误信息保存起来
        }
    }
}
```

(2) 修改 struts.xml 配置文件，在名称(name)为“register”的 action 元素中添加一个名称(name)为“input”的结果视图。有关代码如下：

```
    ...
    <action name="register" class="org.etspace.abc.action.RegisterAction">
```

```
        <result name="success">/register_success.jsp</result>
        <result name="error">/register_error.jsp</result>
        <result name="errorofpassword">/register_errorofpassword.jsp</result>
        <result name="input">/register_s.jsp</result>
        …
    </action>
    …
```

(3) 在项目的 WebRoot 文件夹中添加一个用户注册页面 register_s.jsp，其代码如下：

```
<%@ page language="java" pageEncoding="utf-8" %>
<%@ taglib prefix="s" uri="/struts-tags" %>
<html>
    <head>
        <title>用户注册</title>
    </head>
    <body>
        用户注册<br>
        <s:form action="register.action" method="post">
            <s:textfield label="用户名" name="username"></s:textfield>
            <s:password label="密码" name="password"></s:password>
            <s:submit value="确定"></s:submit>
            <s:reset value="重置"></s:reset>
        </s:form>
    </body>
</html>
```

提示： 在 Action 实现类中，除了 execute()方法以外，往往还会包括其他处理方法，必要时，也可专门针对某个特定的处理方法进行数据校验。为此，只需按规范另外编写相应的校验方法即可。例如，对于处理方法 abc()，其校验方法应为 validateAbc()。

6.6.2 校验框架

通过重写 validate()方法对数据进行校验，需在 Action 实现类中编写相关代码，一方面会使类变得臃肿，另一方面也不利于日后的维护与管理。为避免此弊端，Struts 2 提供了相应的校验框架。采用 Struts 2 校验框架，只需为 Action 实现类添加相应的校验配置文件，即可完成对有关数据的校验功能。

1. 校验配置文件

校验配置文件为 xml 文件，需与对应的 Action 实现类放在一起，其命名规则为 ActionName-validation.xml 或 ActionName-name-validation.xml。其中，ActionName 为相应 Action 实现类的名称，name 为相应 Action 实现类在 struts.xml 中所对应的 action 元素的名称(即属性 name 的值)。例如，在【实例 6-1】中，若要对提交到 Action 实现类 RegisterAction 的数据进行校验，则其所对应的校验配置文件应命名为 RegisterAction-validation.xml 或 RegisterAction-register-validation.xml。

校验配置文件的基本格式如下：

```
<?xml version="1.0" encoding="UTF-8"?>
<!DOCTYPE validators PUBLIC
         "-//Apache Struts//XWork Validator 1.0.2//EN"
         "http://struts.apache.org/dtds/xwork-validator-1.0.2.dtd">
<validators>
    <field name="字段名">
        <field-validator type="校验器名">
        <param name="参数名">参数值</param>
          <message>校验失败信息</message>
        </field-validator>
        …
    </field>
    …
    <validator type="校验器名">
        <param name="fieldName">字段名</param>
        <message>校验失败信息</message>
    </validator>
    …
    <validator type="校验器名">
        <param name="expression">表达式</param>
        <message>校验失败信息</message>
        …
    </validator>
    …
</validators>
```

其中，field 元素用于为指定的字段设置需应用的校验器(一个或多个)，validator 元素用于为指定校验器设置待校验的字段或表达式。

说明： 若使用某些早期的 Struts 2 开发包，则校验配置文件中的<!DOCTYPE>元素应改写为：

```
<!DOCTYPE validators PUBLIC
         "-//OpenSymphony Group//XWork Validator 1.0.2//EN"
         "http://www.opensymphony.com/xwork/xwork-validator-
1.0.2.dtd">
```

2. 数据校验器

Struts 2 提供了大量的数据校验器，用于对指定的字段或表达式进行相应的校验，表 6.18 所示即为 Struts 2 中常用的数据校验器。

表 6.18 Struts 2 的常用数据校验器

校 验 器	说 明
required	检查字段是否为非空
requiredstring	检查字段是否为非空且字符串长度是否大于 0
stringlength	检查字符串长度是否在指定范围内

续表

校 验 器	说　明
email	检查字段是否为 E-mail 格式
url	检查字段是否为 URL 格式
regex	检查字段是否匹配指定的正则表达式
int	检查字段是否为整数且在指定范围内
double	检查字段是否为双精度浮点数且在指定范围内
date	检查字段是否为日期格式且在指定范围内
conversion	检查字段是否发生类型的错误
expression	对指定表达式求值
fieldexpression	字段表达式验证程序
visitor	引用指定对象各属性对应的检验规则

下面，仅对部分数据校验器的基本用法进行简要说明。

(1) required。required 为必填校验器，要求指定的字段必须有值(包括空格字符串)。如以下示例：

```
<field name="username">
    <field-validator type="required">
        <message>用户名是必需的！</message>
    </field-validator>
</field>
```

(2) requiredstring。requiredstring 为必填字符串校验器，要求指定的字段不能为空，即必须有值，且字符串长度大于 0。如以下示例：

```
<field name="username">
    <field-validator type="requiredstring">
        <!--去空格-->
        <param name="trim">true</param>
        <message>用户名不能为空！</message>
    </field-validator>
</field>
```

(3) stringlength 字符串长度校验器。stringlength 为字符串长度校验器，要求字段的长度必须在指定的范围内(通过参数 minLength 与 maxLength 指定)。如以下示例：

```
<field name="password">
    <field-validator type="stringlength">
        <!- -长度最小值-->
        <param name="minLength">6</param>
        <!- -长度最大值-->
        <param name="maxLength">20</param>
        <message>密码长度必须在 6 到 20 之间！</message>
    </field-validator>
</field>
```

(4) int。int 为整数校验器，要求字段的整数值必须在指定范围内(通过参数 min 与 max 指定)。如以下示例：

```
<field name="score">
    <field-validator type="int">
        <!- -最小值-->
        <param name="min">0</param>
        <!- -最大值-->
        <param name="max">100</param>
        <message>成绩必须在 0 至 100 之间！</message>
    </field-validator>
</field>
```

(5) date。date 为日期校验器，要求字段的日期值必须在指定范围内(通过参数 min 与 max 指定)。如以下示例：

```
<field name="date">
    <field-validator type="date">
        <!- -最小值-->
        <param name="min">1980-01-01</param>
        <!- -最大值-->
        <param name="max">2010-12-31</param>
        <message>日期必须在 1980-01-01 至 2010-12-31 之间！</message>
    </field-validator>
</field>
```

(6) email。email 为邮件地址校验器，要求字段非空时必须是合法的邮件地址。如以下示例：

```
<field name="email">
    <field-validator type="email">
        <message>必须输入有效的电子邮件地址！</message>
    </field-validator>
</field>
```

(7) url。url 为网址校验器，要求字段非空时必须是合法的 URL 地址。如以下示例：

```
<field name="url">
    <field-validator type="url">
        <message>必须输入有效的网址！</message>
    </field-validator>
</field>
```

(8) regex。regex 为正则表达式校验器，要求字段值必须匹配指定的正则表达式(通过参数 expression 指定)。如以下示例：

```
<field name="bh">
    <field-validator type="regex">
        <param name="expression"><![CDATA[(\d{7})]]></param>
        <message>编号必须是 7 位的数字！</message>
    </field-validator>
</field>
```

【实例 6-13】Struts 2 数据校验示例。使用 Struts 2 的验证框架实现【实例 6-12】中的数据校验功能，即要求用户名与密码不能为空。

主要步骤(在项目 web_06 中)如下。

(1) 修改 Action 实现类——RegisterAction(其文件名为 RegisterAction.java)，删除或注释掉其中添加校验方法 validate()。

(2) 在 org.etspace.abc.action 包中创建校验文件 RegisterAction-validation.xml，其代码为：

```
<?xml version="1.0" encoding="UTF-8"?>
<!DOCTYPE validators PUBLIC
        "-//Apache Struts//XWork Validator 1.0.2//EN"
        "http://struts.apache.org/dtds/xwork-validator-1.0.2.dtd">
<validators>
    <!-- 需要校验的字段的名称-->
    <field name="username">
        <!--验证字符串不能为空，即必填-->
        <field-validator type="requiredstring">
            <!—删除空格-->
            <param name="trim">true</param>
            <!--错误提示信息-->
            <message>用户名不能为空！</message>
        </field-validator>
    </field>
    <field name="password">
        <field-validator type="requiredstring">
            <param name="trim">true</param>
            <message>用户名不能为空！</message>
        </field-validator>
    </field>
</validators>
```

说明： 校验文件 RegisterAction-validation.xml 的代码也可采用另外一种方式，即：

```
<?xml version="1.0" encoding="UTF-8"?>
<!DOCTYPE validators PUBLIC
        "-//Apache Struts//XWork Validator 1.0.2//EN"
        "http://struts.apache.org/dtds/xwork-validator-
1.0.2.dtd">
<validators>
    <validator type="requiredstring">
        <param name="fieldName">username</param>
        <message>用户名不能为空！</message>
    </validator>
    <validator type="requiredstring">
        <param name="fieldName">password</param>
        <message>密码不能为空！</message>
    </validator>
</validators>
```

6.7 Struts 2 文件上传

文件上传是 Web 应用中的常用功能，分为单文件上传与多文件上传两种情形。为实现文件上传功能，Java 提供了 Commons-FileUpload 框架与 COS 框架。Struts 2 框架对 Commons-FileUpload 框架进行了封装，从而能更容易地实现文件上传功能。

由于 Struts 2 的文件上传默认使用的是 Jakarta 的 Commons-FileUpload 文件上传框架，因此实现文件上传功能时，应注意将 commons-fileupload-1.3.jar 与 commons-io-2.2.jar 这两个 Jar 包加载到项目中(在此假定 Struts 2 开发包为 struts-2.3.16-all.zip)。

6.7.1 单文件上传

所谓单文件上传，就是一次只上传一个文件。在 Struts 2 中实现单文件上传功能较为简单，只需使用普通的 Action 即可。但为了获取上传文件的有关信息(如上传文件的文件名等)，需按照一定的规则为 Action 实现类添加一些属性以及相应的 getter 方法与 setter 方法。

【实例 6-14】 单文件上传示例。图 6.31 所示为文件上传表单，在其中选定一个文件后，再单击“上传”按钮，即可将该文件上传到 Web 服务器的指定路径中(在此为 c:\upload)，并显示图 6.32 所示的结果。

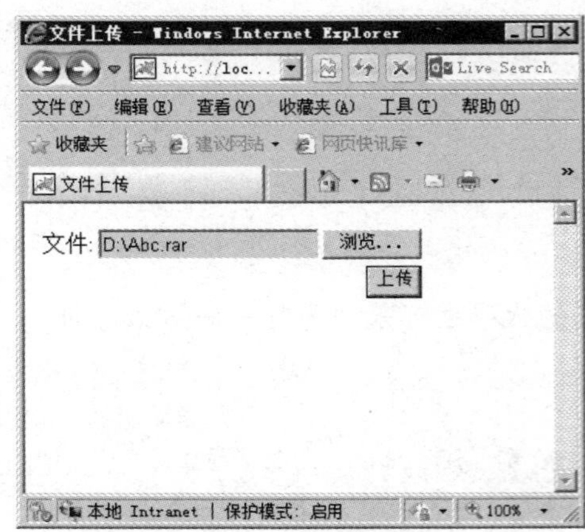

图 6.31 文件上传表单

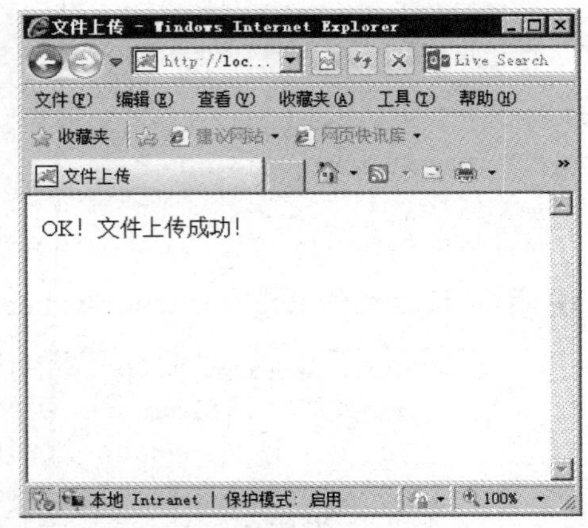

图 6.32 文件上传结果

主要步骤(在项目 web_06 中)如下。

(1) 在项目的 WebRoot 文件夹中添加一个 fileupload.jsp 页面，其代码如下：

```
<%@ page language="java" import="java.util.*" pageEncoding="UTF-8"%>
<%@ taglib uri="/struts-tags" prefix="s" %>
<html>
  <head>
    <title>文件上传</title>
  </head>
```

```
  <body>
    <s:form action="fileupload.action" method="post"
enctype="multipart/form-data">
        <s:file label="文件" name="upload"></s:file>
        <s:submit value="上传"></s:submit>
    </s:form>
  </body>
</html>
```

注意： 为实现文件的上传，在相应表单的<form>标记或<s:form>标签中，除了将 method 属性设置为“post”以外，还应将 enctype 属性设置为“multipart/form-data”(该编码方式表示以二进制流的方法处理表单数据)。

(2) 在 org.etspace.abc.action 包中新建 Action 实现类 FileUploadAction，其文件名为 FileUploadAction.java，代码如下：

```
package org.etspace.abc.action;
import com.opensymphony.xwork2.ActionSupport;
import java.io.InputStream;
import java.io.OutputStream;
import java.io.File;
import java.io.FileInputStream;
import java.io.FileOutputStream;
public class FileUploadAction extends ActionSupport {
    private File upload;
    private String uploadFileName;
    public File getUpload(){
        return upload;
    }
    public void setUpload(File upload){
        this.upload=upload;
    }
    public String getUploadFileName(){
        return uploadFileName;
    }
    public void setUploadFileName(String uploadFileName){
        this.uploadFileName=uploadFileName;
    }
    public String execute() throws Exception{
        InputStream is=new FileInputStream(getUpload());
       OutputStream os=new FileOutputStream("c:\\upload\\"+uploadFileName);
        byte buffer[]=new byte[1024];
        int count=0;
        while((count=is.read(buffer))>0){
            os.write(buffer,0,count);
        }
        os.close();
```

```
        is.close();
        return SUCCESS;
    }
}
```

(3) 在项目的配置文件 struts.xml 中添加 FileUploadAction(Action)的配置代码，具体为：

```
<package name="upload" extends="struts-default">
    <action name="fileupload"
class="org.etspace.abc.action.FileUploadAction">
        <result name="success">/fileupload_success.jsp</result>
    </action>
</package>
```

(4) 在项目的 WebRoot 文件夹中添加一个 fileupload_success.jsp 页面，其代码如下：

```
<%@ page language="java" import="java.util.*" pageEncoding="UTF-8"%>
<%@ taglib uri="/struts-tags" prefix="s" %>
<html>
  <head>
    <title>文件上传</title>
  </head>
  <body>
    OK！文件上传成功！
  </body>
</html>
```

注意： Struts 2 上传文件的默认大小限制为 2M，必要时，可对上传文件的大小限制进行相应修改。为此，只需在项目的 src 文件夹下创建一个 struts.properties (该文件其实就是一个文本文件)文件，并在其中设置好相应的 struts.multipart.maxSize 值(该值表示字节数)即可。例如，在该文件中写入代码“struts.multipart.maxSize=104857600”，即可将上传文件的最大大小设置为 100M。

6.7.2 多文件上传

所谓多文件上传，就是同时上传多个文件。其实，在 Struts 2 中，多文件上传功能的实现并不复杂，只需对单文件上传功能的实现方式稍加修改即可。首先，在文件上传页面中应同时提供多个供用户选择文件的文件框；其次，应将 Action 实现类中的有关属性修改为 List 类型，并相应修改其 getter 方法与 setter 方法；最后，再添加一个循环结构，以便逐一实现各个文件的上传操作。

【实例 6-15】多文件上传示例。图 6.33 所示为文件上传表单，在其中选定相应的文件(在此最多为 3 个文件)后，再单击“上传”按钮，即可同时将其上传到 Web 服务器的指定路径中(在此为 c:\upload)，并显示图 6.34 所示的结果。

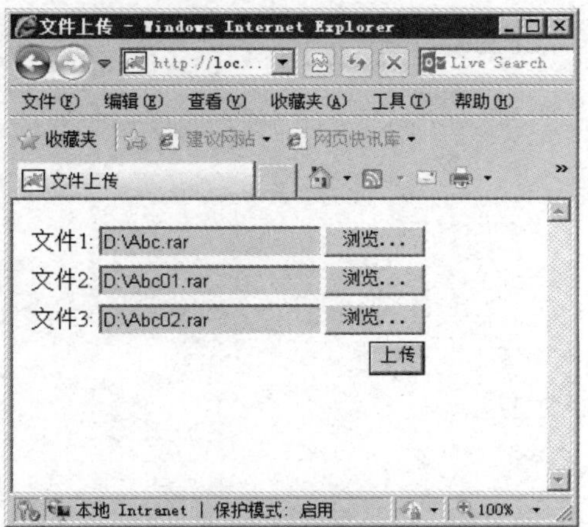

图 6.33　文件上传表单

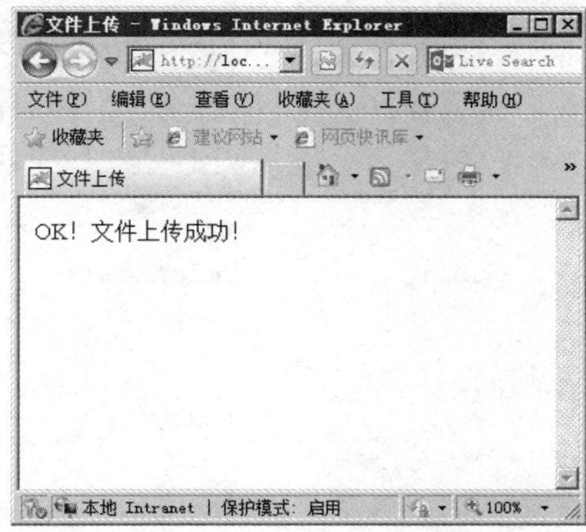

图 6.34　文件上传结果

主要步骤(在项目 web_06 中)如下。

(1) 在项目的 WebRoot 文件夹中添加一个 filesupload.jsp 页面，其代码如下：

```
<%@ page language="java" import="java.util.*" pageEncoding="UTF-8"%>
<%@ taglib uri="/struts-tags" prefix="s" %>
<html>
  <head>
    <title>文件上传</title>
  </head>
  <body>
    <s:form action="filesupload.action" method="post"
enctype="multipart/form-data">
        <s:file label="文件 1" name="upload"></s:file>
        <s:file label="文件 2" name="upload"></s:file>
        <s:file label="文件 3" name="upload"></s:file>
        <s:submit value="上传"></s:submit>
    </s:form>
  </body>
</html>
```

注意： 各文件框的名称(即 name 属性值)应保持一致，以便将其值都封装到相应的 List 集合中。

(2) 在 org.etspace.abc.action 包中新建 Action 实现类 FilesUploadAction，其文件名为 FilesUploadAction.java，代码如下：

```
package org.etspace.abc.action;
import com.opensymphony.xwork2.ActionSupport;
import java.io.InputStream;
import java.io.OutputStream;
import java.io.File;
import java.io.FileInputStream;
import java.io.FileOutputStream;
```

```
import java.util.List;
public class FilesUploadAction extends ActionSupport {
    private List<File> upload;
    private List<String> uploadFileName;
    public List<File> getUpload(){
        return upload;
    }
    public void setUpload(List<File> upload){
        this.upload=upload;
    }
    public List<String> getUploadFileName(){
        return uploadFileName;
    }
    public void setUploadFileName(List<String> uploadFileName){
        this.uploadFileName=uploadFileName;
    }
    public String execute() throws Exception{
        if (upload!=null){
            for (int i=0;i<upload.size();i++){
                InputStream is=new FileInputStream(upload.get(i));
                OutputStream os=new FileOutputStream("c:\\upload\\"+
uploadFileName.get(i));
                byte buffer[]=new byte[1024];
                int count=0;
                while((count=is.read(buffer))>0){
                    os.write(buffer,0,count);
                }
                os.close();
                is.close();
            }
        }
        return SUCCESS;
    }
}
```

(3) 在项目的配置文件 struts.xml 中添加 FilesUploadAction(Action)的配置代码，具体为：

```
<package name="upload" extends="struts-default">
    …
    <action name="filesupload"
class="org.etspace.abc.action.FilesUploadAction">
        <result name="success">/filesupload_success.jsp</result>
    </action>
</package>
```

(4) 在项目的 WebRoot 文件夹中添加一个 filesupload_success.jsp 页面。其代码与【实例 6-15】中的 fileupload_success.jsp 页面相同。

6.8　Struts 2 文件下载

文件下载是 Web 应用中较为常见的功能之一。在 Java Web 应用中，对于一个以西欧字符命名的文件，其下载只需在页面中使用超链接指向即可。如以下代码：

```
<a href="download/example.rar">下载</a>
```

该代码将生成一个“下载”链接，单击该链接后即可顺利下载站点 download 目录下的 example.rar 文件。

不过，如果文件名中包含有非西欧字符(如中文等)，那么这种方法是行不通的。为解决此问题，Struts 2 提供了相应的方案，即在页面生成一个指向 Action 请求的链接，然后在 Action 的实现类中进行相应的属性处理后直接返回“success”，最后在相应 Action 配置的 result 元素中设置好相关的参数。有关参数主要如下。

(1) inputName。该参数用于指定 Action 实现类中作为输入流的属性名。

(2) contentType。该参数用于指定下载文件的类型。若指定该参数，则下载一个图片文件时会直接在浏览器中打开；若不指定该参数，则直接将文件下载到本地。

(3) contentDisposition。该参数用于指定下载文件在客户端上的一些属性。例如，可以设置文件下载后保存到本地时所使用的文件名(该文件名应全部使用西欧字符，否则可能出现乱码)。

(4) bufferSize。该参数用于指定下载文件时缓冲区的大小。

【实例 6-16】Struts 2 文件下载示例：下载站点 download 目录下的“案例.rar”文件。

主要步骤(在项目 web_06 中)如下。

(1) 在项目的 WebRoot 文件夹中新建一个子文件夹 download，并将要下载的文件“案例.rar”置于其中。

(2) 在项目的 WebRoot 文件夹中添加一个 filedownload.jsp 页面，其代码如下：

```
<%@ page language="java" pageEncoding="UTF-8"%>
<html>
  <head>
    <title>文件下载</title>
  </head>
  <body>
    <a href="filedownload.action">下载</a>
  </body>
</html>
```

(3) 在 org.etspace.abc.action 包中新建 Action 实现类 FileDownloadAction，其文件名为 FileDownloadAction.java，代码如下：

```
package org.etspace.abc.action;
import com.opensymphony.xwork2.ActionSupport;
import java.io.InputStream;
```

```
import org.apache.struts2.ServletActionContext;
public class FileDownloadAction extends ActionSupport {
    private String downloadFile;  //需下载的文件
    //生成 setter 方法，参数值由配置文件传递过来
    public void setDownloadFile(String downloadFile){
        this.downloadFile=downloadFile;
    }
    //根据需下载的文件生成其输入流
    public InputStream getTargetFile(){
        return ServletActionContext.getServletContext().getResourceAsStream
(downloadFile);
    }
    public String execute() throws Exception{
        return SUCCESS;
    }
}
```

(4) 在项目的配置文件 struts.xml 中添加 FileDownloadAction(Action)的配置代码，具体为：

```
<package name="download" extends="struts-default">
    <action name="filedownload"
class="org.etspace.abc.action.FileDownloadAction">
        <!-- 传递参数，指定要下载的文件 -->
        <param name="downloadFile">/download/案例.rar</param>
        <result name="success" type="stream">
            <!-- 输入流名称，对应 Action 实现类中的 getTargetFile()方法 -->
            <param name="inputName">targetFile</param>
            <!-- 设置文件下载后所生成的目标文件 -->
            <param name="contentDisposition">
                filename="example.rar"
            </param>
            <!-- 设置下载文件时缓冲区的大小 -->
            <param name="bufferSize">4096</param>
        </result>
    </action>
</package>
```

运行结果如图 6.35 所示。单击“下载”链接后，将弹出图 6.36 所示的“文件下载”对话框，再单击其中的“保存”按钮并完成后续的有关操作，即可将“案例.rar”文件下载到本地指定的路径中。

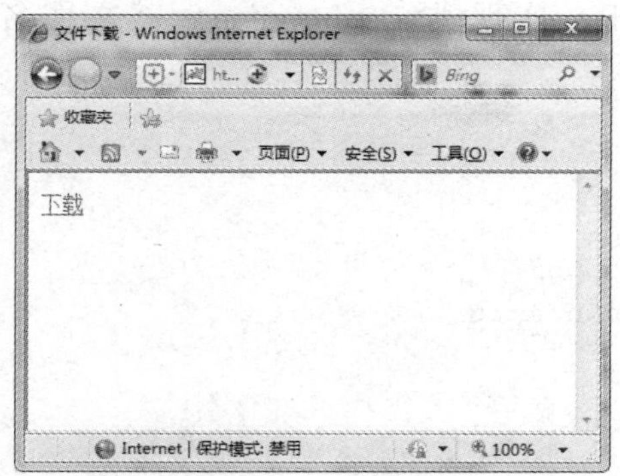

图 6.35 “文件下载”页面

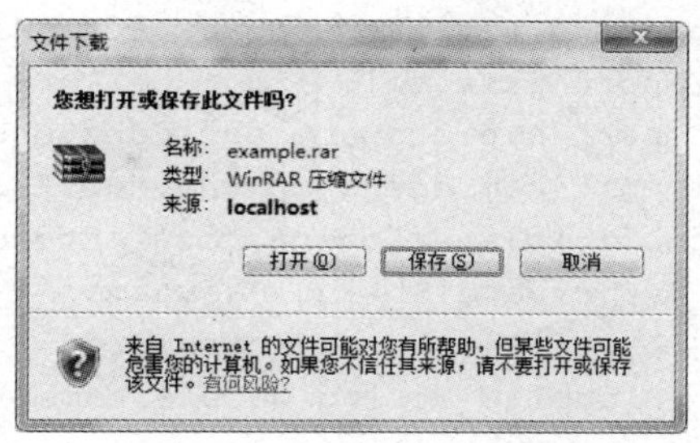

图 6.36 “文件下载”对话框

6.9 Struts 2 应用案例

6.9.1 系统登录

下面，以 JSP+JDBC+JavaBean+Struts 2 模式实现 Web 应用系统中的系统登录功能，其运行结果与第 5 章 5.5.1 小节的系统登录案例相同。

实现步骤(在项目 web_06 中)如下。

(1) 将 SQL Server 2005/2008 的 JDBC 驱动程序 sqljdbc4.jar 添加到项目的 WebRoot\WEB-INF\lib 文件夹中。

(2) 在项目的 src 文件夹中新建一个包 org.etspace.abc.jdbc。

(3) 在 org.etspace.abc.jdbc 包中新建一个 JavaBean——DbBean，其文件名为 DbBean.java，代码与第 5 章 5.5.1 小节系统登录案例中的相同。

(4) 在项目的 WebRoot 文件夹中新建一个子文件夹 syslogin_struts。

(5) 在子文件夹 syslogin_struts 中添加一个新的 JSP 页面 login.jsp，其代码与第 5 章 5.5.1 小节系统登录案例中的类似，只需将其中<form>标记的 action 属性设为“login.action”即可，具体为：

```
<form action="login.action" method="post">
```

说明： 页面 login.jsp 中的表单也可使用 Struts 2 标签实现。为此，应在页面中先添加指令“<%@ taglib prefix="s" uri="/struts-tags" %>”，然后再将其中的表单代码修改为：

```
<s:form action="login.action" method="post">
    <s:textfield label="用户名" name="username"></s:textfield>
    <s:password label="密码" name="password"></s:password>
    <s:submit value="登录"></s:submit>
</s:form>
```

(6) 在 org.etspace.abc.action 包中新建 Action 实现类 LoginAction。其文件名为 LoginAction.java，代码如下：

```
package org.etspace.abc.action;
import java.sql.*;
import org.etspace.abc.jdbc.DbBean;
import com.opensymphony.xwork2.ActionSupport;
public class LoginAction extends ActionSupport{
    private String username;
    private String password;
    public String getUsername(){
        return username;
    }
    public void setUsername(String username){
        this.username=username;
    }
    public String getPassword(){
        return password;
    }
    public void setPassword(String password){
        this.password=password;
    }
    public String execute() throws Exception{
        boolean validated=false;  //验证标识
        String sql="select * from users where UserName='"+username+"'
and password='"+password+ "'";
        DbBean myDbBean=new DbBean();
        myDbBean.openConnection();
        ResultSet rs= myDbBean.executeQuery(sql);
        if(rs.next()){
            validated=true;
        }
        rs.close();
        myDbBean.closeConnection();
        if(validated)
        {
            //验证成功返回"success"
            return "success";
        }
        else
        {
            //验证失败返回"error"
            return "error";
        }
    }
}
```

(7) 在项目的配置文件 struts.xml 中添加 LoginAction(Action)的配置代码，具体为：

```
<package name="login" extends="struts-default">
    <action name="login" class="org.etspace.abc.action.LoginAction">
        <result name="success">/syslogin_struts/welcome.jsp</result>
        <result name="error">/syslogin_struts/error.jsp</result>
    </action>
</package>
```

(8) 在子文件夹 syslogin_struts 中添加一个新的 JSP 页面 welcome.jsp，其代码如下：

```
<%@ page language="java" pageEncoding="utf-8" %>
<%@ taglib prefix="s" uri="/struts-tags" %>
<html>
    <head>
        <title>登录成功</title>
    </head>
    <body>
        <s:property value="username" />，您好！欢迎光临本系统。
    </body>
</html>
```

(9) 在子文件夹 syslogin_struts 中添加一个新的 JSP 页面 error.jsp，其代码与第 5 章 5.5.1 小节系统登录案例中的相同。

6.9.2 数据添加

下面以 JSP+JDBC+JavaBean+Struts 2 模式实现“职工增加”功能。

图 6.37 所示为“职工增加”页面，在其中的表单输入职工的各项信息后，再单击“确定”按钮提交表单。若能成功地将该职工记录添加到职工表中，则显示图 6.38 所示的成功页面，否则图 6.39 所示的失败页面。

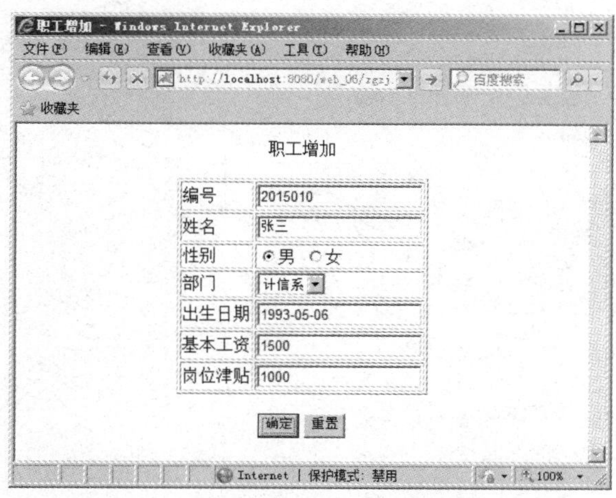

图 6.37　“职工增加”页面

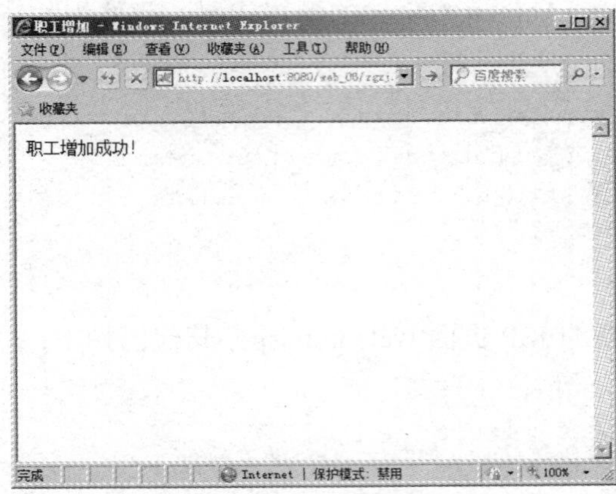

图 6.38　成功页面

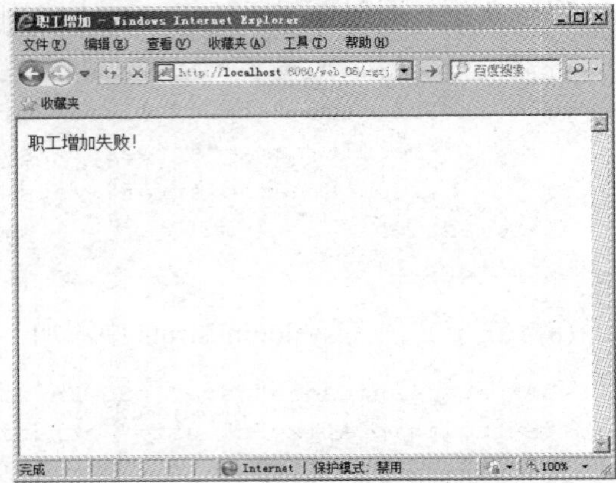

图 6.39　失败页面

主要步骤(在项目 web_06 中)如下。

(1) 在项目的 src 文件夹中新建一个包 org.etspace.abc.bean。

(2) 在 org.etspace.abc.bean 包中新建一个 JavaBean——Zgb，其文件名为 Zgb.java，代码如下：

```
package org.etspace.abc.bean;
import java.sql.Date;
public class Zgb {
    private String bh;
    private String xm;
    private String xb;
    private String bm;
    private Date csrq;
    private Float jbgz;
    private Float gwjt;
    public String getBh() {
        return bh;
    }
    public void setBh(String bh) {
        this.bh=bh;
    }
    public String getXm() {
        return xm;
    }
    public void setXm(String xm) {
        this.xm=xm;
    }
    public String getXb() {
        return xb;
    }
    public void setXb(String xb) {
        this.xb=xb;
```

```
    }
    public String getBm() {
        return bm;
    }
    public void setBm(String bm) {
        this.bm=bm;
    }
    public Date getCsrq() {
        return csrq;
    }
    public void setCsrq(Date csrq) {
        this.csrq=csrq;
    }
    public Float getJbgz() {
        return jbgz;
    }
    public void setJbgz(Float jbgz) {
        this.jbgz=jbgz;
    }
    public Float getGwjt() {
        return gwjt;
    }
    public void setGwjt(Float gwjt) {
        this.gwjt=gwjt;
    }
}
```

(3) 在org.etspace.abc.jdbc包中新建一个JavaBean——DbBean_Zgb，其文件名为DbBean_Zgb.java，代码如下：

```
package org.etspace.abc.jdbc;
import java.sql.*;
import org.etspace.abc.bean.Zgb;
public class DbBean_Zgb extends DbBean {
private PreparedStatement pstmt;
public int insert(Zgb zg){
        int n=0;
        try{
            pstmt=conn.prepareStatement("insert into zgb
values(?,?,?,?,?,?,?)");
            pstmt.setString(1, zg.getBh());
            pstmt.setString(2, zg.getXm());
            pstmt.setString(3, zg.getXb());
            pstmt.setString(4, zg.getBm());
            pstmt.setDate(5, zg.getCsrq());
            pstmt.setFloat(6, zg.getJbgz());
            pstmt.setFloat(7, zg.getGwjt());
            n=pstmt.executeUpdate();
        }catch(Exception e){
```

```
            System.err.println(e.getMessage());
        }
        return n;
    }
}
```

(4) 在项目的 WebRoot 文件夹中添加职工增加页面 zgzj.jsp，其代码如下：

```
<%@ page contentType="text/html;charset=utf-8" language="java" %>
<%@ page import="java.sql.*"%>
<jsp:useBean id="myDbBean" scope="page"
class="org.etspace.abc.jdbc.DbBean"></jsp:useBean>
<%request.setCharacterEncoding("utf-8"); %>
<script language="JavaScript">
function check(theForm)
{
  if (theForm.bh.value.length != 7)
  {
    alert("职工编号必须为 7 位！");
    theForm.bh.focus();
    return (false);
  }
  if (theForm.xm.value == "")
  {
    alert("请输入姓名！");
    theForm.xm.focus();
    return (false);
  }
  if (theForm.csrq.value == "")
  {
    alert("请输入出生日期！");
    theForm.csrq.focus();
    return (false);
  }
  if (theForm.jbgz.value == "")
  {
    alert("请输入基本工资！");
    theForm.jbgz.focus();
    return (false);
  }
  if (theForm.gwjt.value == "")
  {
    alert("请输入岗位津贴！");
    theForm.gwjt.focus();
    return (false);
  }
  return (true);
}
</script>
<html>
<head><title>职工增加</title></head>
```

```
<body>
<div align="center">
<P>职工增加</P>
<form id="form1" name="form1" method="post" action="zgzj.action"
onSubmit="return check(this)">
    <table border="1">
    <tr><td>编号</td><td><input name="zg.bh" type="text" id="bh"
/></td></tr>
    <tr><td>姓名</td><td><input name="zg.xm" type="text" id="xm"
/></td></tr>
    <tr><td>性别</td><td>
    <input type="radio" name="zg.xb" value="男" checked="checked" />男
    <input type="radio" name="zg.xb" value="女" />女
    </td></tr>
<%
        String sql = "select * from bmb order by bmbh";
        myDbBean.openConnection();
        ResultSet rs=myDbBean.executeQuery(sql);
%>
    <tr><td>部门</td>
    <td><select name="zg.bm">
<%
       while(rs.next())
       {
        String bmbh=rs.getString("bmbh");
        String bmmc=rs.getString("bmmc");
%>
    <option value="<%=bmbh%>"><%=bmmc%></option>
<%
       }
       rs.close();
        myDbBean.closeConnection();
%>
    </select>
    </td></tr>
    <tr><td>出生日期</td><td><input name="zg.csrq" type="text" id="csrq"
/></td></tr>
    <tr><td>基本工资</td><td><input name="zg.jbgz" type="text" id="jbgz"
/></td></tr>
    <tr><td>岗位津贴</td><td><input name="zg.gwjt" type="text" id="gwjt"
/></td></tr>
    </table>
    <br>
    <input name="submit" type="submit"  value="确定" />
    <input name="reset" type="reset" value="重置" />
</form>
</div>
</body>
</html>
```

说明： 页面 zgzj.jsp 也可使用 Struts 2 标签实现(其运行结果如图 6.40 所示)，具体代码如下：

```
<%@ page contentType="text/html;charset=utf-8"
language="java" %>
<%@ taglib uri="/struts-tags" prefix="s" %>
<%@ taglib uri="/struts-dojo-tags" prefix="sd" %>
<%@ page import="java.sql.*"%>
<%@ page import="java.util.Map,java.util.HashMap"%>
<jsp:useBean id="myDbBean" scope="page"
class="org.etspace.abc.jdbc.DbBean"></jsp:useBean>
<%request.setCharacterEncoding("utf-8"); %>
<script language="JavaScript">
function check(theForm)
{
  if (theForm.bh.value.length != 7)
  {
    alert("职工编号必须为 7 位！");
    theForm.bh.focus();
    return (false);
  }
  if (theForm.xm.value == "")
  {
    alert("请输入姓名！");
    theForm.xm.focus();
    return (false);
  }
  if (theForm.jbgz.value == "")
  {
    alert("请输入基本工资！");
    theForm.jbgz.focus();
    return (false);
  }
  if (theForm.gwjt.value == "")
  {
    alert("请输入岗位津贴！");
    theForm.gwjt.focus();
    return (false);
  }
  return (true);
}
</script>
<html>
<head><title>职工增加</title></head>
<sd:head/>
<body>
<div align="center">
<P>职工增加</P>
<s:form id="form1" name="form1" method="post"
action="zgzj.action" onSubmit="return check(this)"
theme="simple">
 <table border="1">
 <tr><td>编号</td><td><s:textfield name="zg.bh"
id="bh"></s:textfield></td></tr>
```

```
    <tr><td>姓名</td><td><s:textfield name="zg.xm"
id="xm"></s:textfield></td></tr>
    <tr><td>性别</td><td><s:radio list="{'男','女'}" name="zg.xb"
value="'男'"></s:radio>
    </td></tr>
       <%
        Map<String,String> map=new HashMap<String,String>();
        String sql = "select * from bmb order by bmbh";
        myDbBean.openConnection();
        ResultSet rs=myDbBean.executeQuery(sql);
          while(rs.next())
          {
            String bmbh=rs.getString("bmbh");
            String bmmc=rs.getString("bmmc");
                map.put(bmbh,bmmc);
          }
          rs.close();
        myDbBean.closeConnection();
        request.setAttribute("map",map);
        request.setAttribute("bm","01");
       %>
    <tr><td>部门</td><td>
    <s:select list="#request.map" name="zg.bm" id="bm"
listKey="key" listValue="value" value="#request.bm"></s:select>
    </td></tr>
    <tr><td>出生日期</td><td><sd:datetimepicker name="zg.csrq"
id="csrq" displayFormat="yyyy-MM-dd"></sd:datetimepicker>
    </td></tr>
    <tr><td>基本工资</td><td><s:textfield name="zg.jbgz"
id="jbgz"></s:textfield></td></tr>
    <tr><td>岗位津贴</td><td><s:textfield name="zg.gwjt"
id="gwjt"></s:textfield></td></tr>
    </table>
    <br>
        <s:submit name="sumbit" value="确定"></s:submit>
        <s:reset name="reset" value="重置"></s:reset>
   </s:form>
   </div>
   </body>
   </html>
```

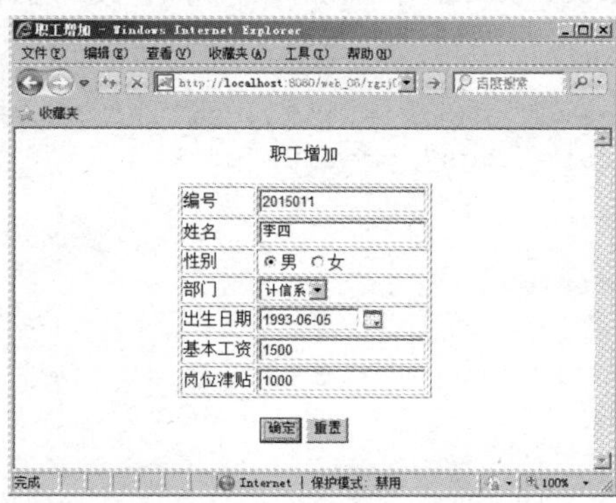

图 6.40　“职工增加”页面

(5) 在 org.etspace.abc.action 包中新建 Action 实现类——ZgzjAction，其文件名为 ZgzjAction.java，代码如下：

```
package org.etspace.abc.action;
import com.opensymphony.xwork2.ActionSupport;
import org.etspace.abc.bean.Zgb;
import org.etspace.abc.jdbc.DbBean_Zgb;
public class ZgzjAction extends ActionSupport {
    private Zgb zg;
    public Zgb getZg(){
        return zg;
    }
    public void setZg(Zgb zg){
        this.zg=zg;
    }
    public String execute() throws Exception{
        DbBean_Zgb myDb=new DbBean_Zgb();
        myDb.openConnection();
        int n=myDb.insert(zg);
        myDb.closeConnection();
        if (n>0){
            return SUCCESS;
        }else{
            return ERROR;
        }
    }
}
```

(6) 在项目的配置文件 struts.xml 中添加 ZgzjAction(Action)的配置代码，具体为：

```
<package name="zg" extends="struts-default">
    <action name="zgzj" class="org.etspace.abc.action.ZgzjAction">
        <result name="success">/zgzj_success.jsp</result>
        <result name="error">/zgzj_error.jsp</result>
    </action>
</package>
```

(7) 在子文件夹 WebRoot_struts 中添加一个新的 JSP 页面 zgzj_success.jsp，其代码如下：

```
<%@ page language="java" import="java.util.*" pageEncoding="UTF-8"%>
<%@ taglib uri="/struts-tags" prefix="s" %>
<html>
  <head>
    <title>职工增加</title>
  </head>
  <body>
    职工增加成功！
  </body>
</html>
```

(8) 在子文件夹 WebRoot_struts 中添加一个新的 JSP 页面 zgzj_error.jsp.jsp。其代码如下：

```
<%@ page language="java" import="java.util.*" pageEncoding="UTF-8"%>
<%@ taglib uri="/struts-tags" prefix="s" %>
<html>
  <head>
    <title>职工增加</title>
  </head>
  <body>
     职工增加失败！
  </body>
</html>
```

说明： 传统的 Java EE 应用开发采用 JSP+JDBC+Servlet+JavaBean 模式(即 Model2 模式)实现 MVC，其控制器(C)为 Servlet，在编程上难度较大。而采用 JSP+JDBC+JavaBean+Struts 2 模式实现 MVC 系统时(见图 6.41)，控制器由 Struts 2 框架本身担任，同时业务逻辑处理功能由用户自定义的 Action 实现。由于 Action 与 Struts 2 的控制核心是互相分离的，因此可进一步降低系统中各部分组件的耦合度与编程难度。

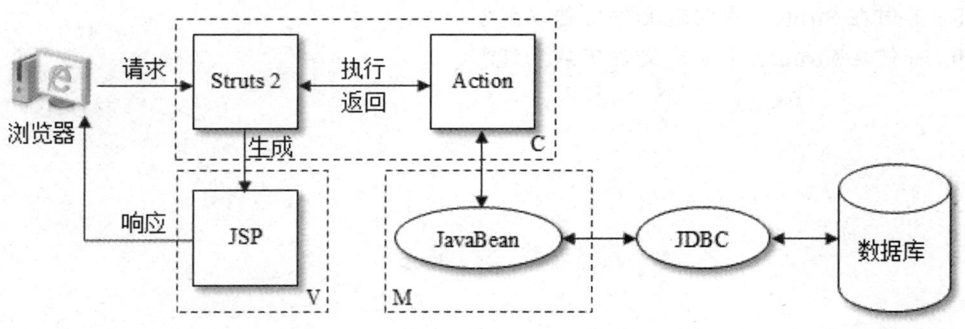

图 6.41　JSP+JDBC+JavaBean+Struts 2 模式

本 章 小 结

本章首先介绍了 Struts 2 框架的基本执行流程，然后通过具体实例讲解了 Struts 2 的基本应用技术、Struts 2 拦截器的实现与配置方法、Struts 2 OGNL 与常用标签的基本用法、Struts 2 的数据验证技术以及 Struts 2 的文件上传与下载方法，最后通过具体案例说明了 Struts 2 的综合应用技术。通过本章的学习，读者应熟练掌握 Struts 2 框架的相关应用技术，并能使用 JSP+JDBC+JavaBean+Struts 2 模式开发相应的 Web 应用系统。

思 考 题

1. 请简述 Struts 2 的基本执行流程。

2. 在开发基于 Struts 2 的 Web 应用时，主要用到 Struts 2 开发包中的哪几个 jar 包？

3. 如何在 Web 项目中应用 Struts 2 框架？

4. Struts 2 的核心过滤器是什么？如何进行配置？

5. Action 的配置文件是什么？其基本格式是什么？package 元素、action 元素与 result 元素的属性有哪些？

6. result 元素常用结果代码与结果类型有哪些？

7. Interceptor 接口定义了哪几个方法？请简述之。

8. 拦截器的常用创建方法有哪些？

9. 如何进行拦截器的配置？

10. 在 OGNL 中，如何创建一个 List 对象与 Map 对象？

11. Struts 2 标签分为哪几类？

12. Struts 2 的数据标签主要有哪些？

13. Struts 2 的控制标签主要有哪些？

14. Struts 2 的表单标签主要有哪些？

15. Struts 2 的非表单标签主要有哪些？

16. 如何在 Action 实现类中实现数据校验？

17. 如何应用 Struts 2 的校验框架完成对有关数据的校验功能？

18. 如何在 Struts 2 中实现文件上传功能？

19. 如何在 Struts 2 中实现文件下载功能？

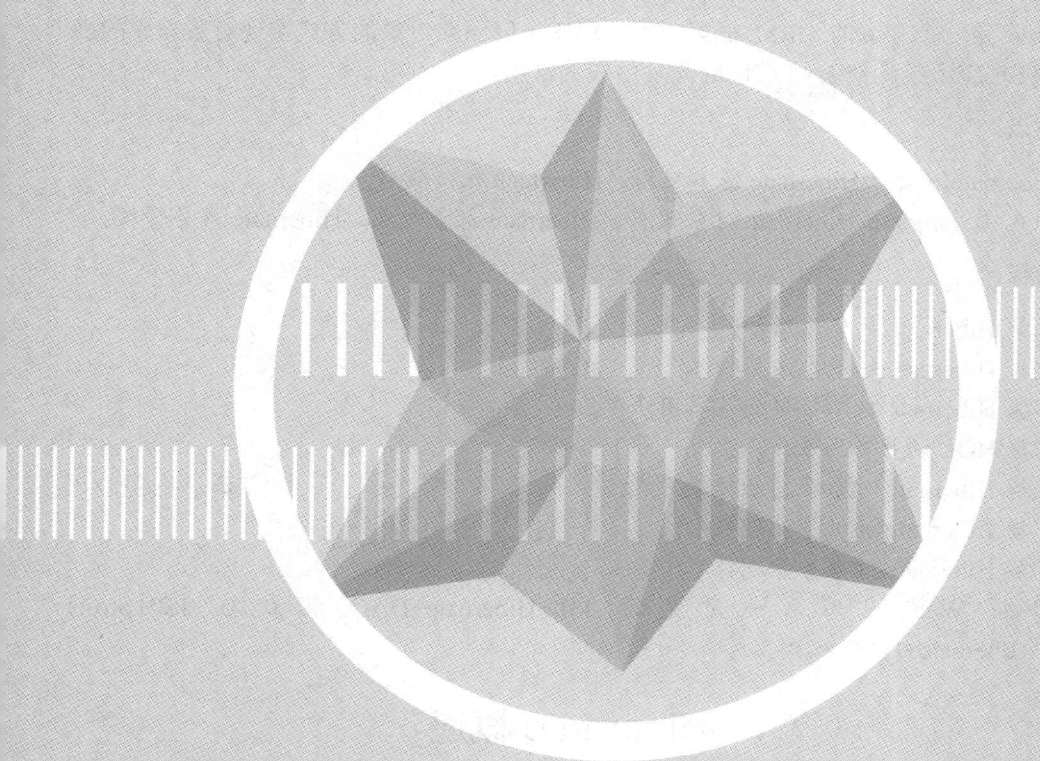

第 7 章

Hibernate 框架

Hibernate 是一个开源的 ORM 框架，可让程序员以面向对象的方式实现对数据库的操作，在 Java EE 应用或其他 Java 应用的开发中较为常用。

本章要点：

- Hibernate 概述、Hibernate 基本应用、Hibernate 核心接口；
- HQL 基本用法、Hibernate 对象状态、Hibernate 批量处理、Hibernate 事务管理。

学习目标：

- 了解 Hibernate 的体系结构；
- 掌握 Hibernate 的基本应用技术；
- 掌握 Hibernate 核心接口的基本用法；
- 掌握 HQL 的基本用法；
- 了解 Hibernate 对象状态的转换方式；
- 掌握 Hibernate 的批量处理技术；
- 了解 Hibernate 的事务管理技术；
- 掌握 Web 应用系统开发的 JSP+Hibernate+DAO 模式与 JSP+Struts 2+Hibernate+DAO 模式。

7.1 Hibernate 概述

Hibernate 是一个开源的 ORM(Object-Relation Mapping，对象关系映射)框架，通过对 JDBC 进行轻量级封装(未完全封装)，让 Java 编程者可以按照面向对象的编程思想来操纵数据库。作为一种 ORM 解决方案，Hibernate 将 Java 中对象与对象之间的关系映射为关系数据库中表与表之间的关系。

7.1.1 ORM 简介

ORM 即对象关系映射，是一种用于将对象与对象之间的关系对应到数据库中表与表之间的关系的模式。基于描述对象与数据库之间映射的元数据，ORM 可将 Java 程序中的对象自动持久化到关系数据库中。

对象与关系数据是业务实体的两种表现形式，业务实体在内存中表现为对象，而在数据库中则表现为关系数据。内存中的对象之间存在着关联与继承关系，而数据库中的关系数据则无法直接表达多对多关联与继承关系。因此，ORM 通常以中间件的形式存在，主要用于实现程序对象到关系数据库数据的映射。

一般来说，ORM 应包括以下几个部分。

(1) 对持久化对象进行 CRUD 操作(即增加、查询、修改与删除操作)的 API。

(2) 与类和类属性查询相关的语言或 API。

(3) 与 mapping metadata 相关的工具。

(4) 让 ORM 实现同事务对象一起进行 dirty checking、lazy association fetching 与其他优化操作的技术。

目前，许多厂商与开源社区均提供了持久层框架的具体实现。在发展过程中，Hibernate 的轻量级 ORM 模型逐步确立了在 Java ORM 架构中的领导地位。

7.1.2　Hibernate 体系结构

Hibernate 的体系结构如图 7.1 所示。其中，Hibernate 的配置文件(hibernate.cfg.xml 或 hibernate.properties)包含了 Hibernate 与数据库的连接信息，而 ORM 映射文件(*.hbm.xml)则用于进行对象关系的映射工作。作为模型层/数据访问层，Hibernate 通过配置文件与映射文件将 Java 对象或 PO(Persistent Object，持久化对象)映射到数据库中的数据表，然后通过操作 PO 实现对数据库中表的各种操作。在此，PO 就是 POJO(Plain Old Java Object，简单的 Java 对象/普通的 Java 对象)加上映射文件。

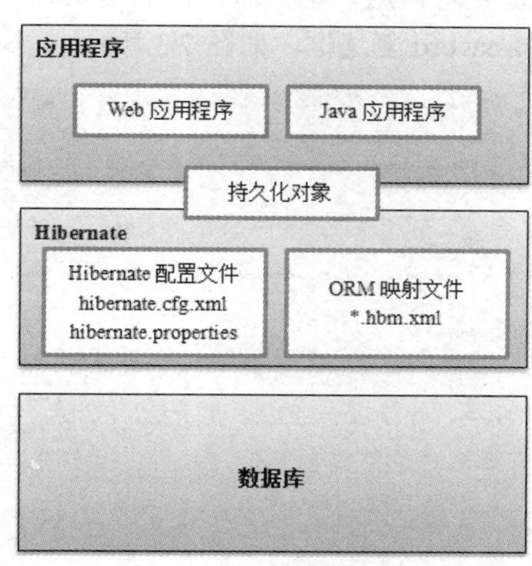

图 7.1　Hibernate 体系结构

Hibernate 是一个优秀的类库或者组件，目前已得到广泛的应用。将 Hibernate 作为企业级应用与关系数据库之间的中间件，即可通过 Hibernate 的 API 实现对数据库的访问，从而极大地减少与对象持久化有关的 JDBC 编程工作量。

7.2　Hibernate 基本应用

由于 MyEclipse 已集成了 Hibernate 功能，因此若要在 MyEclipse 的 Web 项目中使用 Hibernate 框架，只需添加相应的 Hibernate 开发能力即可。

7.2.1　Hibernate 基本用法

下面，通过一个简单的实例说明 Hibernate 框架的基本用法。

【实例 7-1】Hibernate 应用示例：部门增加。

主要步骤如下。

(1) 在 MyEclipse 中创建对数据库(在此为 SQL Server 数据库 rsgl)的连接。为此，可按以下步骤进行操作。

① 在 MyEclipse 中选择 Window→Open Perspective→MyEclipse Database Explorer 菜单项(或者在工具栏中单击 Open Perspective 按钮，并在随之打开的列表中选择 MyEclipse Database Explorer 选项)，进入 MyEclipse 的数据库浏览器(Database Explorer)模式。

② 在 DB Browser 窗格中右击，并在其快捷菜单中选择 New...菜单项，打开图 7.2 所示的 Database Driver 对话框。

③ 在 Driver template 下拉列表框中选择 Microsoft SQL Server 2005 选项(模板)，在

Driver name 文本框中输入欲建连接的名称 rsgl，在 Connection URL 文本框中将欲连接数据库的 URL 修改为“jdbc:sqlserver://localhost:1433;databaseName=rsgl”，在 User name 文本框中输入用户名 sa，在 Password 文本框中输入密码(在此为“abc123!”)，选中 Save password 复选框，如图 7.3 所示。

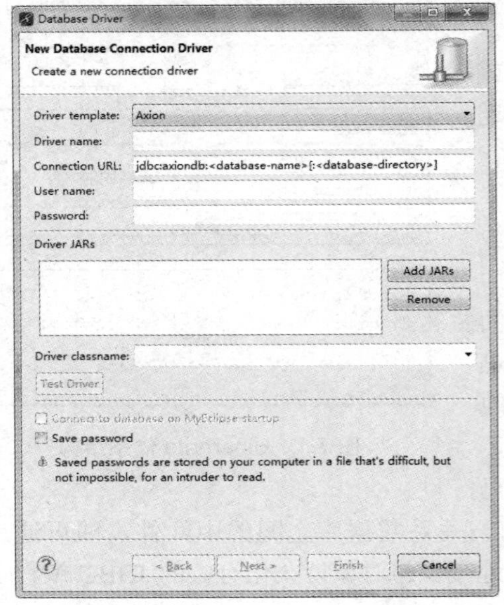

图 7.2 Database Driver 对话框(1)

图 7.3 Database Driver 对话框(2)

④ 单击 Add JARS 按钮，并在随之弹出的图 7.4 所示的“打开”对话框中选择相应的数据库 JDBC 驱动程序(在此为 sqljdbc4.jar)，然后再单击“打开”按钮，关闭“打开”对话框。

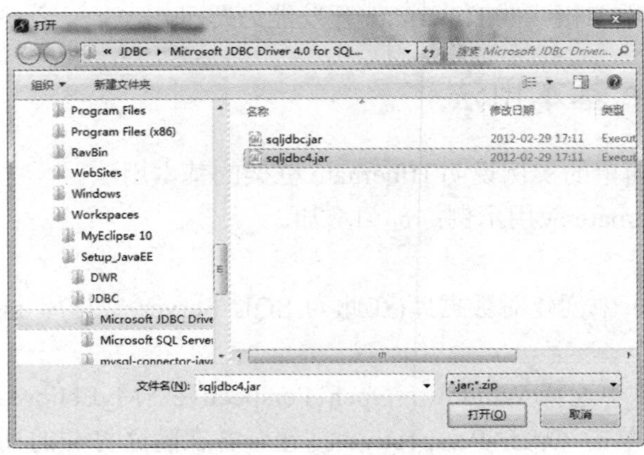

图 7.4 “打开”对话框

⑤ 单击 Test Driver 按钮测试链接，若弹出图 7.5 所示的 Driver Test 提示框，则说明一切正常。

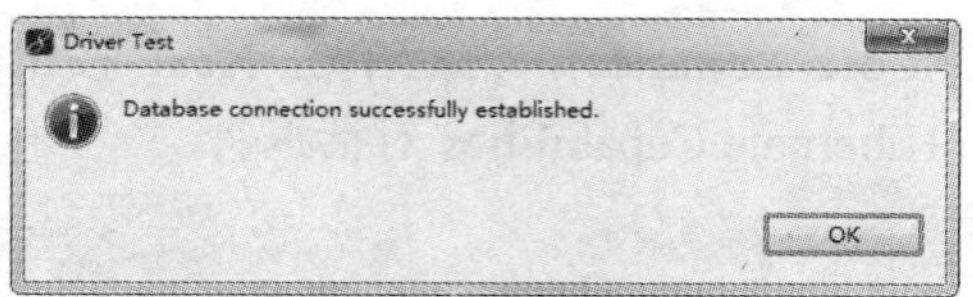

图 7.5　Driver Test 提示框

⑥　单击 Finish 按钮，关闭 Database Driver 对话框。

至此，数据库连接创建成功，在 DB Browser 窗格中将出现相应的数据库连接项(在此为 rsgl)。双击该连接项(或右击，并在其快捷菜单中选择 Open connection…菜单项)，即可打开相应的数据库连接。逐一展开该连接项及其下的有关节点，即可查看到数据库所包含的表。右击某个表，并在其快捷菜单中选择 Edit Data 菜单项，即可查看到表中的数据，或对表中的数据进行编辑，如图 7.6 所示。

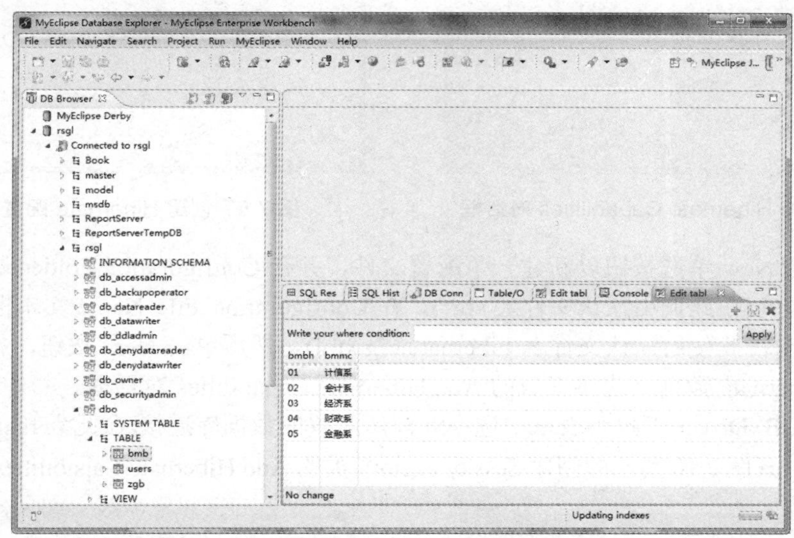

图 7.6　在 MyEclipse 中访问数据库

为便于后续操作，应重新进入 MyEclipse 的 Java 企业版(Java Enterprise)模式。为此，只需在 MyEclipse 中选择 Window→Open Perspective→MyEclipse Java Enterprise 菜单项即可。此外，也可在工具栏中单击 Open Perspective 按钮，并在随之打开的列表中选择 MyEclipse Java Enterprise 选项。

(2)　新建一个 Web 项目 web_07。

(3)　在项目的 src 文件夹中新建一个包 org.etspace.abc.factory(该包用来存放即将创建的 SessionFactory 类)。

(4)　添加 Hibernate 开发能力，可按以下步骤进行操作。

①　右击项目名，并在其快捷菜单中选择 MyEclipse→Add Hibernate Capabilities...菜单项，打开图 7.7 所示的 Add Hibernate Capabilities 窗口。

②　在 Hibernate Specification 选项区选择相应的 Hibernate 框架版本(在此为 Hibernate 3.3)，并在其下的列表框中选中需要的类库(在此为 Hibernate 3.3 Annotations & Entity

Manager 与 Hibernate 3.3 Core Libraries)，然后单击 Next 按钮，打开图 7.8 所示的创建 Hibernate 配置文件的 Add Hibernate Capabilities 对话框。

图 7.7　Add Hibernate Capabilities 对话框

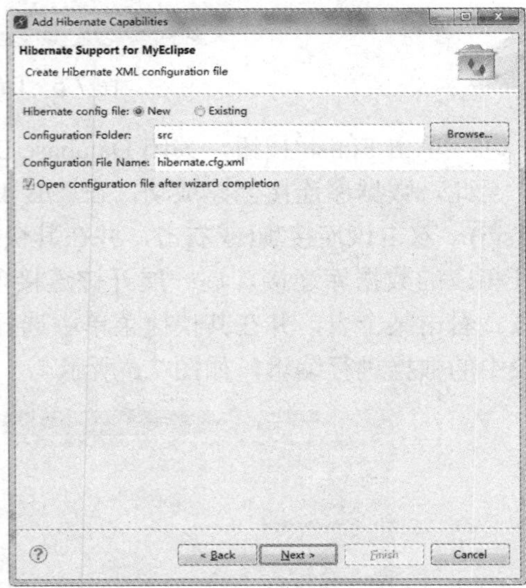

图 7.8　创建 Hibernate 配置文件

③　选中 New 单选按钮以新建一个配置文件，并在 Configuration Folder 处指定存放配置文件的文件夹(在此使用默认文件夹 src)，在 Configuration File Name 文本框中指定配置文件的文件名(在此使用默认文件名 hibernate.cfg.xml)，然后单击 Next 按钮，打开图 7.9 所示的指定 Hibernate 数据库连接细节的 Add Hibernate Capabilities 对话框。

④　在 DB Driver 下拉列表框中选中此前所配置的数据库连接(在此为 rsgl)，然后单击 Next 按钮，打开图 7.10 所示的创建 SessionFactory 类的 Add Hibernate Capabilities 对话框。

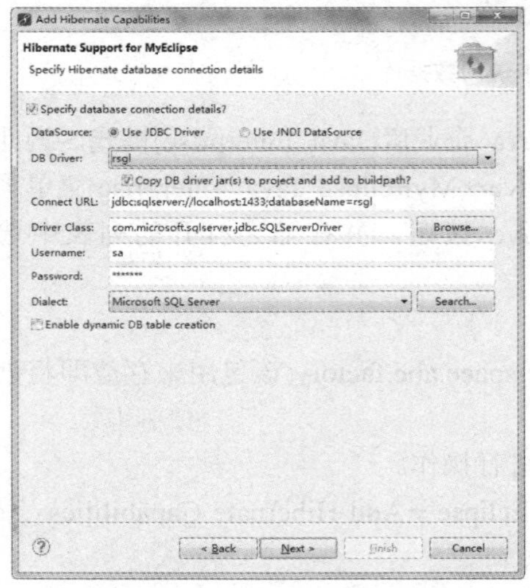

图 7.9　指定 Hibernate 数据库连接细节

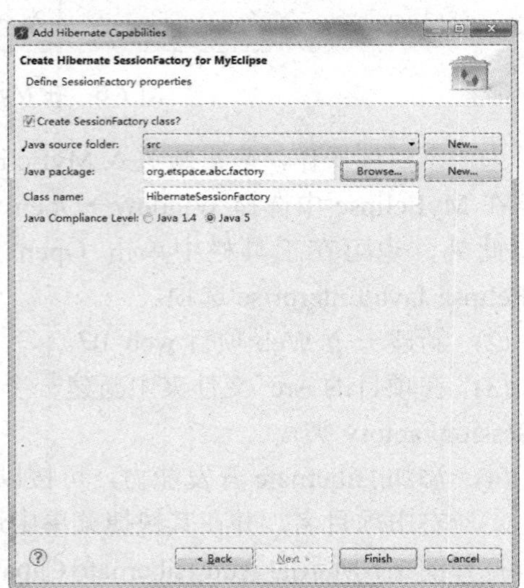

图 7.10　创建 SessionFactory 类

说明： 在 Hibernate 中，有一个与数据库打交道的重要的类 Session，而此类是由 Session 工厂 SessionFactory 创建的。

⑤ 选中 Create SessionFactory class?复选框以创建 SessionFactory 类，并在 Java source folder 下拉列表框中选中 scr，在 Java package 处指定(或单击 Browse...按钮并在随之打开的图 7.11 所示的 Select Package 对话框中选定)相应的包(在此为 org.etspace.abc.factory)，在 Class name 文本框指定相应的类名(在此使用默认类名 HibernateSessionFactory)，然后单击 Finish 按钮关闭对话框。

至此，Hibernate 配置完毕。与此同时，项目中增加了一些 Hibernate 类库(Jar 包)、一个 hibernate.cfg.xml 配置文件、一个 HibernateSessionFactory.java 类文件(见图 7.12)。此外，SQL Server 2005/2008 的 JDBC 驱动程序 sqljdbc4.jar 也会自动添加到项目的 WebRoot\WEB-INF\lib 文件夹中。

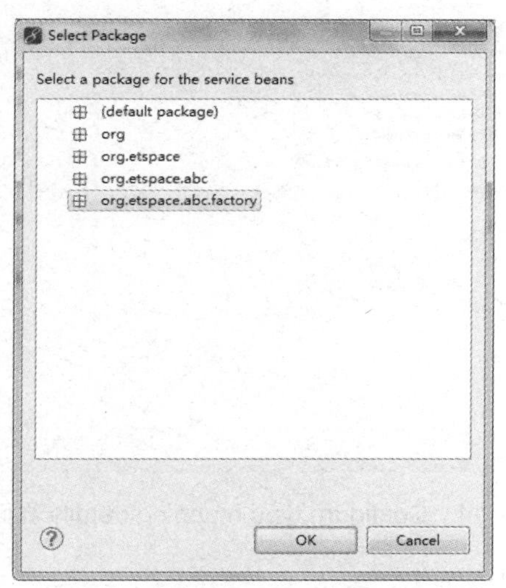

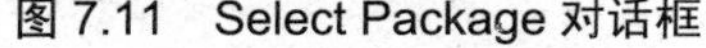

图 7.11　Select Package 对话框

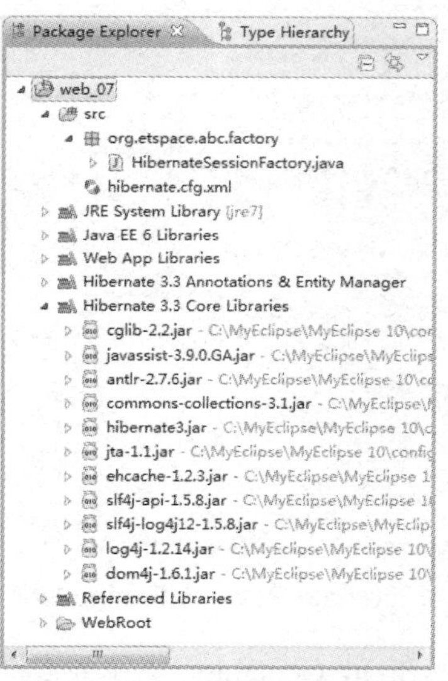

图 7.12　添加了 Hibernate 开发能力的 Web 项目

(5) 在项目的 src 文件夹中新建一个包 org.etspace.abc.vo(该包用来存放与数据库表相对应的 Java 类或 POJO)。

(6) 生成与数据库表相对应的 Java 类与映射文件。为此，可按以下步骤进行操作。

① 在 MyEclipse 中选择 Window→Open Perspective→MyEclipse Database Explorer 菜单项(或者在工具栏中单击 Open Perspective 按钮，并在随之打开的列表中选择 MyEclipse Database Explorer 选项)，进入 MyEclipse 的数据库浏览器(Database Explorer)模式。

② 双击数据库连接项 rsgl，打开相应的数据库连接，然后逐一展开该连接项及其下的有关节点(在此为 Connected to rsgl、rsgl、dbo 与 TABLE 节点)，显示出数据库中所包含的表。

③ 右击某个表(在此为部门表 bmb)，并在其快捷菜单中选择 Hibernate Reverse Engineering...菜单项，启动图 7.13 所示的 Hibernate Reverse Engineering 向导(该向导用于完成将已有的数据库表生成对应的 Java 类与相关映射文件的配置工作)。

④ 在 Java Package 处指定 Java 类与映射文件的存放位置(在此为 org.etspace.abc.vo 包)，选中 Create POJO<>DB Table mapping information 复选框(同时选中其下的 Create a Hibernate mapping file (*.hbm.xml) for each database table 单选按钮与 Update Hibername configuration with mapping resource location 复选框)，选中 Java Data Object (POJO<>DB Table)复选框(同时取消选中其下的 Create abstract class 复选框)，然后单击 Next 按钮，窗口界面切换到图 7.14 所示的 Configure type mapping details。

说明： POJO 即简单的 Java 对象(Plain Old Java Object)，通常也称为 VO(Value Object，值对象)。

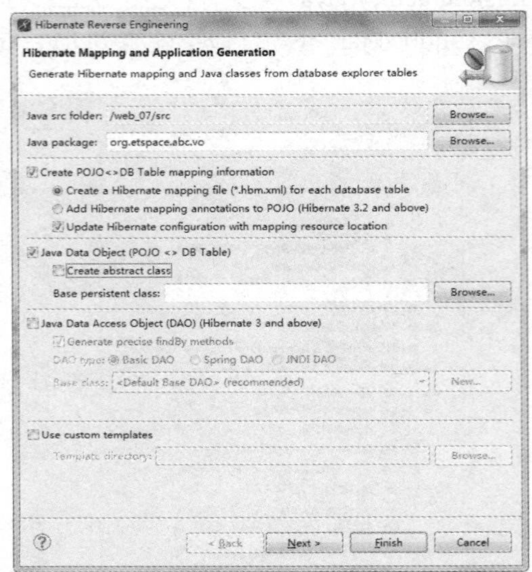

图 7.13 Hibernate Reverse Engineering 对话框

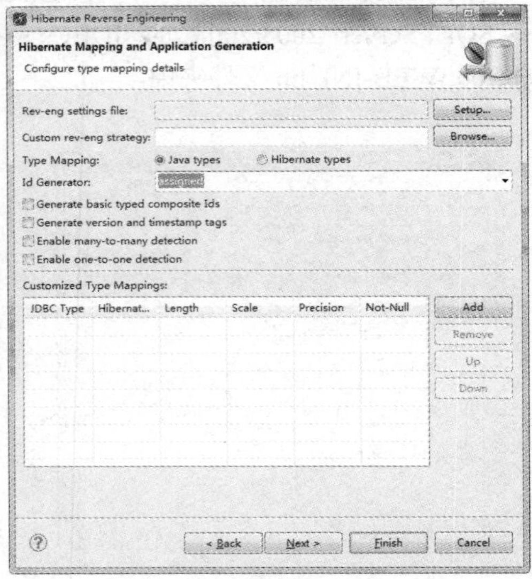

图 7.14 Configure type mapping details 界面

⑤ 在 Id Generator 下拉列表框中选择相应的主键生成方式(在此为 assigned，即由应用程序赋值)，然后单击 Finish 按钮，关闭对话框。

至此，与数据库表相对应的 Java 类与映射文件创建完毕。与此同时，项目中会出现 Bmb.java 类文件与 Bmb.hbm.xml 映射文件(见图 7.15)，并在 hibernate.cfg.xml 完成相应的映射文件配置。

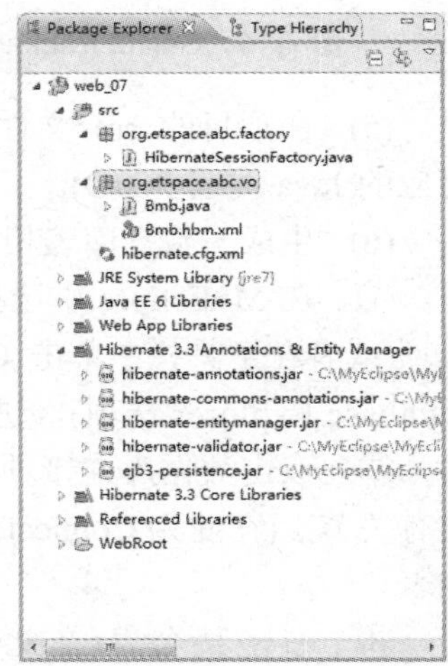

图 7.15 创建了 Java 类与映射文件的 Web 项目

(7) 在项目的 WebRoot 文件夹中新建一个 JSP 页面 bmzj.jsp，其代码如下：

```
<%@ page language="java" import="java.util.*"
pageEncoding="utf-8"%>
<%@ page
import="org.etspace.abc.factory.*,org.etspace
.abc.vo.*,org.hibernate.*" %>
<html>
  <head>
```

```
    <title>部门增加</title>
  </head>
  <body>
    <%
     String bmbh="10";
     String bmmc="学生处";
     Session mySession=HibernateSessionFactory.getSession();
     Transaction myTransaction=mySession.beginTransaction();
     Bmb bmb=new Bmb();
     bmb.setBmbh(bmbh);
     bmb.setBmmc(bmmc);
     mySession.save(bmb);
     myTransaction.commit();
     HibernateSessionFactory.closeSession();
     out.println("OK!");
    %>
  </body>
</html>
```

完成项目的部署后，再启动 Tomcat 服务器，即可打开浏览器访问部门增加页面 bmzj.jsp(地址为 http://localhost:8080/web_07/bmzj.jsp)，从而将指定的部门记录添加到数据库的部门表中。部门增加页面 bmzj.jsp 的运行结果如图 7.16 所示。

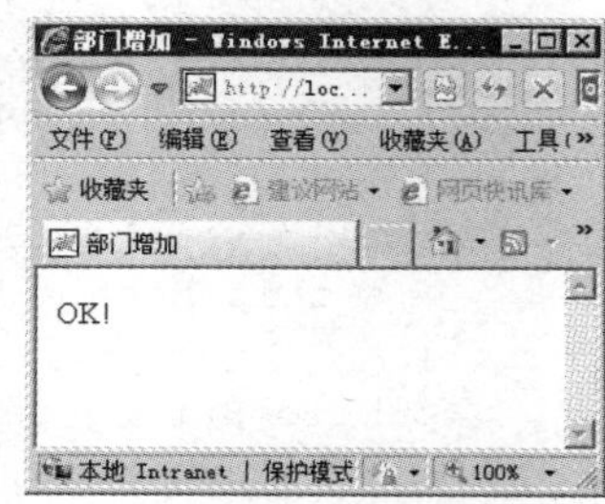

图 7.16　部门增加页面的运行结果

在本实例中，无须编写任何 SQL 语句，而只需保存一个部门表对象，即可在数据库的部门表中添加一条相应的部门记录。可见，借助于 Hibernate 框架，可轻松实现对数据库的有关操作。

7.2.2　Hibernate 相关文件

下面，结合【实例 7-1】对 Hibernate 中的有关文件进行简要说明。

1. POJO 类及其映射文件

POJO 即简单的 Java 对象(Plain Old Java Object)，通常也称为 VO(Value Object，值对象)。使用 POJO 的名称是为了避免与 EJB 相混淆。POJO 包含一些属性以及相应的 getter 与 setter 方法，必要时，也可包含一些简单的运算属性，但不允许存在业务方法。

Hibernate 的映射文件是实体对象与数据库关系表之间相互转换的重要依据。一般而言，一个映射文件对应着数据库中的一个关系表，关系表之间的关联关系也在映射文件中进行配置。

在【实例 7-1】中，部门表所对应的 POJO 类为 Bmb，其文件名为 Bmb.java，代码如下：

```
package org.etspace.abc.vo;
```

```
/**
 * Bmb entity. @author MyEclipse Persistence Tools
 */
public class Bmb implements java.io.Serializable {
    // Fields
    private String bmbh;
    private String bmmc;
    // Constructors
    /** default constructor */
    public Bmb() {
    }
    /** minimal constructor */
    public Bmb(String bmbh) {
        this.bmbh = bmbh;
    }
    /** full constructor */
    public Bmb(String bmbh, String bmmc) {
        this.bmbh = bmbh;
        this.bmmc = bmmc;
    }
    // Property accessors
    public String getBmbh() {
        return this.bmbh;
    }
    public void setBmbh(String bmbh) {
        this.bmbh = bmbh;
    }
    public String getBmmc() {
        return this.bmmc;
    }
    public void setBmmc(String bmmc) {
        this.bmmc = bmmc;
    }
}
```

在 Bmb 类中，定义了两个 String 型属性，即 bmbh 与 bmmc。对于各个属性，均定义了相应的 getter 方法与 setter 方法。例如，对于 bmbh 属性，其 getter 方法为 getBmbh()，setter 方法为 setBmbh()。此外，在 Bmb 类中，还定义了 3 个不同的构造方法，其中包括不带任何参数的默认构造方法 Bmb()。

POJO 类中的属性与表中的字段是一一对应的。实际上，二者是通过相应的映射文件(*.hbm.xml)逐一映射起来的。在此，Bmb 类所对应的映射文件为 Bmb.hbm.xml，其代码如下：

```
<?xml version="1.0" encoding="utf-8"?>
<!DOCTYPE hibernate-mapping PUBLIC "-//Hibernate/Hibernate Mapping DTD
3.0//EN"
"http://hibernate.sourceforge.net/hibernate-mapping-3.0.dtd">
<!--
    Mapping file autogenerated by MyEclipse Persistence Tools
```

```
-->
<hibernate-mapping>
    <class name="org.etspace.abc.vo.Bmb" table="bmb" schema="dbo"
catalog="rsgl">
        <id name="bmbh" type="java.lang.String">
            <column name="bmbh" length="2" />
            <generator class="assigned" />
        </id>
        <property name="bmmc" type="java.lang.String">
            <column name="bmmc" length="20" />
        </property>
    </class>
</hibernate-mapping>
```

映射文件(*.hbm.xml)的根元素为 hibernate-mapping，其他元素均置于该根元素内。从总体来看，映射文件主要分为 3 个部分。

1)　类与表的映射配置

POJO 类与数据库表的映射配置是通过 class 元素实现的。如以下代码：

```
<class name="org.etspace.abc.vo.Bmb" table="bmb" schema="dbo"
catalog="rsgl">
…
</class>
```

其中，name 属性用于指定 POJO 类(在此为 org.etspace.abc.vo.Bmb)，table 属性用于指定当前类所对应的数据库表(在此为 bmb)，catalog 属性用于指定数据库的名称(在此为 rsgl)。

根据 class 元素的具体配置，Hibernate 即可获知类与表的映射关系，即每个类对象对应于相应表中的一条记录。

2)　id(主键)的映射配置

id(主键)的映射配置是通过 class 元素的子元素 id 实现的。如以下代码：

```
<id name="bmbh" type="java.lang.String">
<column name="bmbh" length="2" />
<generator class="assigned" />
</id>
```

其中，name 属性用于指定作为主键的类属性(在此为 bmbh)，type 属性用于指定主键的数据类型(在此为 String)，column 子元素用于指定类主键所对应的数据库表的字段(在此为 bmbh)及其长度(在此为 2)，generator 子元素用于指定主键的生成方式(在此为 assigned，表示由应用程序自身对主键赋值)。

根据 id 元素的具体配置，Hibernate 即可获知类主键与数据库表主键的映射关系，同时确定主键的生成方式。

Hibernate 的主键生成策略可分为三大类，分别为由应用程序自身对主键赋值、由数据库对主键赋值、由 Hibernate 对主键赋值。Hibernate 所支持的各种主键生成方式及其说明(见表 7.1)。

表 7.1 Hibernate 的主键生成方式

方 式	说 明
assigned	由应用程序自身负责对主键赋值
native	由数据库负责对主键赋值。最常见的是整型的自增型主键
hilo	通过 hi/lo 算法实现的主键生成机制，需要额外的数据库表保存主键生成的历史状态
seqhilo	与 hilo 类似，同样为通过 hi/lo 算法实现的主键生成机制，只是主键生成的历史状态保存在 sequence 中，适用于支持 sequence 的数据库，如 Oracle
increment	主键按数值顺序递增。此方式的实现机制为在当前应用实例中维持一个变量，以保存当前的最大值，之后在每次生成主键时便将此值加 1 作为主键。采用这种方式时，在多个实例同时访问同一个数据库时，由于各个实例均各自维护其自身的主键状态，因此不同的实例可能会生成同样的主键，从而造成主键重复异常
identity	采用数据库提供的主键生成机制，如 SQL Server、MySQL 中的自增型主键生成机制
sequence	采用数据库提供的 sequence 机制生成主键，如 Oracle sequence
uuid.hex	由 Hibernate 基于 128 位唯一值产生算法，根据当前设备 IP、时间、JVM 启动时间、内部自增量 4 个参数生成相应的十六进制数值(经编码后以长度为 32 位的字符串表示)作为主键。即使在多实例并发运行的情况下，这种算法也可在最大限度地保证所产生主键的唯一性。不过，重复的概率在理论上是依然存在的，只是概率比较小。一般来说，利用这种方式生成主键可提供最好的数据插入性能与数据平台适应性
uuid.string	与 uuid.hex 类似，只是对生成的主键进行编码(长度为 16)。这种方式在某些数据库中可能出现问题
foreign	使用外部表的字段作为主键
select	Hibernate 3 新引入的主键生成机制，主要针对遗留系统的改造工程

3) 属性与字段的映射配置

属性与字段的映射配置是通过 class 元素的子元素 property 实现的。如以下代码：

```
<property name="bmmc" type="java.lang.String">
<column name="bmmc" length="20" />
</property>
```

其中，name 属性用于指定类的属性(在此为 bmmc)，type 属性用于指定属性的数据类型(在此为 String)，column 子元素用于指定类属性所对应的数据库表的字段(在此为 bmmc)及其长度(在此为 20)。

根据 property 元素的具体配置，Hibernate 即可将类属性与数据库表字段对应或关联起来。

2. hibernate.cfg.xml 配置文件

hibernate.cfg.xml 是 Hibernate 默认的核心配置文件，主要用于配置数据库链接与 Hibernate 运行时所需要的各种基本信息(或基础信息)，包括数据库 URL、数据库用户名、数据库用户密码、数据库 JDBC 驱动类、数据库 dialect(方言)以及数据库表与类对应的映射文件。以下为某个 hibernate.cfg.xml 文件的具体代码：

```
<?xml version='1.0' encoding='UTF-8'?>
```

```
<!DOCTYPE hibernate-configuration PUBLIC
        "-//Hibernate/Hibernate Configuration DTD 3.0//EN"
        "http://hibernate.sourceforge.net/hibernate-configuration-
3.0.dtd">
<!-- Generated by MyEclipse Hibernate Tools.                   -->
<hibernate-configuration>
    <session-factory>
        <property name="dialect">
            org.hibernate.dialect.SQLServerDialect
        </property>
        <property name="connection.url">
            jdbc:sqlserver://localhost:1433;databaseName=rsgl
        </property>
        <property name="connection.username">sa</property>
        <property name="connection.password">abc123!</property>
        <property name="connection.driver_class">
            com.microsoft.sqlserver.jdbc.SQLServerDriver
        </property>
        <property name="myeclipse.connection.profile">rsgl</property>
        <mapping resource="org/etspace/abc/vo/Users.hbm.xml" />
        <mapping resource="org/etspace/abc/vo/Bmb.hbm.xml" />
        <mapping resource="org/etspace/abc/vo/Zgb.hbm.xml" />
    </session-factory>
</hibernate-configuration>
```

Hibernate 初始化期间，会自动在 CLASSPATH 中查找 hibernate.cfg.xml 配置文件，并读取其中的配置信息，为后续数据库操作做好准备。

3. HibernateSessionFactory 类文件

HibernateSessionFactory 是默认生成的 SessionFactory 类(其名字可根据自己的喜好来决定，通常为 HibernateSessionFactory)，具体代码如下：

```
package org.etspace.abc.factory;
import org.hibernate.HibernateException;
import org.hibernate.Session;
import org.hibernate.cfg.Configuration;
/**
 * Configures and provides access to Hibernate sessions, tied to the
 * current thread of execution.  Follows the Thread Local Session
 * pattern, see {@link http://hibernate.org/42.html }.
 */
public class HibernateSessionFactory {
    /**
     * Location of hibernate.cfg.xml file.
     * Location should be on the classpath as Hibernate uses
     * #resourceAsStream style lookup for its configuration file.
     * The default classpath location of the hibernate config file is
     * in the default package. Use #setConfigFile() to update
     * the location of the configuration file for the current session.
```

```
     */
    private static final ThreadLocal<Session> threadLocal = new
ThreadLocal<Session>();
    private static org.hibernate.SessionFactory sessionFactory;
    private static Configuration configuration = new Configuration();
    private static String CONFIG_FILE_LOCATION = "/hibernate.cfg.xml";
    private static String configFile = CONFIG_FILE_LOCATION;
    static {
    try {
            configuration.configure(configFile);
            sessionFactory = configuration.buildSessionFactory();
        } catch (Exception e) {
           System.err.println("%%%% Error Creating SessionFactory %%%%");
            e.printStackTrace();
        }
    }
    private HibernateSessionFactory() {
    }
    /**
     * Returns the ThreadLocal Session instance.  Lazy initialize
     * the <code>SessionFactory</code> if needed.
     *
     *  @return Session
     *  @throws HibernateException
     */
    public static Session getSession() throws HibernateException {
        Session session = (Session) threadLocal.get();
         if (session == null || !session.isOpen()) {
            if (sessionFactory == null) {
                rebuildSessionFactory();
            }
            session = (sessionFactory != null) ?
            sessionFactory.openSession(): null;
            threadLocal.set(session);
        }
        return session;
    }
    /**
     *  Rebuild hibernate session factory
     *
     */
    public static void rebuildSessionFactory() {
        try {
            configuration.configure(configFile);
            sessionFactory = configuration.buildSessionFactory();
        } catch (Exception e) {
           System.err.println("%%%% Error Creating SessionFactory %%%%");
            e.printStackTrace();
```

```
        }
    }
    /**
     *  Close the single hibernate session instance.
     *
     *  @throws HibernateException
     */
    public static void closeSession() throws HibernateException {
        Session session = (Session) threadLocal.get();
        threadLocal.set(null);
        if (session != null) {
            session.close();
        }
    }
    /**
     *  return session factory
     *
     */
    public static org.hibernate.SessionFactory getSessionFactory() {
        return sessionFactory;
    }
    /**
     *  return session factory
     *
     *   session factory will be rebuilded in the next call
     */
    public static void setConfigFile(String configFile) {
        HibernateSessionFactory.configFile = configFile;
        sessionFactory = null;
    }
    /**
     *  return hibernate configuration
     *
     */
    public static Configuration getConfiguration() {
        return configuration;
    }
}
```

在 Hibernate 中，由 Session 负责完成对象的持久化操作。至于 Session 对象的创建与关闭，则是由 HibernateSessionFactory 类负责的。由 HibernateSessionFactory 类文件可知，Session 对象的创建大致需要以下 3 个步骤。

(1) 初始化 Hibernate 配置管理类 Configuration。

(2) 通过 Configuration 类实例创建 Session 工厂类 SessionFactory。

(3) 通过 SessionFactory 类得到 Session 实例。

7.3 Hibernate 核心接口

Hibernate 所提供的核心接口主要有 5 个，即 Configuration、SessionFactory、Session、Transaction 与 Query 接口。配合使用相应的接口，即可实现面向对象的数据库操作，包括持久化对象的存取与事务的控制等。

7.3.1 Configuration 接口

Configuration 负责读取 Hibernate 的配置信息(保存在 hibernate.cfg.xml 文件中)，并完成相应的初始化工作，从而为后续操作做好准备。

其实，要让 Hibernate 完成相关的初始化操作，只需创建一个 Configuration 实例即可。如以下示例：

```
Configuration config=new Configuration().configure();
```

在此，所创建的 Configuration 实例为 config。

7.3.2 SessionFactory 接口

SessionFactory 负责创建 Session 实例，而 SessionFactory 实例则是通过 Configuration 实例创建的。Configuration 实例会根据当前的数据库配置信息，构造 SessionFacory 实例并返回之。如以下示例：

```
Configuration config=new Configuration().configure();
SessionFactory sessionFactory=config.buildSessionFactory();
```

在此，通过调用 Configuration 实例 config 的 buildSessionFactory()方法，创建了一个 SessionFacory 实例 sessionFactory。

注意： SessionFactory 实例一旦构造完毕，即被赋予特定的配置信息，此后，相应 Configuration 实例的任何变更都不会影响到已创建好的 SessionFactory 实例。因此，如果 Configuration 实例已发生变更，往往要重新创建 SessionFactory 实例。此外，如果要同时访问多个数据库，那么针对每一个数据库，应分别为其创建相应的 SessionFactory 实例。

SessionFactory 实例保存了相应数据库配置的所有映射关系，同时也负责维护当前的二级数据缓存与 Statement Pool，其创建过程是非常复杂的，代价也较为高昂。因此，在系统设计中，应充分考虑 SessionFactory 的重用策略。另外，由于 SessionFactory 采用了线程安全的设计，可由多个线程并发调用，因此在大多数情况下，对于同一个数据库，只需共享一个 SessionFactory 实例即可。

7.3.3 Session 接口

Session 是 Hibernate 持久化操作的基础，提供了众多持久化方法，如 save()、

update()、delete()、query()、get()、load()等。通过这些方法，可透明地完成持久化对象的增加、修改、删除、查找等操作。

Session 实例的创建可通过调用 SessionFactory 实例的 openSession()方法实现。如以下示例：

```
Configuration config=new Configuration().configure();
SessionFactory sessionFactory=config.buldSessionFactory();
Session session=sessionFactory.openSession();
```

在此，所创建的 Session 实例为 session。此后，通过调用 session 的相关方法即可完成相应的持久层操作。

注意： 在 Hibernate 中，Session 的设计是非线程安全的，即一个 Session 实例同时只能由一个线程使用，同一个 Session 实例的多线程并发调用将导致难以预知的错误。

7.3.4　Transaction 接口

Transaction 是 Hibernate 中进行事务操作的接口，其实例(即事务对象)的创建可通过调用 Session 实例的 beginTransaction ()方法实现。如以下示例：

```
Transaction transaction =session.beginTransaction();
```

在此，所创建的 Transaction 实例为 transaction。

说明： Transaction 接口是对实际事务实现(包括 JDBC 的事务、JTA 中的 UserTransaction 与 CORBA 事务)的一个抽象。如此设计，可让开发者能够使用一个统一的操作界面，确保自己的项目可以在不同的环境与容器之间方便地进行移植。

7.3.5　Query 接口

Query 接口是 Hibernate 的查询接口，用于执行 HQL(Hibernate Query Language，Hibernate 查询语言)语句。

Query 与 HQL 是密不可分的，其实例(即查询对象)需通过调用 Session 实例的 createQuery()方法并根据指定的 HQL 语句来创建。如以下示例：

```
Query query=session.createQuery("from Bmb where bmbh='01'");
```

在此，所创建的 Query 实例为 query，相应的 HQL 语句为“from Bmb where bmbh='01'”。

7.4　HQL 基本用法

HQL 即 Hibernate 查询语言(Hibernate Query Language)，是 Hibernate 官方推荐使用的标准数据库查询语言，采用了新的面向对象的查询方式。从语法及用法上看，HQL 类似于

SQL，不过，HQL 的操作对象是类、实例(对象)、属性等，而 SQL 的操作对象则是数据库、表、列(字段)等。其实，这正是二者之间的本质区别。

下面以部门表类 Bmb 为例，简要说明 HQL 的基本功能及其相关用法。

7.4.1 HQL 查询

HQL 查询用于从指定的类中查询满足相应条件的对象、对象属性或其他相关信息，通常使用 from 语句或 select 语句实现。在此，仅介绍 HQL 查询最基本的一些用法。

1. 基本查询

基本查询是 HQL 查询中最简单的一种方式，用于从某个类中查询所有的对象或对象的属性，其基本格式为：

```
[select <属性名列表>] from <类名>
```

如以下示例：

(1) from Bmb。

(2) select bmbh,bmmc from Bmb。

其中，第一个语句用于查询所有的部门，第二个语句用于查询所有部门的编号与名称。

2. 条件查询

条件查询需使用 where 子句指定相应的查询条件。如以下示例：

(1) from Bmb where bmbh='01'。

(2) from Bmb where bmbh>='10' and bmbh<'20'。

(3) from Bmb where bmbh between'10' and '19'。

(4) from Bmb where bmbh in ('10','20')。

(5) from Bmb where bmmc is null。

(6) from Bmb where bmmc is not null。

(7) from Bmb where bmmc like '%计%'。

其中，语句(1)的条件为部门编号为“01”，语句(2)的条件为部门编号大于等于“10”且小于“20”，语句(3)的条件为部门编号在“10”与“19”之间(包括“10”与“19”)，语句(4)的条件为部门编号为“10”或“20”，语句(5)的条件为部门名称为空，语句(6)的条件为部门名称为非空，语句(7)的条件为部门名称中包含有一个“计”字。

3. 结果排序

必要时，可对查询结果进行排序。为此，需使用 order by 子句指定相应的排序方式。如以下示例：

(1) from Bmb order by bmbh。

(2) from Bmb order by bmbh desc。

其中，第一个语句按部门编号的升序进行排序，第二个语句按部门编号的降序进行排序。

其实，HQL 查询的功能是十分强大的，其用法也相当灵活，除了只涉及一个类的简单查询以外，还可以同时对多个类进行链接查询，或者在查询的 where 子句中嵌套相应的子查询。关于 HQL 查询的各种详细用法，请查阅相关的使用手册或技术资料，在此不作详述。

7.4.2　HQL 更新

HQL 更新包括修改与删除两个方面。

1. HQL 修改

HQL 修改用于对类中的某些对象的有关属性进行修改，需使用 update 语句实现，其基本格式为：

```
update <类名> set <属性名>=<属性值> [,<属性名>=<属性值> …] [where <条件>]
```

如以下示例：

```
update Bmb set bmmc='学工部' where bmbh='10'
```

该语句用于将部门编号为“10”的部门的名称修改为“学工部”。

2. HQL 删除

HQL 删除用于删除类中的某些对象，需使用 delete 语句实现，其基本格式为：

```
delete <类名> [where <条件>]
```

如以下示例：

(1)　delete Bmb where bmbh='10'。

(2)　delete Bmb。

其中，第一个语句用于删除部门编号为“10”的部门，第二个语句用于删除所有的部门。

7.4.3　HQL 语句的执行

HQL 语句的执行依赖于 Query 实例。其实，每个 Query 实例均为一个完成特定功能的查询对象。

在创建 Query 实例时，需指定相应的 HQL 语句。例如：

```
Query query11=session.createQuery("from Bmb where bmbh='01'");
Query query12=session.createQuery("update Bmb set bmmc='学工部' where
bmbh='10'");
Query query13=session.createQuery("delete Bmb where bmbh='10'");
```

必要时，可在 HQL 语句中指定相应的位置参数(以“?”表示)或命名参数(以“:参数名”的方式表示)。例如：

```
Query query21=session.createQuery("from Bmb where bmbh=?");
Query query22=session.createQuery("from Bmb where bmbh=:bmbh");
```

对于查询类 HQL 语句，只需调用相应 Query 实例的 list()方法即可执行之，并返回一个 List 实例(Object 集合或 Object 数组集合)。对于更新类 HQL 语句，只需调用相应 Query 实例的 executeUpdate()方法即可执行之，并返回受影响的对象个数。例如：

```
List list=query11.list();
Int n= query12.executeUpdate();
Int m= query13.executeUpdate();
```

如果在 HQL 语句中定义有位置参数或命名参数，那么在执行之前应先为其设置好相应的参数值。为此，可根据具体情况调用相应 Query 实例的有关参数设置方法。参数设置方法及其基本用法见表 7.2。

表 7.2　Query 实例的参数设置方法

方　法	说　明
setXxx(index, value)	设置指定序号的位置参数的值。位置参数的序号从 0 开始，“Xxx”表示数据类型(如 Int、String 等)
setParameter(index, value)	设置指定序号的位置参数的值。适用于任何数据类型的参数
setXxx(name, value)	设置指定名称的命名参数的值。“Xxx”表示数据类型(如 Int、String 等)
setParameter(name, value)	设置指定名称的命名参数的值。适用于任何数据类型的参数

例如：

```
Query query=session.createQuery("from Bmb where bmbh=?");
query.setString(0,"01");
List list=query.list();
```

其中，“query.setString(0,"01");”也可改写为“query.settParameter(0,"01");”。

又如：

```
Query query=session.createQuery("from Bmb where bmbh=:bmbh");
query.setString("bmbh","01");
List list=query.list();
```

其中，“query.setString("bmbh","01");”也可改写为“query.settParameter("bmbh","01");”。

【实例 7-2】Hibernate 应用示例：部门查询。图 7.17 所示为“部门查询”页面，其中显示了所有部门的编号与名称。

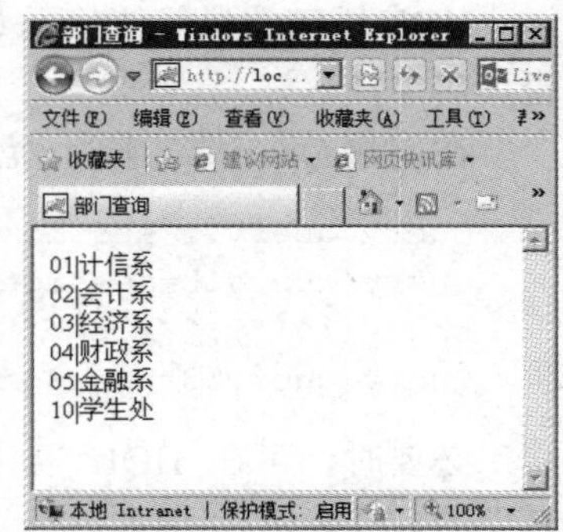

图 7.17　“部门查询”页面

主要步骤(在项目 web_07 中)如下。

在项目的 WebRoot 文件夹中新建一个 JSP 页面 bmcx.jsp，其代码如下：

```
<%@ page language="java" import="java.util.*"
pageEncoding="utf-8"%>
<%@ page
```

```
import="org.etspace.abc.factory.*,org.etspace.abc.vo.*,org.hibernate.*"
%>
<%@ page import="java.sql.*" %>
<html>
  <head>
    <title>部门查询</title>
  </head>
  <body>
    <%
     Session mySession=HibernateSessionFactory.getSession();
     Query query=mySession.createQuery("from Bmb order by bmbh");
     List list=query.list();
     Iterator iterator=list.iterator();
     while(iterator.hasNext()){
         Bmb bmb=(Bmb)iterator.next();
         out.println(bmb.getBmbh()+"|"+bmb.getBmmc()+"<br>");
     }
     HibernateSessionFactory.closeSession();
    %>
  </body>
</html>
```

说明： bmcx.jsp 页面中<body>与</body>之间的代码也可修改为：

```
<%
 Session mySession=HibernateSessionFactory.getSession();
 Query query=mySession.createQuery("select bmbh,bmmc from Bmb
order by bmbh");
 List list=query.list();
 Iterator iterator=list.iterator();
 while(iterator.hasNext()){
     Object[] bmb=(Object[])iterator.next();
     out.println(bmb[0]+"|"+bmb[1]+"<br>");
 }
 HibernateSessionFactory.closeSession();
%>
```

【实例 7-3】Hibernate 应用示例：部门修改。图 7.18 所示为“部门修改”页面，其功能为将编号为“10”的部门的名称修改为“学工部”。

图 7.18　“部门修改”页面

主要步骤(在项目 web_07 中)如下。

在项目的 WebRoot 文件夹中新建一个 JSP 页面 bmxg.jsp，其代码如下：

```
<%@ page language="java" import="java.util.*"
pageEncoding="utf-8"%>
<%@ page
import="org.etspace.abc.factory.*,org.etspace.abc
.vo.*,org.hibernate.*" %>
<%@ page import="java.sql.*" %>
```

```
<html>
  <head>
    <title>部门修改</title>
  </head>
  <body>
  <%
     String bmbh="10";
     String bmmc="学工部";
     Session mySession=HibernateSessionFactory.getSession();
     Transaction transaction=mySession.beginTransaction();
     Query query=mySession.createQuery("update Bmb set bmmc=? where
bmbh=?");
     query.setString(0, bmmc);
     query.setString(1, bmbh);
     query.executeUpdate();
     transaction.commit();
     HibernateSessionFactory.closeSession();
     out.println("OK!");
  %>
 </body>
</html>
```

【实例 7-4】Hibernate 应用示例：部门删除。图 7.19 所示为“部门删除”页面，其功能为删除编号为“10”的部门。

主要步骤(在项目 web_07 中)如下。

在项目的 WebRoot 文件夹中新建一个 JSP 页面 bmsc.jsp，其代码如下：

图 7.19 “部门删除”页面

```
<%@ page language="java" import="java.util.*"
pageEncoding="utf-8"%>
<%@ page
import="org.etspace.abc.factory.*,org.etspace.abc
.vo.*,org.hibernate.*" %>
<%@ page import="java.sql.*" %>
<html>
  <head>
    <title>部门删除</title>
  </head>
  <body>
  <%
     String bmbh="10";
     Session mySession=HibernateSessionFactory.getSession();
     Transaction transaction=mySession.beginTransaction();
     Query query=mySession.createQuery("delete Bmb where bmbh=:bmbh");
     query.setParameter("bmbh", bmbh);
     query.executeUpdate();
     transaction.commit();
     HibernateSessionFactory.closeSession();
     out.println("OK!");
   %>
```

```
  </body>
</html>
```

7.5 Hibernate 对象状态

实体对象的生命周期是 Hibernate 应用中的一个关键概念。在此，实体对象特指 Hibernate O/R 映射关系中的域对象(即 O/R 中的 O)。

实体对象在其生命周期中共有 3 种状态，即 transient(瞬时态)、persisent(持久态)与 detached(脱管态)。

7.5.1 瞬时态

瞬时态(transient)是指实体对象只在内存中存在，尚未被 Session 实例关联(即不处于 Session 缓存中)，且与数据库中的记录无关。

使用 new 创建一个对象时，该对象即处于瞬时态。如以下示例：

```
Bmb bmb=new Bmb();
bmb.setBmbh("10");
bmb.setBmmc("学生处");
```

此处 bmb 即为一个处于瞬时态的对象。

7.5.2 持久态

持久态(persisent)是指实体对象已被 Session 实例关联(即处于 Session 缓存中)，对应着数据库中的一条记录，其变更将被保存到数据库中。持久态对象正处于 Hibernate 框架的管理之中，其引用也已被纳入 Hibernate 实体容器之中加以管理。

可通过调用 Session 实例的 save()方法，将处于瞬时态的对象转换为持久态。如以下示例：

```
Bmb bmb=new Bmb();
bmb.setBmbh("10");
bmb.setBmmc("学生处");
Bmb bmb0=new Bmb();
bmb0.setBmbh("11");
bmb0.setBmmc("教务处");  //至此，bmb 与 bmb0 均处于瞬时态
Transaction transaction =session.beginTransaction();  //开始一个事务
session.save(bmb);  //至此，bmb 转换为持久态，bmb0 仍处于瞬时态
transaction.commit();  //提交事务，bmb 被持久化到数据库中
Transaction transaction0=session.beginTransaction();
bmb.setBmmc ("学工部");
bmb0.setBmmc ("教学部");
transaction0.commit();
```

以上示例包含有两个事务，即 transaction 与 transaction0。在 transaction0 事务中，虽然没有显式地调用 session.save()方法保存 bmb 对象，但由于其正处于持久态，因此会自动将

其更改保存到数据库中。而 bmb0 对象仍处于瞬时态，因此不会对数据库产生任何影响。

如果一个实体对象是由 Hibernate 所加载的，那么该对象将自动处于持久态。如以下示例：

```
Bmb bmb=(Bmb)session.load(Bmb.class,"01");  // bmb 处于持久态
```

说明： 对瞬时态对象执行 session.save()方法时，Hibernate 将对其进行持久化，并赋予其相应的主键值，从而与数据库表中具有相同主键值的记录相关联。由于持久态对象对应着数据库中的一条记录，因此可将其看作是数据库记录的对象化操作接口，其状态的变更将对数据库中的记录产生影响。

7.5.3 脱管态

持久态对象在其对应的 Session 实例关闭后，即处于脱管态(detached)。如以下示例：

```
Bmb bmb=new Bmb();
bmb.setBmbh("10");
bmb.setBmmc("学生处");  // bmb 处于瞬时态
Transaction transaction =session.beginTransaction();
session.save(bmb);  //此时，bmb 转换为持久态
transaction.commit();
session.close();  //此时，bmb 转换为脱管态
```

可见，脱管态对象是由持久态对象转换过来的，已不再被 Session 实例关联(即不再处于 Session 缓存中)，但数据库中可能存在着与其相对应的记录。

Session 实例可以看作是持久态对象的宿主，一旦宿主失效，其从属的持久态对象将进入脱管态。必要时，可将脱管态对象再次与某个 Session 实例相关联而成为持久态对象。

说明： 瞬时态对象与数据库表中的记录缺乏对应关系，而脱管态对象却可能在数据库表中存在着对应的记录。与持久态对象不同，由于脱管态对象脱离了 Session 这个数据操作平台，因此其状态的改变将无法更新到数据库表中。

为方便起见，通常将处于瞬时态与脱管态的对象统称为值对象(Value Object，VO)，将处于持久态的对象称为持久对象(Persistent Object，PO)。其实，这是从“实体对象是否被纳入 Hibernate 实体管理容器”的角度加以区别的，不被管理的实体对象统称为 VO，被管理的实体对象则称为 PO。

7.6 Hibernate 批量处理

在 Hibernate 应用中，有时会遇到批量处理的情况。批量处理主要包括批量插入、批量修改与批量删除。

7.6.1 批量插入

Hibernate 应用的批量插入有两种实现方法，其一是通过 Hibernate 缓存，其二是绕过

Hibernate 直接调用 JDBC API。

1. 通过 Hibernate 缓存进行批量插入

为实现基于 Hibernate 缓存的批量插入，需先在 Hibernate 的配置文件(通常为 hibernate.cfg.xml)中设置好批量尺寸属性 hibernate.jdbc.batch_size，且最好关闭 Hibernate 的二级缓存以提高效率。有关代码如下：

```
<hibernate-configuration>
  <session-factory>
    …
    <property name="hibernate.jdbc.batch_size">5</property>
    <property name="hibernate.cache.use_second_level_cache">false</property>
    …
  </session-factory>
</hibernate-configuration>
```

在此，将批量尺寸的大小设置为 5，并关闭 Hibernate 二级缓存。其中，hibernate.jdbc.batch_size 与 hibernate.cache.use_ second_level_cache 也可分别改写为 jdbc.batch_size 与 cache.use_second_level_cache。

图 7.20　“部门增加”页面

【实例 7-5】批量插入示例。图 7.20 所示为“部门增加”页面，其功能为批量插入 50 个部门到数据库的部门表中(每 5 个部门为一批)。

主要步骤(在项目 web_07 中)如下。

在项目的 WebRoot 文件夹中新建一个 JSP 页面 bmzj_batch.jsp，其代码如下：

```
<%@ page language="java" import="java.util.*" pageEncoding="utf-8"%>
<%@ page
import="org.etspace.abc.factory.*,org.etspace.abc.vo.*,org.hibernate.*"
%>
<html>
  <head>
    <title>部门增加</title>
  </head>
  <body>
<%
    Session mySession=HibernateSessionFactory.getSession();
    Transaction myTransaction=mySession.beginTransaction();
    for (int i=0;i<30;i++){
        Bmb bmb=new Bmb();
        bmb.setBmbh(String.valueOf(50+i));
        bmb.setBmmc("部门"+String.valueOf(50+i));
        mySession.save(bmb);
        //以 5 个部门为一批向数据库提交(该值应与已配置的批量尺寸保持一致)
        if ((i+1)%5==0){
```

```
                mySession.flush();  //将当前批的数据立即插入数据库中
                mySession.clear();  //清空缓存区，释放内存供下一批数据使用
            }
        }
        myTransaction.commit();
        HibernateSessionFactory.closeSession();
        out.println("OK!");
%>
  </body>
</html>
```

2. 绕过 Hibernate 直接调用 JDBC API 进行批量插入

由于 Hibernate 只对 JDBC 进行了轻量级的封装，因此完全可以绕过 Hibernate 直接调用 JDBC API 进行批量插入。为此，应在 bmzj_batch.jsp 页面中添加一条指令“<%@ page import="java.sql.*"%>”，然后再将其中的 Java 脚本代码修改为：

```
<%
    Session mySession=HibernateSessionFactory.getSession();
    Transaction myTransaction=mySession.beginTransaction();
    Connection conn=mySession.connection();
    try {
        PreparedStatement stmt=conn.prepareStatement("insert into
bmb(bmbh,bmmc) values(?,?)");
        for (int i=0;i<30;i++) {
            stmt.setString(1,String.valueOf(50+i));
            stmt.setString(2,"部门"+String.valueOf(50+i));
            stmt.addBatch();  // 添加到批处理中
        }
        stmt.executeBatch();  // 执行批处理任务
    } catch (SQLException e) {
        e.printStackTrace();
    }
    myTransaction.commit();
    HibernateSessionFactory.closeSession();
    out.println("OK!");
%>
```

说明： 自 Hibernate 3.3.2 版本开始，通过调用 Session 实例的 connection()方法来获取 Connection 实例的方法已经被弃用。按照 Hibernate 官方的计划，Hibernate 4 推荐以调用 doWork()方法实现 org.hibernate.jdbc.Work 接口的新方式来使用 JDBC API。为此，应在 bmzj_batch.jsp 页面中再添加一条指令“<%@ page import="org.hibernate.jdbc.Work"%>”，同时将其中的 Java 脚本代码修改为：

```
<%
    Session mySession=HibernateSessionFactory.getSession();
    Transaction myTransaction=mySession.beginTransaction();
    mySession.doWork(
        new Work(){  //定义一个实现了 Work 接口的匿名类
```

```
            public void execute(Connection conn)throws
SQLException{
                try {
                    PreparedStatement stmt=conn.prepareStatement
("insert into bmb(bmbh,bmmc) values(?,?)");
                    for (int i=0;i<30;i++) {
                        stmt.setString(1,String.valueOf(50+i));
                        stmt.setString(2,"部门"+String.valueOf
(50+i));
                        stmt.addBatch();  // 添加到批处理中
                    }
                    stmt.executeBatch();  // 执行批处理任务
                } catch (SQLException e) {
                    e.printStackTrace();
                }
            }
        });
    myTransaction.commit();
    HibernateSessionFactory.closeSession();
    out.println("OK!");
%>
```

注意：在用 Hibernate 进行操作时，操作的是类、对象及其属性；而用 JDBC 进行操作时，操作的是表、记录及其字段。

7.6.2　批量修改

与批量插入类似，Hibernate 的批量修改也有两种实现方法，其一是由 Hibernate 直接处理，其二是绕过 Hibernate 直接调用 JDBC API 进行处理。

1. 由 Hibernate 直接进行批量修改

为实现基于 Hibernate 的批量修改，需先在 Hibernate 的配置文件(通常为 hibernate.cfg.xml)中设置好 HQL/SQL 查询翻译器属性 hibernate.query.factory_class(也可将其改写为 query.factory_class)，以使 Hibernate 的 HQL 直接支持 update/delete 的批量更新语法。有关代码如下：

```
<hibernate-configuration>
   <session-factory>
   …
   <property name="hibernate.
query.factory_class">

    org.hibernate.hql.ast.ASTQuery
TranslatorFactory
    </property>
   …
   </session-factory>
</hibernate-configuration>
```

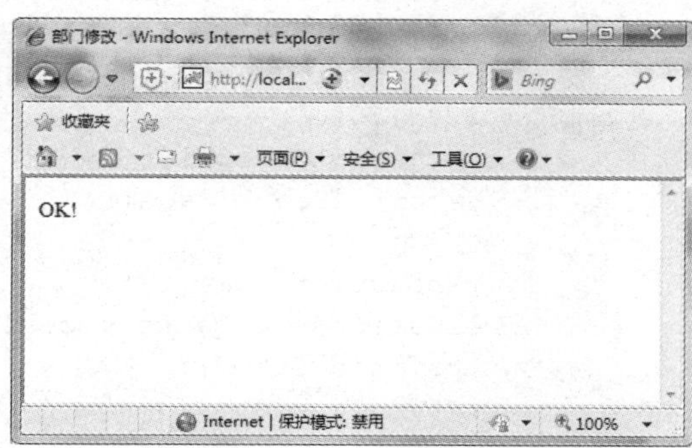

图 7.21　“部门修改”页面

【**实例 7-6**】批量修改示例。图 7.21 所示为“部门修改”页面，其功能为修改部门表中编号大于等于“50”的部门，在其名称后加上一个“!”。

主要步骤(在项目 web_07 中)如下。

在项目的 WebRoot 文件夹中新建一个 JSP 页面 bmxg_batch.jsp，其代码如下：

```
<%@ page language="java" import="java.util.*" pageEncoding="utf-8"%>
<%@ page
import="org.etspace.abc.factory.*,org.etspace.abc.vo.*,org.hibernate.*"
%>
<html>
  <head>
    <title>部门修改</title>
  </head>
  <body>
<%
    Session mySession=HibernateSessionFactory.getSession();
    Transaction myTransaction=mySession.beginTransaction();
    Query query=mySession.createQuery("update Bmb set bmmc=bmmc+'!'
where bmbh>='50'");
    query.executeUpdate();
    myTransaction.commit();
    HibernateSessionFactory.closeSession();
    out.println("OK!");
%>
  </body>
</html>
```

2. 绕过 Hibernate 直接调用 JDBC API 进行批量修改

为绕过 Hibernate 直接调用 JDBC API 进行批量修改，应在 bmxg_batch.jsp 页面中添加一条指令“<%@ page import="java.sql.*"%>”，然后再将其中的 Java 脚本代码修改为：

```
<%
    Session mySession=HibernateSessionFactory.getSession();
    Transaction myTransaction=mySession.beginTransaction();
    Connection conn=mySession.connection();
    try {
        Statement stmt=conn.createStatement();
        //调用 JDBC 的 update 进行批量修改
        stmt.executeUpdate("update bmb set bmmc=bmmc+'!' where
bmbh>='50'");
    } catch (SQLException e) {
        e.printStackTrace();
    }
    myTransaction.commit();
    HibernateSessionFactory.closeSession();
    out.println("OK!");
%>
```

7.6.3　批量删除

与批量修改类似，Hibernate 的批量删除也有两种实现方法，其一是由 Hibernate 直接处理，其二是绕过 Hibernate 直接调用 JDBC API 进行处理。

1. 由 Hibernate 直接进行批量删除

为实现基于 Hibernate 的批量删除，需先在 Hibernate 的配置文件(通常为 hibernate.cfg.xml)中设置好 HQL/SQL 查询翻译器属性 hibernate.query.factory_class(也可将其改写为 query.factory_class)，以便使 Hibernate 的 HQL 直接支持 update/delete 的批量更新语法。有关代码与批量修改的相同，在此不再赘述。

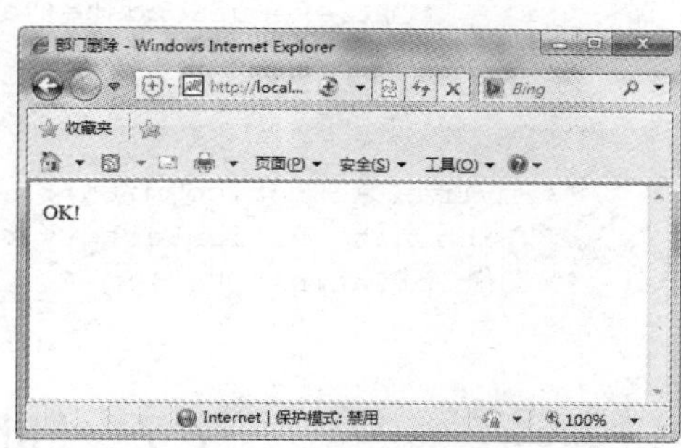

图 7.22　“部门删除”页面

【实例 7-7】批量删除示例。图 7.22 所示为“部门删除”页面，其功能为删除部门表中编号大于等于“50”的部门。

主要步骤(在项目 web_07 中)如下。

在项目的 WebRoot 文件夹中新建一个 JSP 页面 bmsc_batch.jsp，其代码如下：

```
<%@ page language="java" import="java.util.*" pageEncoding="utf-8"%>
<%@ page
import="org.etspace.abc.factory.*,org.etspace.abc.vo.*,org.hibernate.*"
%>
<html>
  <head>
    <title>部门删除</title>
  </head>
  <body>
<%
    Session mySession=HibernateSessionFactory.getSession();
    Transaction myTransaction=mySession.beginTransaction();
    Query query=mySession.createQuery("delete Bmb where bmbh>='50'");
    query.executeUpdate();
    myTransaction.commit();
    HibernateSessionFactory.closeSession();
    out.println("OK!");
%>
  </body>
</html>
```

2. 绕过 Hibernate 直接调用 JDBC API 进行批量删除

为绕过 Hibernate 直接调用 JDBC API 进行批量删除，应在 bmsc_batch.jsp 页面中添加一条指令“<%@ page import="java.sql.*"%>”，然后再将其中的 Java 脚本代码修改为：

```
<%
```

```
    Session mySession=HibernateSessionFactory.getSession();
    Transaction myTransaction=mySession.beginTransaction();
    Connection conn=mySession.connection();
    try {
        Statement stmt=conn.createStatement();
        //调用 JDBC 的 delete 进行批量删除
        stmt.executeUpdate("delete from bmb where bmbh>='50'");
    } catch (SQLException e) {
        e.printStackTrace();
    }
    myTransaction.commit();
    HibernateSessionFactory.closeSession();
    out.println("OK!");
%>
```

7.7 Hibernate 事务管理

在各种管理系统中，事务的应用是非常广泛、十分重要的。从本质上看，Hibernate 只是 JDBC 的轻量级封装，因此其本身并不具备事务管理的能力。事实上，Hibernate 的事务管理与调度功能是通过委托给底层的 JDBC 或 JTA(Java Transaction API)来实现的。

7.7.1 事务的基本概念

事务(Transaction)是数据库系统所执行的基本逻辑单元(或工作单元)，是一系列相关数据库操作的集合(或序列)。一个事务所包含的所有操作均执行成功之后，方可提交(commit)，从而将所有相关操作的结果都同步到数据库中。反之，若事务中的某个操作执行失败，则应回滚(rollback)该事务，以便撤销事务中已执行的有关操作，让数据库返回到事务开始时的状态。可见，事务主要用于确保数据库能够被正确修改，避免数据只修改了一部分而导致数据不完整或不一致。

事务的重要特性有 4 个，即原子性(Atomicity)、一致性(Consistency)、隔离性(Isolation)与持久性(Durability)。事务的这 4 个特性通常又称为事务的 ACID 准则。

(1) 原子性。事务是一个不可分割的整体，其中的操作要么全做，要么一个都不做。只有全部操作都成功执行了，才能提交事务以便确认；否则，若其中的任何一个操作执行失败，则要回滚整个事务以便撤销。

(2) 一致性。事务不能破坏数据库的完整性与一致性。不管执行是否成功，事务结束时数据库内部的数据都是正确的。

(3) 隔离性。不同事务的执行互不干扰、互不影响。在并发操作中，一个事务不会查看到或访问到另外一个事务正在修改的数据。

(4) 持久性。事务被成功提交后，其所做的所有更改就被永久地保存到数据库中。

7.7.2 基于 JDBC 的事务管理

Hibernate 的默认事务处理机制是基于 JDBC 事务的。在 JDBC 的数据库操作中，一个

事务就是由一条或多条 SQL 语句所组成的一个不可分割的工作单元，通过提交或回滚事务，即可结束一个事务的执行。

默认情况下，JDBC 事务是自动提交的。换言之，一条对数据库进行更新(修改或删除)的 SQL 语句就代表着一个事务，执行成功后系统将自动调用 commit()方法提交之，否则就调用 rollback()方法回滚。必要时，可通过调用 Connection 对象实例的 setAutoCommit(false)方法禁止自动提交，从而将多条 SQL 语句作为一个事务，最后再整体提交或回滚。

将事务管理委托给 JDBC 进行处理是最简单易行的实现方式。事实上，Hibernate 对于 JDBC 事务的封装也较为简单。如下面的代码：

```
Session session=sessionFactory.openSession();
Transaction transaction =session.beginTransaction();
…
transaction.commit();
```

从 JDBC 层面而言，以上代码对应着：

```
Connection cnnection =getConnection;
cnnection.setAutoCommit(false);
… //JDBC 执行相关的 SQL 语句
cnnection.commit();
```

注意： 在调用 sessionFactory.openSession()方法时，Hibernate 会初始化数据库链接，并将其 AutoCommit 设置为关闭状态(false)。也就是说，一开始从 sessionFactory 处获取的 session 的自动提交属性是已被关闭的，因此，以下代码不会对数据库产生任何效果：

```
Session session=sessionFactory.openSession();
session.save(bmb);
session.close();
```

这实际上相当于 JDBC Connection 的 AutoCommit 属性被设置为 false 时，执行了若干 SQL 语句之后却并没有执行提交操作。若要使代码真正作用到数据库，则必须显式地调用 Transaction 的有关指令。如以下代码：

```
Session session = sessionFactory.openSession();
Transaction transaction = session.beginTransaction();
session.save(bmb);
transaction.commit();
session.close();
```

说明： 基于 Hibernate 的事务应用通常可分为以下几个步骤。

(1) 通过 sessionFactory 获取 Session 实例。

(2) 通过 Session 实例开始一个事务。

(3) 执行相关的数据操作。

(4) 若事务处理一切正常，则提交。

(5) 若事务处理出现异常，则回滚。

(6) 关闭 Session，结束操作。

如以下代码：

```
…
Configuration myConfiguration = new Configuration().configure();
SessionFactory mySessionFactory =
myConfiguration.buildSessionFactory();
Session mySession = mySessionFactory.openSession();
Transaction my Transaction = mySession.beginTransaction();
try{
Bmb bmb=new Bmb();
bmb.setBmbh("10");
bmb.setBmmc("学生处");
mySession.save(bmb);
    myTransaction.commit();
}catch(Exception e){
    if(myTransaction!=null){
        myTransaction.rollback();
    }
    e.printStackTrace();
}finally{
    mySession.close();
}
…
```

7.7.3 基于 JTA 的事务管理

JTA(Java Transaction API)是由 Java EE Transaction Manager 管理的事务，其事务范围的界定、事务的提交与事务的回滚是通过调用 UserTransaction 接口的 begin()、commit()与 rollback()方法来完成的。基于 JTA，可以实现同一事务对应不同的数据库。

与 JDBC 的 Connection 只提供单个数据源的事务不同，JTA 主要用于分布式的多个数据源的两阶段提交的事务。JDBC 事务由于只涉及单个数据源，因此可由数据库自己单独实现。而 JTA 事务因其分布式与多数据源的特性，只能由事务管理器实现(而无法由任何一个数据源单独实现)。事务管理器通过使用两阶段提交的技术，可在多个数据源之间统筹事务。

JTA 与 JDBC 事务最大的差异是提供了跨 Session 的事务管理能力。JDBC 事务由 Connection 管理，即事务管理实际上是在 JDBC Connection 中实现的，事务周期仅限于 Connection 的生命周期之内。同样，对于基于 JDBC 事务的 Hibernate 事务管理机制而言，事务管理是在 Session 所依托的 JDBC Connection 中实现的，事务周期仅限于 Session 的生命周期之内。与 JDBC 事务不同，JTA 事务管理是由 JTA 容器实现的，JTA 容器对当前加入事务的各个 Connection 进行调度并实现事务性要求，因此 JTA 的事务周期可横跨多个 JDBC Connection 生命周期。同样，对于基于 JTA 事务的 Hibernate 事务管理机制而言，JTA 事务可横跨多个 Session。

7.8　Hibernate 应用案例

下面，通过一些典型的应用案例进一步说明 Hibernate 框架的有关应用技术。

7.8.1　数据查询

【实例 7-8】分页查询。图 7.23 所示为“部门查询”表单，在其中输入欲查询的页数与每页记录的个数后，再单击“确定”按钮，即可显示相应的查询结果，如图 7.24 所示。

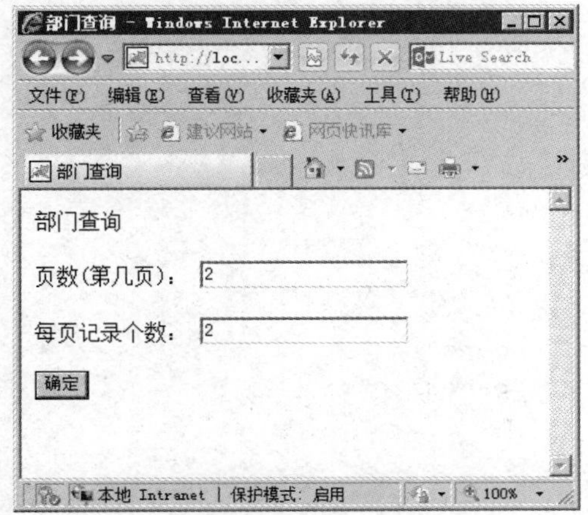

图 7.23　“部门查询”表单

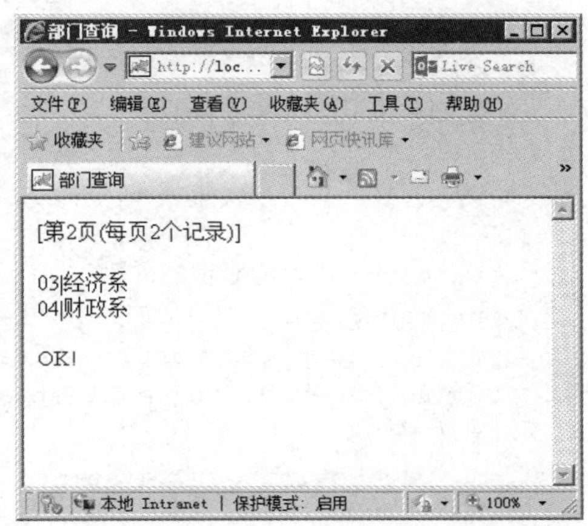

图 7.24　“部门查询”结果

主要步骤(在项目 web_07 中)如下。

(1) 在项目的 WebRoot 文件夹中新建一个 HTML 页面 bmcx_page_form.htm，其代码如下：

```
<html>
    <head>
        <title>部门查询</title>
    </head>
    <body>
        <p>
        部门查询
        <p>
        <form method="post" action="bmcx_page_result.jsp">
            <p>
                页数(第几页)：
                <input type="text" name="pageno">
            </p>
            <p>
                每页记录个数：
                <input type="text" name="pagesize">
            </p>
            <p>
```

```
                <input type="submit" value="确定" name="ok">
            </p>
        </form>
    </body>
</html>
```

(2) 在项目的 WebRoot 文件夹中新建一个 JSP 页面 bmcx_page_result.jsp，其代码如下：

```
<%@ page language="java" import="java.util.*" pageEncoding="utf-8"%>
<%@ page
import="org.etspace.abc.factory.*,org.etspace.abc.vo.*,org.hibernate.*" %>
<html>
  <head>
    <title>部门查询</title>
  </head>
  <body>
<%
    int pageNo;  //查询的页数
    int pageSize;  //页面的大小(记录个数)
    String pageno=request.getParameter("pageno");
    String pagesize=request.getParameter("pagesize");
    try{
        pageNo=Integer.parseInt(pageno);
    }
    catch(Exception e){
        pageNo=1;
    }
    try{
        pageSize=Integer.parseInt(pagesize);
    }
    catch(Exception e){
        pageSize=10;
    }
    out.println("[第"+pageNo+"页(每页"+pageSize+"个记录)]<br><br>");
    Session mySession=HibernateSessionFactory.getSession();
    Query query=mySession.createQuery("from Bmb order by bmbh");
    query.setFirstResult((pageNo-1)*pageSize); //指定查询的开始位置(对象序号)
    query.setMaxResults(pageSize); //指定结果集的最大对象数目
    List list=query.list();
    Iterator iterator=list.iterator();
    while(iterator.hasNext()){
        Bmb bmb=(Bmb)iterator.next();
        out.println(bmb.getBmbh()+"|"+bmb.getBmmc()+"<br>");
    }
    HibernateSessionFactory.closeSession();
    out.println("<br>OK!");
```

```
%>
  </body>
</html>
```

解析：

(1) bmcx_page_form.htm 为一静态页面，用于生成一个“部门查询”表单，让用户自行指定所要查询的页数(pageno)以及每页所能显示的记录个数(pagesize)。该表单提交后，将由 JSP 页面 bmcx_page_result.jsp 进行处理。

(2) 在 bmcx_page_result.jsp 页面中，先获取通过表单提交的页数(pageno)与每页记录个数(pagesize)，并进行相应的类型转换与异常处理，确定要查询的页数(pageNo)与每页的所能显示的最大记录数(pageSize)。然后创建用于实现部门查询功能的 Query 实例 query，并调用该实例的 setFirstResult()与 setMaxResults()方法设置好开始查询的对象序号与结果集所能包含的最大对象数目，最后再调用其 list()方法即可获取相应的查询结果集。

说明： 为满足分页查询的需要，Hibernate 的 Query 实例提供了以下两个相关的方法。

(1) setFirstResult(int firstResult)。该方法用于指定查询的开始位置，即从哪一个对象开始进行查询。开始位置通过对象的序号表示，第 1 个对象的序号为 0，以此类推。

(2) setMaxResults(int maxResult)。该方法用于指定查询所能返回的对象的最大数目，即结果集的最大对象数目。

设置好查询的开始位置与查询所能返回的对象的最大数目后，再调用 Query 实例的 list()方法，即可获取在指定范围内由相应对象所构成的结果集。而默认情况下，直接调用 Query 实例的 list()方法，将从第 1 个对象开始进行查询，并返回所有满足条件的对象。

【实例 7-9】分页查询。图 7.25 所示为“部门查询”页面，其功能为以分页的方式显示部门的信息。

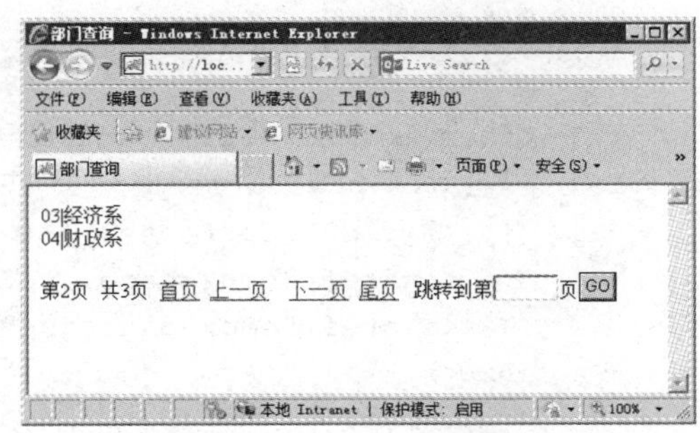

图 7.25　“部门查询”页面

主要步骤(在项目 web_07 中)如下。

在项目的 WebRoot 文件夹中新建一个 HTML 页面 bmcx_page.jsp，其代码如下：

```
<%@ page language="java"
import="java.util.*"
pageEncoding="utf-8"%>
<%@ page
import="org.etspace.abc.factory.*,org.etspace.abc.vo.*,org.hibernate.*"
%>
<html>
  <head>
    <title>部门查询</title>
  </head>
```

```
  <body>
    <%
        int pageSize=2;  //每页的大小
        int pageCount;  //总页数
        int pageCurrent;  //当前页
        String pageNo=request.getParameter("pageno");
        try{
            pageCurrent=Integer.parseInt(pageNo);
        }
        catch(Exception e){
            pageCurrent=1;
        }
        Session mySession=HibernateSessionFactory.getSession();
        Query query=mySession.createQuery("from Bmb order by bmbh");
        List list=query.list();
        pageCount = (list.size() + pageSize - 1)/pageSize;
        if(pageCurrent>pageCount)
            pageCurrent=pageCount;
        if(pageCurrent<1)
            pageCurrent=1;
        query.setFirstResult((pageCurrent-1)*pageSize);
        query.setMaxResults(pageSize);
        list=query.list();
        Iterator iterator=list.iterator();
        while(iterator.hasNext()){
            Bmb bmb=(Bmb)iterator.next();
            out.println(bmb.getBmbh()+"|"+bmb.getBmmc()+"<br>");
        }
        HibernateSessionFactory.closeSession();
    %>
    <form method="POST" action="bmcx_page.jsp">
        第<%=pageCurrent %>页 共<%=pageCount %>页 
        <%if(pageCurrent>1){ %>
        <a href="bmcx_page.jsp?pageno=1">首页</a>
        <a href="bmcx_page.jsp?pageno=<%=pageCurrent-1 %>">上一页</a>
        <%} %>
         
        <%if(pageCurrent<pageCount){ %>
        <a href="bmcx_page.jsp?pageno=<%=pageCurrent+1 %>">下一页</a>
        <a href="bmcx_page.jsp?pageno=<%=pageCount %>">尾页</a>
        <%} %>
         跳转到第<input type="text" name="pageno" size="3"
maxlength="5">页<input name="submit" type="submit" value="GO">
    </form>
  </body>
</html>
```

解析：

(1) 在本页面中，pageSize 表示每页的大小(在此将其设置为 2)，pageCount 表示总页数，pageCurrent 表示当前页。

(2) 在本页面中，包含一个用于提供“页链接”与“页跳转”的表单，该表单被提交后，将由页面自身进行处理。

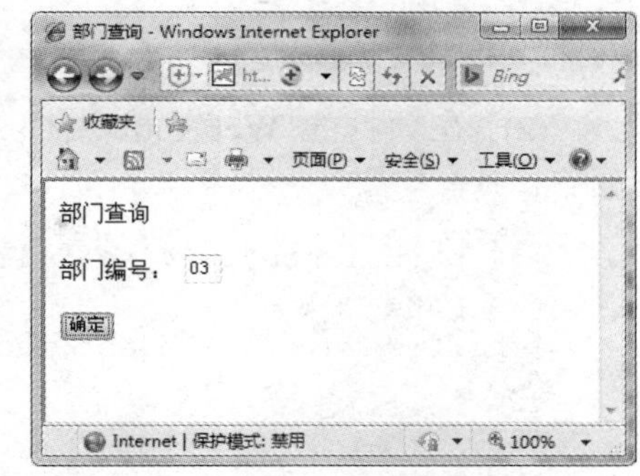

图 7.26　“部门查询”表单

【实例 7-10】唯一查询。图 7.26 所示为“部门查询”表单，在其中输入欲查询的部门的编号后，再单击“确定”按钮，即可显示相应的查询结果。图 7.27(a)所示为所查部门存在时的结果页面；图 7.27(b)所示则为所查部门不存在时的结果页面。

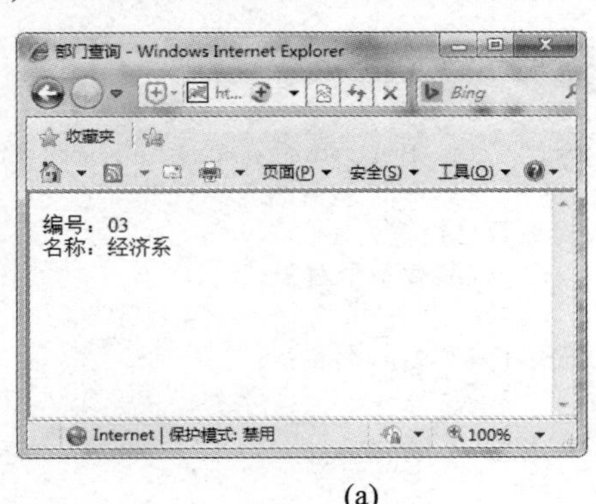

(a)

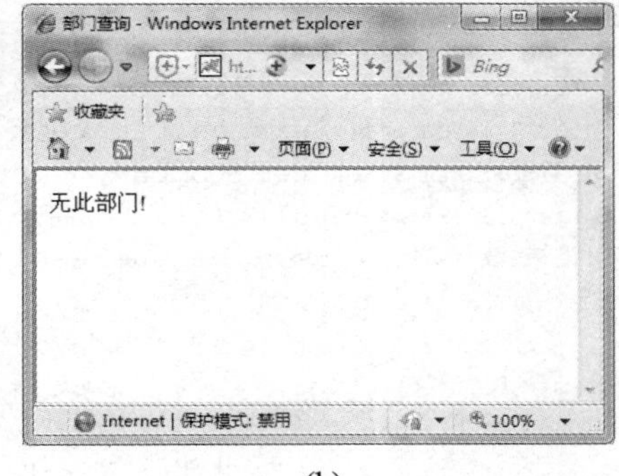

(b)

图 7.27　“部门查询”结果

主要步骤(在项目 web_07 中)如下。

(1) 在项目的 WebRoot 文件夹中新建一个 HTML 页面 bmcx_bh_form.htm，其代码如下：

```
<html>
    <head>
        <title>部门查询</title>
    </head>
    <body>
        <p>
        部门查询
        <p>
        <form method="post" action="bmcx_bh_result.jsp">
            <p>
                部门编号：
                <input type="text" name="bmbh" size="2" maxlength="2">
            </p>
            <p>
                <input type="submit" value="确定" name="ok">
            </p>
```

```
        </form>
    </body>
</html>
```

(2) 在项目的 WebRoot 文件夹中新建一个 JSP 页面 bmcx_bh_result.jsp，其代码如下：

```
<%@ page language="java" import="java.util.*" pageEncoding="utf-8"%>
<%@ page
import="org.etspace.abc.factory.*,org.etspace.abc.vo.*,org.hibernate.*"
%>
<html>
  <head>
    <title>部门查询</title>
  </head>
  <body>
<%
    String bmbh=request.getParameter("bmbh");
    Session mySession=HibernateSessionFactory.getSession();
    Query query=mySession.createQuery("from Bmb where bmbh=:bmbh");
    query.setParameter("bmbh", bmbh);
    query.setMaxResults(1);  //将最大对象数目设置为 1
    Bmb bmb=(Bmb)query.uniqueResult();  //装载单个对象
    if (bmb != null) {
        out.println("编号: "+bmb.getBmbh()+"<br>名称:
"+bmb.getBmmc()+"<br>");
    } else {
        out.println("无此部门!<br>");
    }
    HibernateSessionFactory.closeSession();
%>
  </body>
</html>
```

解析：由于部门的编号是唯一的，因此在 bmcx_bh_result.jsp 页面中，先调用 Query 实例 query 的 setMaxResults()方法将查询的最大对象数目设置为 1，然后再调用 uniqueResult()方法返回结果集中的单个对象(若结果集为空，则返回 null)。

说明： Query 接口的 uniqueResult()方法用于返回结果集中的唯一对象(当结果集中有且只有一个对象时)或 null(当结果集为空时)。若结果集中的对象有多个，该方法将抛出异常(NonUniqueResultException)。

7.8.2 系统登录

【实例 7-11】系统登录。以 JSP+ Hibernate 模式实现 Web 应用系统中的系统登录功能，其运行结果与第 3 章 3.4.1 小节的系统登录案例相同。

实现步骤(在项目 web_07 中)如下。

(1) 在 org.etspace.abc.vo 包中生成与数据库表 users 相对应的 Java 类 Users(其文件名

为 Users.java)与映射文件 Users.hbm.xml。其操作步骤与本章【实例 7-1】中的类似。

(2) 在项目的 WebRoot 文件夹中新建一个子文件夹 syslogin_hibernate。

(3) 在子文件夹 syslogin_hibernate 中添加一个新的 JSP 页面 login.jsp。其代码与第 3 章 3.4.1 小节系统登录案例中的相同。

(4) 在子文件夹 syslogin_hibernate 中添加一个新的 JSP 页面 validate.jsp，其代码如下：

```
<%@ page language="java" pageEncoding="utf-8" import="java.util.*" %>
<%@ page
import="org.etspace.abc.factory.*,org.etspace.abc.vo.*,org.hibernate.*"
%>
<%request.setCharacterEncoding("utf-8"); %>
<html>
    <head>
        <title>验证页面</title>
        <meta http-equiv="Content-Type" content="text/html;charset=utf-
8">
    </head>
    <body>
        <%
        String username=request.getParameter("username");
        String password=request.getParameter("password");
        boolean validated=false;  //验证标识
        Session mySession=HibernateSessionFactory.getSession();
        Query query=mySession.createQuery("from Users where username=?
and password=?");
        query.setParameter(0, username);
        query.setParameter(1, password);
        List list=query.list();
        if(list.size()!=0)
        {
            validated=true;
        }
        HibernateSessionFactory.closeSession();

        if(validated)  //验证成功跳转到成功页面
        {
        %>
            <jsp:forward page="welcome.jsp"></jsp:forward>
        <%
        }
        else  //验证失败跳转到失败页面
        {
        %>
            <jsp:forward page="error.jsp"></jsp:forward>
        <%
        }
        %>
    </body>
</html>
```

(5) 在子文件夹 syslogin_hibernate 中添加一个新的 JSP 页面 welcome.jsp。其代码与第 3 章 3.4.1 小节系统登录案例中的相同。

(6) 在子文件夹 syslogin_hibernate 中添加一个新的 JSP 页面 error.jsp。其代码与第 3 章 3.4.1 小节系统登录案例中的相同。

【实例 7-12】系统登录。以 JSP+ Hibernate+DAO 模式实现 Web 应用系统中的系统登录功能，其运行结果与第 3 章 3.4.1 节的系统登录案例相同。

实现步骤(在项目 web_07 中，在【实例 7-11】的基础上)如下。

(1) 在项目的 WebRoot 文件夹中新建一个子文件夹 syslogin_hibernate_dao。

(2) 在子文件夹 syslogin_hibernate_dao 中添加一个新的 JSP 页面 login.jsp。其代码与第 3 章 3.4.1 小节系统登录案例中的相同。

(3) 在项目的 src 文件夹中新建一个包 org.etspace.abc.dao(该包用于存放有关的 DAO 接口)。

(4) 在 org.etspace.abc.dao 包中新建一个接口 IUsersDAO。其文件名为 IUsersDAO.java，代码如下：

```
package org.etspace.abc.dao;
import org.etspace.abc.vo.Users;
public interface IUsersDAO {
    public Users validateUsers(String username,String password);
}
```

(5) 在项目的 src 文件夹中新建一个包 org.etspace.abc.dao.impl(该包用于存放有关的 DAO 接口的实现类)。

(6) 在 org.etspace.abc.dao.impl 包中新建一个 IUsersDAO 接口的实现类 UsersDAO。其文件名为 UsersDAO.java，代码如下：

```
package org.etspace.abc.dao.impl;
import org.etspace.abc.dao.IUsersDAO;
import org.etspace.abc.vo.Users;
import java.util.*;
import org.etspace.abc.factory.*;
import org.hibernate.*;
public class UsersDAO implements IUsersDAO {
    public Users validateUsers(String username, String password) {
        Session mySession=HibernateSessionFactory.getSession();
        Query query=mySession.createQuery("from Users where username=?
and password=?");
        query.setParameter(0, username);
        query.setParameter(1, password);
        List list=query.list();
        if(list.size()!=0)
        {
            Users users=(Users)list.get(0);
            return users;
        }
        HibernateSessionFactory.closeSession();
```

```
        return null;
    }
}
```

(7) 在子文件夹 syslogin_hibernate_dao 中添加一个新的 JSP 页面 validate.jsp，其代码如下：

```
<%@ page language="java" pageEncoding="utf-8" %>
<%@ page import="org.etspace.abc.dao.*,org.etspace.abc.dao.impl.*" %>
<%request.setCharacterEncoding("utf-8"); %>
<html>
    <head>
        <title>验证页面</title>
        <meta http-equiv="Content-Type" content="text/html;charset=utf-
8">
    </head>
    <body>
        <%
        String username=request.getParameter("username");
        String password=request.getParameter("password");
        boolean validated=false;  //验证标识
        IUsersDAO usersDAO=new UsersDAO();
        if(usersDAO.validateUsers(username, password)!=null)
        {
            validated=true;
        }
        if(validated)  //验证成功跳转到成功页面
        {
        %>
            <jsp:forward page="welcome.jsp"></jsp:forward>
        <%
        }
        else  //验证失败跳转到失败页面
        {
        %>
            <jsp:forward page="error.jsp"></jsp:forward>
        <%
        }
        %>
    </body>
</html>
```

(8) 在子文件夹 syslogin_hibernate_dao 中添加一个新的 JSP 页面 welcome.jsp。其代码与第 3 章 3.4.1 小节系统登录案例中的相同。

(9) 在子文件夹 syslogin_hibernate_dao 中添加一个新的 JSP 页面 error.jsp。其代码与第 3 章 3.4.1 小节系统登录案例中的相同。

说明： DAO 即数据访问对象(Data Access Object)，介于数据源与业务逻辑之间，用于将底层的数据访问操作与高层的业务逻辑相分离。一般来说，数据源不同，其访问方式也会有所不同。为建立一个健全的应用系统，应将所有对数据源的访问操作抽象封装在一个公共 API 中。换言之，就是建立一个接口，并在接口中定义应用程序中将会用到的所有业务的抽象方法。在应用程序中，当需要与特定的数据源进行交互的时候就使用该接口，并编写一个相应的类来实现。这就是通常所说的DAO技术或模式。

在开发 Java Web 应用系统时，JSP+ Hibernate+DAO 是一种较为常用的模式(见图 7.28)。在此模式中，用 Hibernate 取代 JDBC 实现对数据库的访问，同时使用 DAO 接口技术实现系统的持久层功能。

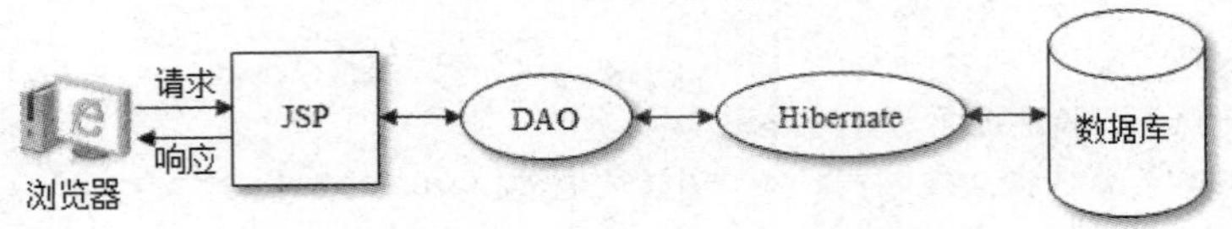

图 7.28　JSP+ Hibernate+DAO 模式

7.9　Hibernate 与 Struts 2 整合应用

在 Java EE 应用中，控制层通常用 Struts 2 框架实现，而持久层则通常用 Hibernate 框架实现，并采用 DAO 技术。图 7.29 所示即为 Java EE 应用开发的 JSP+Struts 2+Hibernate+DAO 模式，也表明了 Hibernate 框架与 Struts 2 框架的整合原理。在此模式中，Hibernate 取代了 JDBC，可利用逆向工程生成数据库表所对应的 POJO 类与映射文件，并允许以面向对象(OO)的方式访问数据库。而 Struts 2 中的 Action 模块则不再直接执行 JDBC 操作，而是通过 DAO 接口的方式间接地访问数据库。可见，在应用开发中整合 Struts 2 与 Hibernate 框架，并采用相应的 DAO 技术，即可彻底解耦前端的程序模块与后台的数据库操作。这样，在更换后台数据库管理系统(DBMS)时，只需改写相应的 DAO 实现即可，而无须对 Action 模块的代码进行任何修改。

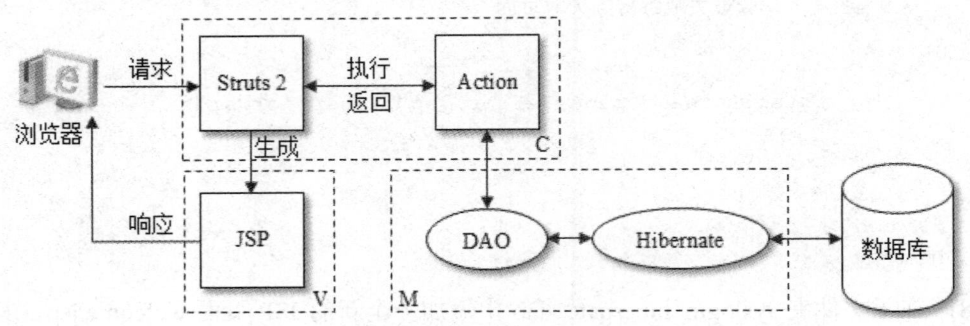

图 7.29　JSP+Struts 2+Hibernate+DAO 模式

下面以 JSP+Struts 2+Hibernate+DAO 模式实现 Web 应用系统中的系统登录功能，其运行结果与第 6 章 6.9.1 小节的系统登录案例相同。

实现步骤(在项目 web_07 中)如下。

(1) 加载 Struts 2 类库。为此，只需将 Struts 2 的 5 个基本类库与 4 个附加类库(共 9

个 jar 包)复制到项目的 WebRoot\WEB-INF\lib 目录下即可。

(2) 配置 Struts 2 框架的核心过滤器。为此，只需对 Web 项目的配置文件 web.xml 进行相应的修改即可。具体代码如下：

```
<?xml version="1.0" encoding="UTF-8"?>
<web-app …>
    …
    <filter>
        <filter-name>struts2</filter-name>
        <filter-
class>org.apache.struts2.dispatcher.FilterDispatcher</filter-class>
    </filter>
    <filter-mapping>
        <filter-name>struts2</filter-name>
        <url-pattern>/*</url-pattern>
    </filter-mapping>
    …
</web-app>
```

(3) 在 MyEclipse 中创建对数据库(在此为 SQL Server 数据库 rsgl)的连接(在此将其命名为 rsgl)，具体方法请参见本章的【实例 7-1】。

(4) 在项目的 src 文件夹中新建一个包 org.etspace.abc.factory(该包用来存放即将创建的 SessionFactory 类)。

(5) 添加 Hibernate 开发能力，具体方法请参见本章的【实例 7-1】。

(6) 在项目的 src 文件夹中新建一个包 org.etspace.abc.vo(该包用来存放与数据库表相对应的 Java 类或 POJO)。

(7) 在 org.etspace.abc.vo 包中生成与数据库表 users 相对应的 Java 类 Users(其文件名为 Users.java)与映射文件 Users.hbm.xml。

(8) 在项目的 WebRoot 文件夹中新建一个子文件夹 syslogin_struts_hibernate_dao。

(9) 在子文件夹 syslogin_struts_hibernate_dao 中添加一个新的 JSP 页面 login.jsp，其代码与第 6 章 6.9.1 小节系统登录案例中的相同。

(10) 在项目的 src 文件夹中新建一个包 org.etspace.abc.dao(该包用于存放有关的 DAO 接口)。

(11) 在 org.etspace.abc.dao 包中新建一个接口 IUsersDAO，其文件名为 IUsersDAO.java，代码与本章【实例 7-12】中的相同。

(12) 在项目的 src 文件夹中新建一个包 org.etspace.abc.dao.impl(该包用于存放有关的 DAO 接口的实现类)。

(13) 在 org.etspace.abc.dao.impl 包中新建一个 IUsersDAO 接口的实现类 UsersDAO，其文件名为 UsersDAO.java，代码与本章【实例 7-12】中的相同。

(14) 在项目的 src 文件夹中新建一个包 org.etspace.abc.action(该包用于存放有关的 Action 实现类)。

(15) 在 org.etspace.abc.action 包中新建 Action 实现类 LoginAction，其文件名为 LoginAction.java，代码如下：

```
package org.etspace.abc.action;
import org.etspace.abc.dao.*;
import org.etspace.abc.dao.impl.*;
import com.opensymphony.xwork2.ActionSupport;
public class LoginAction extends ActionSupport{
    private String username;
    private String password;
    public String getUsername(){
        return username;
    }
    public void setUsername(String username){
        this.username=username;
    }
    public String getPassword(){
        return password;
    }
    public void setPassword(String password){
        this.password=password;
    }
    public String execute() throws Exception{
        boolean validated=false;  //验证标识
        IUsersDAO usersDAO=new UsersDAO();
        if(usersDAO.validateUsers(username, password)!=null)
        {
            validated=true;
        }
        if(validated)
        {
            //验证成功返回"success"
            return "success";
        }
        else
        {
            //验证失败返回"error"
            return "error";
        }
    }
}
```

(16) 在项目的 src 文件夹中创建配置文件 struts.xml，并在其中完成 LoginAction 的配置。具体代码如下：

```
<?xml version="1.0" encoding="UTF-8"?>
<!DOCTYPE struts PUBLIC
    "-//Apache Software Foundation//DTD Struts Configuration 2.0//EN"
    "http://struts.apache.org/dtds/struts-2.0.dtd">
<struts>
    <package name="login" extends="struts-default">
        <action name="login" class="org.etspace.abc.action.LoginAction">
            <result
name="success">/syslogin_struts_hibernate_dao/welcome.jsp</result>
```

```
            <result
name="error">/syslogin_struts_hibernate_dao/error.jsp</result>
        </action>
    </package>
</struts>
```

(17) 在子文件夹 syslogin_struts_hibernate_dao 中添加一个新的 JSP 页面 welcome.jsp，其代码与第 6 章 6.9.1 小节系统登录案例中的相同。

(18) 在子文件夹 syslogin_struts_hibernate_dao 中添加一个新的 JSP 页面 error.jsp，其代码与第 6 章 6.9.1 小节系统登录案例中的相同。

本 章 小 结

本章首先介绍了 Hibernate 框架的体系结构，然后通过具体实例讲解了 Hibernate 的基本应用技术、Hibernate 核心接口的基本用法、HQL 的基本用法、Hibernate 对象状态的转换方式、Hibernate 的批量处理技术以及 Hibernate 的事务管理技术，最后再通过具体案例说明了 Hibernate 的综合应用技术。通过本章的学习，读者应熟练掌握 Hibernate 框架的相关应用技术，并能使用 JSP+ Hibernate+DAO 模式与 JSP+Struts 2+Hibernate+DAO 模式开发相应的 Web 应用系统。

思 考 题

1. 请简述 Hibernate 的体系结构。
2. 如何在 Web 项目中添加 Hibernate 开发能力？
3. Hibernate 默认的核心配置文件是什么？有何作用？
4. 如何生成与数据库表相对应的 Java 类与映射文件？
5. Hibernate 映射文件的作用是什么？可分为哪几个部分？
6. Hibernate 的主键生成策略可分为哪几大类？具体有哪些主键生成方式？
7. Hibernate 的核心接口主要有哪几个？各有何作用？如何创建各接口的实例？
8. HQL 查询、修改与删除语句的基本格式是什么？
9. 如何执行 HQL 语句？
10. 在 Hibernate 中，实体对象在其生命周期中共有哪几种状态？
11. 如何实现 Hibernate 的批量处理？
12. 请简述 Hibernate 的事务管理功能。
13. 请简述 Hibernate 与 Struts 2 框架的整合方法。

第 8 章

Spring 框架

Spring 是一个由 Rod Johnson(罗德·约翰逊)发明的开源框架，其初衷是为了解决企业级应用开发的复杂性问题。由于其本身已完成了大量开发中的通用步骤，因此可以大大提高企业级应用的开发效率，在当前 Java EE 的应用开发中十分流行。

本章要点：

- Spring 概述；
- Spring 基本应用；
- Spring 关键配置；
- Spring 核心接口；
- Spring AOP；
- Spring 事务支持。

学习目标：

- 了解 Spring 框架的分层架构；
- 掌握 Spring 的基本应用技术；
- 掌握 Spring 关键配置的设置方法；
- 掌握 Spring 核心接口的基本用法；
- 掌握 Spring AOP 的基本应用技术；
- 掌握 Spring 事务支持的主要配置方法；
- 掌握 Web 应用系统开发的 JSP+JDBC+JavaBean+Struts2+ Spring 模式、JSP+Spring+Hibernate+DAO 模式与 JSP+Struts 2+Spring+Hibernate+DAO 模式(即 SSH2 模式)。

8.1 Spring 概述

Spring 是一个由世界著名的 Java EE 大师 Rod Johnson(罗德·约翰逊)所创建的开源框架，其初衷是为了解决经典企业级 Java EE 应用开发中 EJB 的臃肿、低效与复杂性问题。对于以前只能由 EJB 完成的事情，Spring 只需使用基本的 JavaBean 即可胜任。Spring 1.0 正式版于 2004 年 3 月 24 日发布，随后便引发了 Java EE 应用框架的轻量化革命。实际上，Spring 是一个从实际开发中抽取出来的轻量级的应用软件框架，其本身已完成了大量开发中的通用步骤，留给开发人员的仅仅是与特定应用相关的部分，因此可大大提高企业级应用的开发效率。

Spring 框架具有简单、可测试与松耦合的特征，其核心是一个轻量级的容器，提供了为数众多的服务，包括事务管理服务、消息服务、JMS 服务与持久化服务等。借助于 Spring 的 AOP(Aspect-Oriented Programming，面向切面编程)技术，可在应用软件中轻松实现诸如权限拦截、运行监控等功能。对于主流的其他应用框架(如 Struts、Hibernate 等)，Spring 也提供了极其良好的集成支持。

Spring 框架的主要优势之一是其分层架构，如图 8.1 所示。该分层架构由 7 个定义良好的模块(子框架)组成，各个模块可单独存在，也可与其他一个或多个模块联合使用，具有极强的灵活性。

Spring 分层架构中各个模块的主要功能如下。

(1) Spring Core(核心容器)。提供 Spring 框架的基本功能，定义了创建、配置与管理 Bean 的方式。其主要组件为 BeanFactory，是工厂模式的实现。Spring Core 通过控制反转模式，将应用程序配置和依赖性规范与实际应用程序代码分开。Spring 的其他模块均构建在该核心容器之上。

(2) Spring Context(上下文)。向 Spring 框架提供上下文信息，包括企业级服务，如 JNDI、EJB、电子邮件、国际化、校验与调度等。

(3) Spring AOP。通过配置管理特性，可以很容易地使 Spring 框架管理的任何对象支持 AOP。Spring AOP 模块直接将面向切面编程的功能集成到 Spring 框架中，为基于 Spring 的应用程序对象提供了事务管理服务。通过 Spring AOP，无须依赖 EJB 组件，即可将声明性事务管理集成到应用程序中。

(4) Spring DAO。提供了 JDBC DAO 抽象层，可有效简化代码的编写与数据库厂商的异常错误处理。

(5) Spring ORM。Spring 框架插入了若干 ORM 框架，提供 ORM 的对象关系工具，其中包括 JDO、Hibernate 与 iBatis SQL Map，并且都遵从 Spring 的通用事务与 DAO 异常层次结构。

(6) Spring Web。为基于 Web 的应用程序提供上下文。Spring Web 建立在应用程序上下文模块之上，简化了处理多份请求及将请求参数绑定到域对象的工作。Spring 框架支持与 Jakarta Struts 的集成。

(7) Spring Web MVC。Spring Web MVC(简称 Spring MVC)是一个全功能构建 Web 应用程序的 MVC 实现。通过策略接口实现高度可配置，并容纳了大量的视图技术，其中包括 JSP、Velocity、Tiles、iText 与 POI 等。

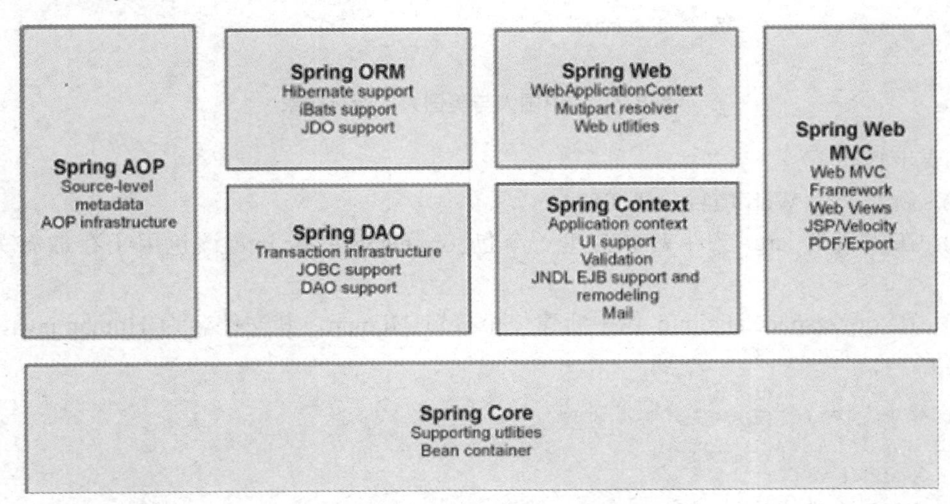

图 8.1　Spring 框架的分层架构

Spring 框架的功能可用于任何 Java EE 服务器中，其中大多数功能也适用于不受管理的环境。对于 Spring 来说，其核心要点是支持不绑定到特定 Java EE 服务的可重用业务与数据访问对象，而这样的对象方可在不同的 Java EE 环境(Web 或 EJB)、独立应用程序与测试环境之间进行重用。

8.2 Spring 基本应用

Spring 的核心机制为依赖注入(Dependency Inversion，DI)。所谓依赖注入，是指在程序运行过程中，当需要调用另一个对象协助时，无须在代码中直接创建该对象，而是依赖于外部容器的注入。换言之，依赖注入机制将对象间的依赖关系交由外部容器管理，而程序内部则只负责接口的控制，从而实现控制权从内部代码到外部容器的转移。因此，依赖注入通常又称为控制反转(Inversion of Control，IoC)。

为便于 Spring 核心机制与基本用法的理解与掌握，在此先简要介绍一下工厂模式，然后再通过具体实例说明 Spring 的基本用法以及常用的依赖注入方式。

8.2.1 工厂模式

工厂模式是一种常用的设计模式，是指在应用程序中，当甲组件需要乙组件的协助时，无须直接创建乙组件的实例对象，而是通过乙组件的工厂生成(某种类型的工厂可以生成相应类型组件的各种实例对象)。可见，在工厂模式下，甲组件无须与乙组件以硬编码的方式耦合在一起，而只需与乙组件的工厂进行耦合即可。

【实例 8-1】工厂模式应用示例：“中国人与美国人”，如图 8.2 所示。

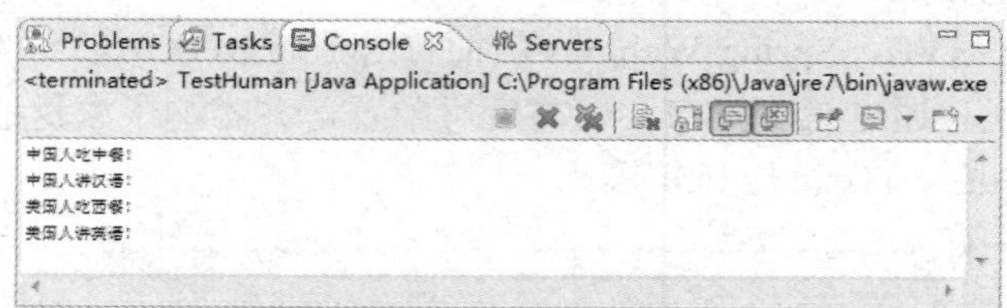

图 8.2 “中国人与美国人”运行结果

主要步骤如下。

(1) 新建一个 Web 项目 web_08。

(2) 在项目的 src 文件夹中新建一个包 org.etspace.abc.inte(该包用于存放有关的接口)。

(3) 在 org.etspace.abc.inte 包中新建一个接口 Human，其文件名为 Human.java，代码如下：

```
package org.etspace.abc.inte;
public interface Human {
    void eat();
    void speak();
}
```

(4) 在项目的 src 文件夹中新建一个包 org.etspace.abc.inte.impl(该包用于存放有关的接口实现类)。

(5) 在 org.etspace.abc.inte.impl 包中新建一个 Human 接口的实现类 Chinese，其文件名为 Chinese.java，代码如下：

```
package org.etspace.abc.inte.impl;
import org.etspace.abc.inte.Human;
public class Chinese implements Human {
    public void eat() {
        System.out.println("中国人吃中餐！");
    }
    public void speak() {
        System.out.println("中国人讲汉语！");
    }
}
```

(6) 在 org.etspace.abc.inte.impl 包中新建一个 Human 接口的实现类 American，其文件名为 American.java，代码如下：

```
package org.etspace.abc.inte.impl;
import org.etspace.abc.inte.Human;
public class American implements Human {
    public void eat() {
        System.out.println("美国人吃西餐！");
    }
    public void speak() {
        System.out.println("美国人讲英语！");
    }
}
```

(7) 在项目的 src 文件夹中新建一个包 org.etspace.abc.factory(该包用于存放有关的工厂类)。

(8) 在 org.etspace.abc.factory 包中新建一个工厂类 HumanFactory，其文件名为 HumanFactory.java，代码如下：

```
package org.etspace.abc.factory;
import org.etspace.abc.inte.Human;
import org.etspace.abc.inte.impl.Chinese;
import org.etspace.abc.inte.impl.American;
public class HumanFactory {
    public Human getHuman(String name){
        if(name.equals("Chinese")){
            return new Chinese();
        }else if(name.equals("American")){
            return new American();
        }else{
            throw new IllegalArgumentException("参数不正确");
        }
    }
}
```

(9) 在项目的 src 文件夹中新建一个包 org.etspace.abc.test(该包用于存放有关的测

试类)。

(10) 在 org.etspace.abc.test 包中新建一个测试类 TestHuman。其文件名为 TestHuman.java，代码如下：

```
package org.etspace.abc.test;
import org.etspace.abc.inte.Human;
import org.etspace.abc.factory.HumanFactory;
public class TestHuman {
    public static void main(String[] args) {
        Human myHuman=null;
        HumanFactory myHumanFactory=new HumanFactory();
        myHuman=myHumanFactory.getHuman("Chinese");
        myHuman.eat();
        myHuman.speak();
        myHuman=myHumanFactory.getHuman("American");
        myHuman.eat();
        myHuman.speak();
    }
}
```

运行结果如图 8.2 所示。TestHuman.java 为 Java 应用程序，因此可直接运行之。为此，只需在 MyEclipse 中右击，并在其快捷菜单中选择 Run As→1 Java Application 菜单项即可。

解析：

在本实例的测试类 TestHuman 中，要用到 Chinese 类与 American 类的对象，传统的方法是直接创建，但在此并没有这样做，而是通过工厂来获得，从而大大地降低了程序的耦合性。

8.2.2 Spring 基本用法

使用工厂模式，当甲组件需要乙组件的实例对象时，无须直接创建，而是通过工厂获得，因此需要创建一个乙组件的工厂。相对于工厂模式，使用 Spring 框架则更加省事，开发人员无须创建任何工厂，而只需直接利用 Spring 所提供的依赖注入方式即可。

由于 MyEclipse 已集成了 Spring 功能，因此若要在 MyEclipse 的 Web 项目中使用 Spring 框架，只需添加相应的 Spring 开发能力即可。下面通过一个简单的实例说明 Spring 框架的基本用法。

【实例 8-2】Spring 应用示例：“中国人与美国人”，如图 8.2 所示。

主要步骤(在项目 web_08 中、【实例 8-1】的基础上)如下。

(1) 添加 Spring 开发能力，可按以下步骤进行操作。

① 右击项目名，并在其快捷菜单中选择 MyEclipse→Add Spring Capabilities...菜单项，打开图 8.3 所示的 Add Spring Capabilities 对话框。

② 在 Spring version 选项区选择相应的 Spring 框架版本(在此为 Spring 3.1)，然后在其下的列表框中选中需要的类库(在此为 Spring 3.1 Core Libraries、Spring 3.1 Persistence Libraries 与 Spring 3.1 Web Libraries)，然后单击 Next 按钮，界面切换到 Add Spring bean configuration file(见图 8.4)，以便创建 Spring 的配置文件。

③　选中 New 单选按钮以新建一个配置文件，并在 Folder 文本框中指定存放配置文件的文件夹(在此使用默认文件夹 src)，在 File 文本框中指定配置文件的文件名(在此使用默认文件名 applicationContext.xml)，然后单击 Finish 按钮关闭对话框。

说明： applicationContext.xml 为 Spring 默认的核心配置文件。

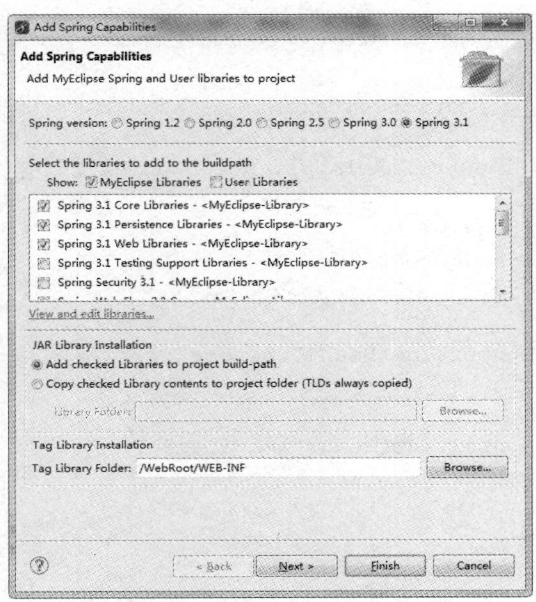

图 8.3　Add Spring Capabilities 对话框

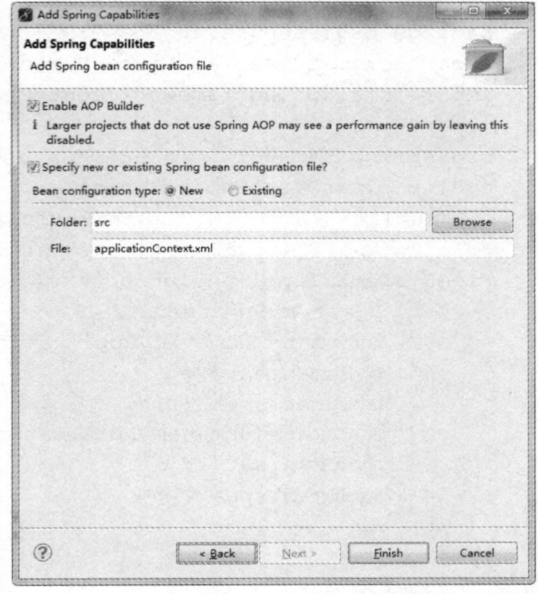

图 8.4　Add Spring bean configuration file 界面

至此，Spring 配置完毕。与此同时，项目中增加了一些 Spring 类库(Jar 包)、一个 applicationContext.xml 配置文件(在 src 文件夹中)以及其他一些文件，如图 8.5 所示。

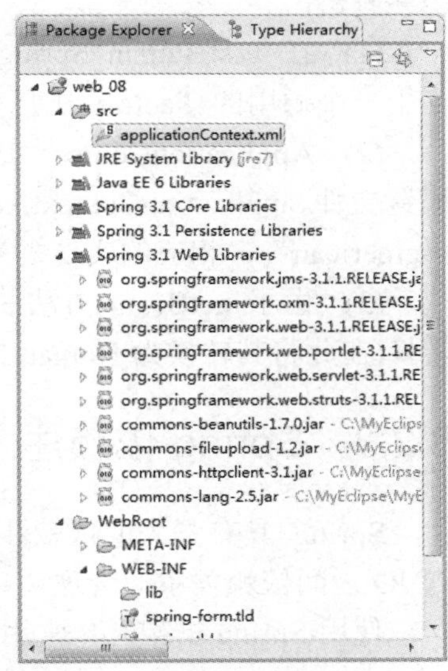

图 8.5　添加了 Spring 开发能力的 Web 项目

(2) 修改配置文件 applicationContext.xml。在此，主要是使用 bean 元素完成 Chinese 与 American 这两个实现类所对应的 Bean 的配置，其 id 分别为“chinese”与“american”。具体代码如下：

```
<?xml version="1.0" encoding="UTF-8"?>
<beans
    xmlns="http://www.springframework.org/sc
hema/beans"
    xmlns:xsi="http://www.w3.org/2001/XMLSch
ema-instance"
    xmlns:p="http://www.springframework.org/
schema/p"
    xsi:schemaLocation="http://www.springfra
mework.org/schema/beans
```

```
http://www.springframework.org/schema/beans/spring-beans-3.1.xsd">
    <bean id="chinese" class="org.etspace.abc.inte.impl.Chinese"></bean>
    <bean id="american"
class="org.etspace.abc.inte.impl.American"></bean>
</beans>
```

(3) 在 org.etspace.abc.test 包中新建一个测试类 TestHuman_Spring，其文件名为 TestHuman_Spring.java，代码如下：

```
package org.etspace.abc.test;
import org.etspace.abc.inte.Human;
import org.springframework.context.ApplicationContext;
import
org.springframework.context.support.FileSystemXmlApplicationContext;
public class TestHuman_Spring {
    public static void main(String[] args) {
        ApplicationContext myApplicationContext=new
FileSystemXmlApplicationContext("src/applicationContext.xml");
        Human myHuman=null;
        myHuman=(Human)myApplicationContext.getBean("chinese");
        myHuman.eat();
        myHuman.speak();
        myHuman=(Human)myApplicationContext.getBean("american");
        myHuman.eat();
        myHuman.speak();
    }
}
```

运行结果如图 8.2 所示。TestHuman_Spring.java 为 Java 应用程序，因此可以直接运行。

解析：

(1) 在 TestHuman_Spring.java 中，ApplicationContext 的实例 myApplicationContext 相当于一个通用的 Factory(工厂)。

(2) ApplicationContext 实例的 getBean()方法用于创建指定类的实例，其参数为 Spring 配置文件 applicationContext.xml 中相应 Bean 的 id(String 型，在此分别为“chinese”与“american”，相应的实现类为 org.etspace.abc.inte.impl 包中的 Chinese 与 American)。

(3) 由于 getBean()方法的返回值类型为 Object，因此在具体应用时要进行相应的类型转换(在此将其转换为 Human 接口类型)。

8.2.3 Spring 依赖注入

Spring 具有强大的依赖注入功能，对调用者与被调用者几乎没有任何要求，完全支持 POJO 之间依赖关系的管理。

使用 Spring 框架时，常用的依赖注入方式主要有两种，即设置注入与构造注入。

1. 设置注入

设置注入是在调用者中通过 setter 方法来注入被调用者实例的方式。具体地说，设置注入方式在调用无参构造函数(方法)创建默认的 Bean 实例后，再调用相应的 setter 方法设置属性值(即注入依赖关系)。由于设置注入方式较为简单、直观，且易于理解、掌握，因而被广为使用。

【实例 8-3】设置注入应用示例：“中国人说英语”，如图 8.6 所示。

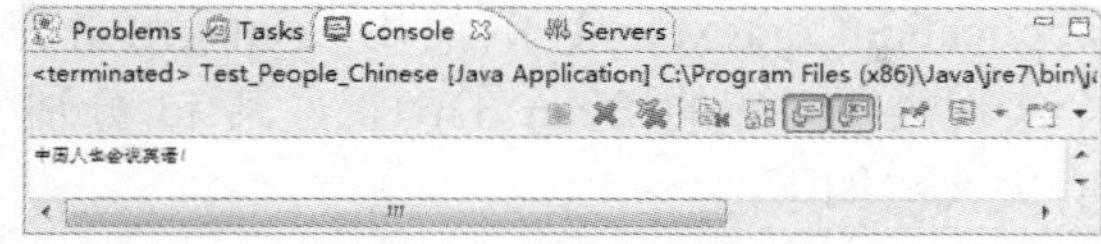

图 8.6　“中国人说英语”运行结果

主要步骤(在项目 web_08 中)如下。

(1)　在 org.etspace.abc.inte 包中新建一个接口 People，其文件名为 People.java，代码如下：

```
package org.etspace.abc.inte;
public interface People {
    void speak();
}
```

(2)　在 org.etspace.abc.inte 包中新建一个接口 Language，其文件名为 Language.java，代码如下：

```
package org.etspace.abc.inte;
public interface Language {
    public String type();
}
```

(3)　在 org.etspace.abc.inte.impl 包中新建一个 Language 接口的实现类 Language_English，其文件名为 Language_English.java，代码如下：

```
package org.etspace.abc.inte.impl;
import org.etspace.abc.inte.Language;
public class Language_English implements Language {
    public String type() {
        return "中国人也会说英语！";
    }
}
```

(4)　在 org.etspace.abc.inte.impl 包中新建一个 People 接口的实现类 People_Chinese，其文件名为 People_Chinese.java，代码如下：

```
package org.etspace.abc.inte.impl;
import org.etspace.abc.inte.People;
import org.etspace.abc.inte.Language;
public class People_Chinese implements People {
    private Language language;
    public People_Chinese(){};
    public void speak() {
        System.out.println(language.type());
```

```
    }
    public void setLanguage(Language language) {
        this.language = language;
    }
}
```

(5) 在配置文件 applicationContext.xml 中使用 bean 元素完成 People_Chinese 与 Language_English 这两个实现类所对应的 Bean 的配置，其 id 分别为“people_chinese”与“language_english”。有关代码如下：

```
…
<beans …>
    …
    <bean id="people_chinese"
class="org.etspace.abc.inte.impl.People_Chinese">
        <!-- property 元素用于指定需要容器注入的属性 -->
        <property name="language" ref="language_english"></property>
    </bean>
    <bean id="language_english"
class="org.etspace.abc.inte.impl.Language_English">
</bean>
    …
</beans>
```

其中，People_Chinese 类所对应的 bean 元素包含有一个子元素 property，该元素用于指定需要容器注入的属性及其取值。在此，属性名为“language”，其取值是 id 为“language_english”的 Bean 的实例(即 Language_English 类的实例)。

(6) 在 org.etspace.abc.test 包中新建一个测试类 Test_People_Chinese，其文件名为 Test_People_Chinese.java，代码如下：

```
package org.etspace.abc.test;
import org.etspace.abc.inte.People;
import org.springframework.context.ApplicationContext;
import
org.springframework.context.support.FileSystemXmlApplicationContext;
public class Test_People_Chinese {
    public static void main(String[] args) {
        ApplicationContext myApplicationContext = new
FileSystemXmlApplicationContext("src/applicationContext.xml");
        People people = null;
        people = (People) myApplicationContext.getBean("people_chinese");
        people.speak();
    }
}
```

运行结果如图 8.6 所示。Test_People_Chinese.java 为 Java 应用程序，因此可以直接运行。

2. 构造注入

构造注入是在调用者中利用构造函数(方法)来注入被调用者实例的方式。与设置注入方式不同，构造注入方式在创建 Bean 实例时，就已通过构造函数完成了相应属性的初始化(即已完成了依赖关系的注入)。显然，构造注入方式对于依赖关系的注入顺序的控制是较为方便的。

【实例 8-4】构造注入应用示例：“中国人说英语”。

主要步骤(在项目 web_08 中、在【实例 8-3】的基础上)如下。

(1) 将 People_Chinese 类的代码修改为：

```
package org.etspace.abc.inte.impl;
import org.etspace.abc.inte.People;
import org.etspace.abc.inte.Language;
public class People_Chinese implements People {
    private Language language;
    public People_Chinese(){};
    public void speak() {
        System.out.println(language.type());
    }
    //构造注入所需要的带参数的构造函数(方法)
    public People_Chinese(Language language){
        this.language=language;
    }
}
```

(2) 将配置文件 applicationContext.xml 中 People_Chinese 类所对应的 bean 元素修改为：

```
<bean id="people_chinese"
class="org.etspace.abc.inte.impl.People_Chinese">
    <!-- constructor-arg 元素用于指定需要容器注入的参数 -->
    <constructor-arg ref="language_english"></constructor-arg>
</bean>
```

其中，constructor-arg 子元素用于指定需要容器注入的构造函数的参数及其取值，在此，参数只有一个，其取值是 id 为“language_english”的 Bean 的实例(即 Language_English 类的实例)。

8.3　Spring 关键配置

Spring 的关键配置体现在其核心配置文件中，而 applicationContext.xml 即为 Spring 默认的核心配置文件。在 Spring 中，各个 Bean 及其相互之间的依赖关系都是在配置文件中进行设置的。通过配置文件，Spring 容器能精确地获知各个 Bean 的基本信息，并在运行过程中将有关的 Bean 装配起来。应用 Spring 框架后，程序中的各种组件即可作为 Spring 的 Bean 在配置文件中加以定义与配置(也就是进行注册)。

8.3.1 Bean 的基本定义

Spring 配置文件的根元素为<beans>，其中可包含多个<bean>子元素。<bean>元素用于定义一个 Bean 如何被装配到 Spring 容器中(任何一个 Java 对象均可作为一个 Bean 进行定义)。

对于一个 Bean 来说，最基本的定义包括 Bean 的 id 及其全称类名，分别通过<bean>元素的属性 id 与 class 指定。其中，id 属性用于指定 Bean 的唯一性标识符，class 属性用于指定 Bean 的具体实现类。例如：

```
<bean id="chinese" class="org.etspace.abc.inte.impl.Chinese"></bean>
<bean id="american" class="org.etspace.abc.inte.impl.American"></bean>
```

注意：(1) Spring 容器对 Bean 的管理、访问以及有关 Bean 之间的依赖关系，都是通过 Bean 的 id 属性完成的。因此，各个 Bean 的 id 属性在 Spring 容器中必须是唯一的。

(2) 通常情况下，Spring 会直接使用 new 关键字创建 Bean 的实例，因此各个 Bean 的 class 属性不能是接口，而必须是真正的实现类。

8.3.2 Bean 的依赖配置

在具体应用中，有关的 Bean 之间往往具有一定的依赖关系。对于 Bean 的依赖注入，常用的方式有设置注入与构造注入两种。注入方式不同，其配置方法也有所差异。

在使用设置注入时，应为<bean>元素添加相应的<property>子元素。例如：

```
<bean id="people_chinese"
class="org.etspace.abc.inte.impl.People_Chinese">
<property name="language" ref="language_english"></property>
</bean>
```

在使用构造注入时，应为<bean>元素添加相应的<constructor-arg>子元素。在<constructor-arg>子元素中，应设置相应的 index 属性以表明该元素应对应到构造函数(方法)中的哪一个参数(“index="0"”表示第 1 个参数，“index="1"”表示第 2 个参数，以此类推。若构造函数只有一个参数，则相应的<constructor-arg>子元素可省略 index 属性)。例如：

```
<bean id="people_chinese"
class="org.etspace.abc.inte.impl.People_Chinese">
    <constructor-arg ref="language_english"></constructor-arg>
</bean>
```

其中的<constructor-arg>子元素也可改写为：

```
<constructor-arg index="0" ref="language_english"></constructor-arg>
```

在进行依赖配置时，<property>元素与<constructor-arg>元素所指定的依赖关系的值可以是 Spring 容器中其他 Bean 的引用，也可以是一个确定的值。不过，在大多数情况下，通常只使用配置文件管理 Bean 实例间的依赖关系，而不使用配置文件管理普通的属性值。

特别地，如果需要注入的属性的类型为 List、Set 或 Map，那么在进行依赖配置时，还应进一步使用 list、set 或 map 子元素以指定相应 List、Set 或 Map 所包含的各个元素。

【实例 8-5】Bean 的依赖配置示例：“Hello World!”，如图 8.7 所示。

图 8.7　“Hello World!”运行结果

主要步骤(在项目 web_08 中)如下。

(1) 在项目的 src 文件夹中新建一个包 org.etspace.abc.bean(该包用于存放有关的类或 JavaBean)。

(2) 在 org.etspace.abc.bean 包中新建一个类 SomeMessage，其文件名为 SomeMessage.java，代码如下：

```
package org.etspace.abc.bean;
public class SomeMessage {
    private String message;
    public String getMessage() {
        return message;
    }
    public void setMessage(String message) {
        this.message = message;
    }
}
```

(3) 在配置文件 applicationContext.xml 中使用 bean 元素完成 SomeMessage 类所对应的 Bean 的配置，其 id 为“someMessage”。有关代码如下：

```
<bean id="someMessage" class="org.etspace.abc.bean.SomeMessage">
    <property name="message">
        <value>Hello World!</value>
    </property>
</bean>
```

(4) 在 org.etspace.abc.test 包中新建一个测试类 TestSomeMessage，其文件名为 TestSomeMessage.java，代码如下：

```
package org.etspace.abc.test;
import org.etspace.abc.bean.SomeMessage;
import org.springframework.context.ApplicationContext;
import
org.springframework.context.support.FileSystemXmlApplicationContext;
public class TestSomeMessage {
    public static void main(String[] args) {
```

```
        ApplicationContext myApplicationContext=new
FileSystemXmlApplicationContext("src/applicationContext.xml");
        SomeMessage mySomeMessage=(SomeMessage)myApplicationContext.getBean
("someMessage");
        System.out.println(mySomeMessage.getMessage());
    }
}
```

运行结果如图 8.7 所示。TestSomeMessage.java 为 Java 应用程序，因此可以直接运行。

【**实例 8-6**】List 类型属性注入示例：“工程系列职称”，如图 8.8 所示。

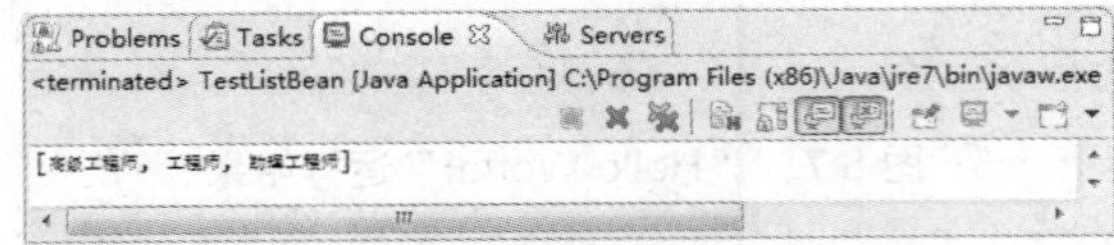

图 8.8 “工程系列职称”运行结果(1)

主要步骤(在项目 web_08 中)如下。

(1) 在 org.etspace.abc.bean 包中新建一个类 ListBean，其文件名为 ListBean.java，代码如下：

```
package org.etspace.abc.bean;
import java.util.List;
public class ListBean {
    private List list;
    public List getList() {
        return list;
    }
    public void setList(List list) {
        this.list = list;
    }
}
```

(2) 在配置文件 applicationContext.xml 中使用 bean 元素完成 ListBean 类所对应的 Bean 的配置，其 id 为“listBean”。有关代码如下：

```
<bean id="listBean" class="org.etspace.abc.bean.ListBean">
    <property name="list">
        <list>
            <value>高级工程师</value>
            <value>工程师</value>
            <value>助理工程师</value>
        </list>
    </property>
</bean>
```

(3) 在 org.etspace.abc.test 包中新建一个测试类 TestListBean，其文件名为 TestListBean.java，代码如下：

```
package org.etspace.abc.test;
import java.util.List;
```

```
import org.etspace.abc.bean.ListBean;
import org.springframework.context.ApplicationContext;
import
org.springframework.context.support.FileSystemXmlApplicationContext;
public class TestListBean {
    public static void main(String[] args) {
        ApplicationContext myApplicationContext=new
FileSystemXmlApplicationContext("src/applicationContext.xml");
        ListBean
myListBean=(ListBean)myApplicationContext.getBean("listBean");
        List myList=myListBean.getList();
        System.out.println(myList);
    }
}
```

运行结果如图 8.8 所示。TestListBean.java 为 Java 应用程序，因此可以直接运行。

【实例 8-7】Set 类型属性注入示例：“工程系列职称”，如图 8.9 所示。

主要步骤(在项目 web_08 中)如下。

图 8.9　“工程系列职称”运行结果(2)

(1) 在 org.etspace.abc.bean 包中新建一个类 SetBean，其文件名为 SetBean.java，代码如下：

```
package org.etspace.abc.bean;
import java.util.Set;
public class SetBean {
    private Set set;
    public Set getSet() {
        return set;
    }
    public void setSet(Set set) {
        this.set = set;
    }
}
```

(2) 在配置文件 applicationContext.xml 中使用 bean 元素完成 SetBean 类所对应的 Bean 的配置，其 id 为“setBean”。有关代码如下：

```
<bean id="setBean" class="org.etspace.abc.bean.SetBean">
    <property name="set">
        <set>
            <value>高级工程师</value>
            <value>工程师</value>
            <value>助理工程师</value>
        </set>
```

```
    </property>
</bean>
```

(3) 在 org.etspace.abc.test 包中新建一个测试类 TestSetBean，其文件名为 TestSetBean.java，代码如下：

```
package org.etspace.abc.test;
import java.util.Set;
import org.etspace.abc.bean.SetBean;
import org.springframework.context.ApplicationContext;
import
org.springframework.context.support.FileSystemXmlApplicationContext;
public class TestSetBean {
    public static void main(String[] args) {
        ApplicationContext myApplicationContext=new
FileSystemXmlApplicationContext("src/applicationContext.xml");
        SetBean
mySetBean=(SetBean)myApplicationContext.getBean("setBean");
        Set mySet=mySetBean.getSet();
        System.out.println(mySet);
    }
}
```

运行结果如图 8.9 所示。TestSetBean.java 为 Java 应用程序，因此可以直接运行。

【实例 8-8】Map 类型属性注入示例：“工程系列职称”，如图 8.10 所示。

图 8.10 “工程系列职称”运行结果(3)

主要步骤(在项目 web_08 中)如下。

(1) 在 org.etspace.abc.bean 包中新建一个类 MapBean，其文件名为 MapBean.java，代码如下：

```
package org.etspace.abc.bean;
import java.util.Map;
public class MapBean {
    private Map map;
    public Map getMap() {
        return map;
    }
    public void setMap(Map map) {
        this.map = map;
    }
}
```

(2) 在配置文件 applicationContext.xml 中使用 bean 元素完成 MapBean 类所对应的 Bean 的配置，其 id 为“mapBean”。有关代码如下：

```
<bean id="mapBean" class="org.etspace.abc.bean.MapBean">
    <property name="map">
        <map>
            <entry key="1">
                <value>高级工程师</value>
            </entry>
            <entry key="2">
                <value>工程师</value>
            </entry>
            <entry key="3">
                <value>助理工程师</value>
            </entry>
        </map>
    </property>
</bean>
```

(3) 在 org.etspace.abc.test 包中新建一个测试类 TestMapBean，其文件名为 TestMapBean.java，代码如下：

```
package org.etspace.abc.test;
import java.util.Map;
import org.etspace.abc.bean.MapBean;
import org.springframework.context.ApplicationContext;
import
org.springframework.context.support.FileSystemXmlApplicationContext;
public class TestMapBean {
    public static void main(String[] args) {
        ApplicationContext myApplicationContext=new
FileSystemXmlApplicationContext("src/applicationContext.xml");
        MapBean
myMapBean=(MapBean)myApplicationContext.getBean("mapBean");
        Map myMap=myMapBean.getMap();
        System.out.println(myMap);
    }
}
```

运行结果如图 8.10 所示。TestMapBean.java 为 Java 应用程序，因此可以直接运行。

8.3.3　Bean 的别名设置

必要时，可为 Bean 指定别名。为此，应使用<bean>元素的 name 属性。例如：

```
<bean id="chinese" name="zhongguoren"
class="org.etspace.abc.inte.impl.Chinese"></bean>
```

在此，将 id 为 chinese 的 Bean 的别名指定为 zhongguoren。

必要时，还可以为 Bean 指定多个别名。为此，只需在 name 属性中将多个别名用逗号(,)、分号(;)或空格隔开即可。例如：

```
<bean id="chinese" name="zhongguoren;zgr;chinaren"
class="org.etspace.abc.inte.impl.Chinese"></bean>
```

在此，id 为 chinese 的 Bean 的别名有 3 个，分别为 zhongguoren、zgr 与 chinaren。

为 Bean 指定别名后，即可通过别名来获取或访问相应的 Bean 实例。例如，将 id 为 chinese 的 Bean 的别名指定为 zhongguoren、zgr 与 chinaren 后，以下 4 个语句的作用是等价的：

```
Human myHuman=(Human)myApplicationContext.getBean("chinese");
Human myHuman=(Human)myApplicationContext.getBean("zhongguoren");
Human myHuman=(Human)myApplicationContext.getBean("zgr");
Human myHuman=(Human)myApplicationContext.getBean("chinaren");
```

8.3.4 Bean 的作用域设置

在对 Bean 进行定义时，还可以根据需要设置其作用域。为此，应使用<bean>元素的 scope 属性。该属性共有 5 种不同的取值，分别为 singleton、prototype、request、session 与 global session。相应地，Bean 的作用域也分为 5 种。

1. singleton 作用域

当一个 Bean 的作用域设置为 singleton 时，则该 Bean 处于单例模式。在这种情况下，Spring 容器只会创建该 Bean 的唯一实例，并将其存储到单例缓存(singleton cache)中，此后所有针对该 Bean 的请求与引用均将返回被缓存的同一个对象实例。换言之，对于具有 singleton 作用域的 Bean，Spring 容器中只会存在该 Bean 的一个共享的实例。例如：

```
<bean id="chinese" class="org.etspace.abc.inte.impl.Chinese"
scope="singleton"></bean>
```

在此，id 为 chinese 的 Bean 的作用域被设置为 singleton。该代码也可改写为：

```
<bean id="chinese" class="org.etspace.abc.inte.impl.Chinese"
singleton="true"></bean>
```

在此，<bean>元素的 singleton 属性值为 true(该属性的默认值也为 true)，表明相应的 Bean 为单例 Bean。

2. prototype 作用域

当一个 Bean 的作用域设置为 prototype 时，则该 Bean 处于原型模式。在这种情况下，对于该 Bean 每次的请求与引用，Spring 容器均会产生一个新的实例(相当于执行一次 new 操作)。换言之，对于具有 prototype 作用域的 Bean，Spring 容器可保证每次返回的该 Bean 的实例都是不同的。例如：

```
<bean id="chinese" class="org.etspace.abc.inte.impl.Chinese"
scope="prototype"></bean>
```

在此，id 为 chinese 的 Bean 的作用域被设置为 prototype。该代码也可改写为：

```
<bean id="chinese" class="org.etspace.abc.inte.impl.Chinese"
singleton="false"></bean>
```

在此，<bean>元素的 singleton 属性值为 false，表明相应的 Bean 不是单例 Bean，而是原型 Bean。

【实例 8-9】Bean 的作用域示例。

主要步骤(在项目 web_08 中、【实例 8-5】的基础上)如下。

(1) 在配置文件 applicationContext.xml 中添加两个 bean 元素，其实现类均为 SomeMessage，但 id 分别为“someMessage1”与“someMessage2”，作用域分别为 singleton 与 prototype。有关代码如下：

```
<bean id="someMessage1" class="org.etspace.abc.bean.SomeMessage"></bean>
<bean id="someMessage2" class="org.etspace.abc.bean.SomeMessage"
scope="prototype"></bean>
```

(2) 在 org.etspace.abc.test 包中新建一个测试类 TestSomeMessageScope，其文件名为 TestSomeMessageScope.java，代码如下：

```
package org.etspace.abc.test;
import org.etspace.abc.bean.SomeMessage;
import org.springframework.context.ApplicationContext;
import
org.springframework.context.support.FileSystemXmlApplicationContext;
public class TestSomeMessageScope {
    public static void main(String[] args) {
        ApplicationContext myApplicationContext = new
FileSystemXmlApplicationContext("src/applicationContext.xml");
        SomeMessage mySomeMessage =
(SomeMessage)myApplicationContext.getBean("someMessage");
        System.out.println(mySomeMessage.getMessage());
        SomeMessage mySomeMessage11 =
(SomeMessage)myApplicationContext.getBean("someMessage1");
        SomeMessage mySomeMessage12 =
(SomeMessage)myApplicationContext.getBean("someMessage1");
        System.out.println(mySomeMessage11 == mySomeMessage12);
        SomeMessage mySomeMessage21 =
(SomeMessage)myApplicationContext.getBean("someMessage2");
        SomeMessage mySomeMessage22 =
(SomeMessage)myApplicationContext.getBean("someMessage2");
        System.out.println(mySomeMessage21 == mySomeMessage22);
    }
}
```

运行结果如图 8.11 所示。TestSomeMessageScope.java 为 Java 应用程序，因此可直接运行。

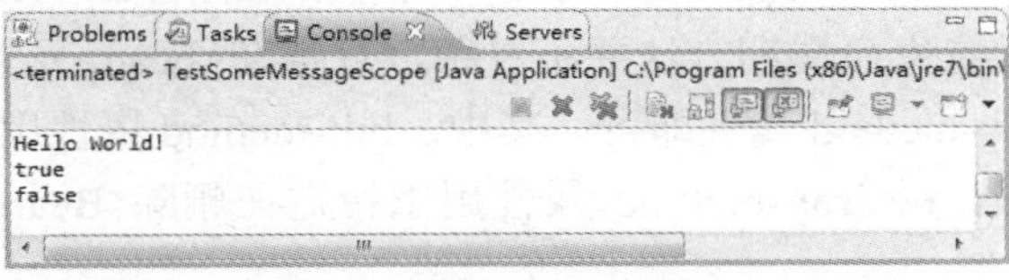

图 8.11　【实例 8-9】运行结果

解析：

(1) id 为“someMessage1”的 Bean 的作用域为 singleton，即该 Bean 处于单例模式，因此该 Bean 的两个实例 mySomeMessage11 与 mySomeMessage12 是相同的，表达式“mySomeMessage11 == mySomeMessage12”的结果为 true。

(2) id 为“someMessage2”的 Bean 的作用域为 prototype，即该 Bean 处于原型模式，因此该 Bean 的两个实例 mySomeMessage21 与 mySomeMessage22 是不同的，表达式“mySomeMessage21 == mySomeMessage22”的结果为 false。

3. request 作用域

当一个 Bean 的作用域设置为 request 时，则针对该 Bean 的每个 HTTP 请求，Spring 容器均会产生一个新的实例，同时该实例仅在当前的 HTTP request 内有效。request 作用域只有在 Web 应用中使用 Spring 时才有效。

4. session 作用域

当一个 Bean 的作用域设置为 session 时，则针对该 Bean 的每个 HTTP 请求，Spring 容器均会产生一个新的实例，同时该实例仅在当前的 HTTP session 内有效。session 作用域只有在 Web 应用中使用 Spring 时才有效。

5. global session 作用域

global session(全局 session)作用域类似于标准的 HTTP Session 作用域，不过仅在基于 portlet 的 Web 应用中才有意义。在 portlet 规范中，global session 被所有构成某个 portlet Web 应用的各种不同的 portlet 所共享。作用域为 global session 的 Bean 被限定在全局 portlet session 的生命周期范围内，只对应于一个 Bean 实例。若在 Web 中使用 global session 作用域来定义 Bean，则 Web 会自动将其当成 session 作用域。

说明： 在 Spring 2.0 以前，Bean 只有 singleton 与 prototype 两种作用域。而在 Spring 2.0 之后，Bean 的作用域又增加了 3 种，分别为 request、session 与 global session。默认情况下，Spring 中的 Bean 均为单例模式，即作用域为 singleton。

8.3.5 Bean 的生命周期方法设置

必要时，可让 Spring 容器在创建一个 Bean 实例后或删除一个 Bean 实例前，自动调用该 Bean 实例中的某个方法以完成相应的处理过程(如 Bean 实例创建后的初始化工作或删除前的清理工作)。为此，需在定义 Bean 时为其设置相应的生命周期方法。

其实，为 Bean 设置生命周期方法并不复杂，只需在<bean>元素中根据需要添加相应的 init-method 或 destroy-method 属性即可。其中，init-method 属性用于指定在创建 Bean 实例后需立即调用的方法，destroy-method 属性用于指定在删除 Bean 实例前需先行调用的方法。

现假定有一个数据库连接池实现类 DbConnectionPool，其部分代码如下：

```
package org.etspace.abc.pool;
```

```
…
public class DbConnectionPool{
    …
    public void initialize(){
    //初始化连接池
}
    public void cleanup(){
    //释放连接
}
    …
}
```

与此数据库连接池实现类相对应的 Bean 可定义为：

```
<bean id="dbConnectionPool"
class="org.etspace.abc.pool.DbConnectionPool"
   init-method="initialize" destroy-method="cleanup"></bean>
```

按此定义，DbConnectionPool 被实例化并载入容器后，将立即调用 initalize()方法进行初始化。反之，当 DbConnectionPool 的实例要从容器中删除时，将先行调用 cleanup()方法释放数据库连接。

【实例 8-10】Bean 的生命周期示例。

主要步骤(在项目 web_08 中、【实例 8-9】的基础上)如下。

(1) 修改 org.etspace.abc.bean 包中的 SomeMessage 类，在其中添加 initialize()与 cleanup()方法。有关代码如下：

```
package org.etspace.abc.bean;
public class SomeMessage {
    private String message;
    …
    public void initialize() {
        this.message = "How are you?";
        System.out.println("初始化完毕！");
    }
    public void cleanup() {
        this.message = "";
        System.out.println("OK！清理完毕！");
    }
}
```

(2) 在配置文件 applicationContext.xml 中添加一个 bean 元素，其实现类为 SomeMessage ，id 为“someMessageWithOther”。有关代码如下：

```
<bean id="someMessageWithOther" class="org.etspace.abc.bean.SomeMessage"
init-method="initialize" destroy-method="cleanup">
    <property name="message">
        <value>Hello World!</value>
    </property>
</bean>
```

(3) 在 org.etspace.abc.test 包中新建一个测试类 TestSomeMessageWithOther，其文件名为 TestSomeMessageWithOther.java，代码如下：

```
package org.etspace.abc.test;
import org.etspace.abc.bean.SomeMessage;
import org.springframework.context.support.AbstractApplicationContext;
import
org.springframework.context.support.FileSystemXmlApplicationContext;
public class TestSomeMessageWithOther {
    public static void main(String[] args) {
        AbstractApplicationContext myAbstractApplicationContext = new
FileSystemXmlApplicationContext("src/applicationContext.xml");
        SomeMessage mySomeMessageWithOther =
(SomeMessage)myAbstractApplicationContext.getBean("someMessageWithOther");
        System.out.println(mySomeMessageWithOther.getMessage());
        myAbstractApplicationContext.registerShutdownHook();
    }
}
```

运行结果如图 8.12 所示。TestSomeMessageWithOther.java 为 Java 应用程序，因此可以直接运行。

图 8.12 【实例 8-10】运行结果

解析：

(1) 在测试类 TestSomeMessageWithOther 中，先创建一个 AbstractApplicationContext 类的实例 myAbstractApplicationContext，然后调用该实例的 getBean()方法创建一个 id 为“someMessageWithOther”的 Bean(其实现类为 SomeMessage)的实例 mySomeMessageWithOther。在输出 mySomeMessageWithOther 实例的 message 属性值后，再调用 myAbstractApplicationContext 实例的 registerShutdownHook()方法为 Spring 容器注册关闭，让程序在退出 JVM 之前先关闭 Spring 容器。

(2) 根据 SomeMessage 类所对应的 bean 元素的具体设置，在测试类 TestSomeMessageWithOther 中创建其实例时，会先调用其 initialize()方法，从而将其 message 属性的值由最初注入的“Hello World!”修改为“How are you?”，并输出“初始化完毕！”的信息。反之，在容器删除其实例时，也会先调用其 cleanup()方法，从而将其 message 属性的值修改为空串，并输出“OK！清理完毕！”的信息。

8.4 Spring 核心接口

Spring 的核心接口有两个，分别为 BeanFactory(Bean 工厂)与 ApplicationContext(应用

上下文)。实际上，Spring 的核心接口相当于 Spring 容器。

8.4.1 BeanFactory

BeanFactory 由 org.springframework.beans.factory.BeanFactory 接口定义，是 Spring 最简单的容器，负责创建和分发 Bean，并提供了基本的依赖注入支持。对于一般的应用来说，使用 BeanFactory 就可以很好地满足各种需求。

顾名思义，BeanFactory 采用工厂设计模式，而且是一个通用的工厂，可创建与分发各种类型的 Bean。这一点，显然不同于其他只创建与分发一种类型的对象的工厂模式的实现。

在 Spring 中，BeanFactory 的具体实现有多种，其中最为常用的实现是 org.springframework.beans.factory.xml.XmlBeanFactory。该实现(XmlBeanFactory)根据 XML 配置文件中 Bean 的定义装载 Bean。

可以通过传递一个 java.io.InputStream 对象给 XmlBeanFactory 的构造函数(即使用 InputStream 对象提供 XML 配置文件给工厂)，创建一个 XmlBeanFactory 的实例。如以下代码：

```
InputStream myInputStream=new FileInputStream("applicationContext.xml");
BeanFactory myBeanFactory=new XmlBeanFactory(myInputStream);
```

在此，BeanFactory 只从 XML 配置文件中读取 Bean 的定义信息，但 Bean 并未被实例化。换言之，BeanFactory 会立即加载 Bean 的定义信息，但只在需要时才实例化相应的 Bean。因此，相对于 Bean 的定义信息来说，Bean 实例是被延迟加载到 BeanFactory 之中的。

此外，通过传递一个 ClassPathResource 对象给 XmlBeanFactory 的构造函数，也可创建一个 XmlBeanFactory 的实例。如以下代码：

```
ClassPathResource myClassPathResource=new
ClassPathResource("applicationContext.xml");
BeanFactory myBeanFactory=new XmlBeanFactory(myClassPathResource);
```

对于 BeanFactory 来说，最常用的方法为 getBean()。该方法用于根据 Bean 的 id 返回相应的 Bean 实例，其基本格式为：

```
Object getBean(String id)
```

其中，String 型参数 id 用于指定相应 Bean 在定义时所使用的 id。由于 getBean()方法的返回值类型为 Object，因此在具体应用时需进行强制类型转换。例如：

```
MyBean myBean = (MyBean)myBeanFactory.getBean("myBean");
```

在此，MyBean 为某个 Bean 的实现类，在 Bean 定义中其 id 为“myBean”。

在调用 getBean()方法时，BeanFactory 才实例化相应的 Bean，由此而产生的 Bean 实例便在 Spring 容器中开启了其自身的生命周期。

8.4.2 ApplicationContext

ApplicationContext 由 org.springframework.context.ApplicationContext 接口定义(该接口是 BeanFactory 的子接口)，是 Spring 更加高级的容器。

从表面上看，ApplicationContext 与 BeanFactory 差不多，均可载入 Bean 的定义信息、装配与分发相应的 Bean 实例。但实际上，ApplicationContext 建立在 BeanFactory 的基础之上，并进行了适当的扩充，从而提供了更多的功能，具体如下。

(1) 提供消息解析方法，包括国际化支持消息。

(2) 提供资源访问方法，如装载文件资源(包括图片等)。

(3) 可以向注册为监听器的 Bean 发送事件。

(4) 可以载入多个(有继承关系)上下文类，使得每个上下文类都专注于一个特定的层次，如应用的 Web 层。

由于 ApplicationContext 提供了诸多附加的功能，因此大多数应用系统都会选用 ApplicationContext，而不是 BeanFactory。

在 Spring 中，ApplicationContext 有多种不同的具体实现，其中最常用实现为以下 3 个。

(1) ClassPathXmlApplicationContext。从类路径中的 XML 配置文件载入上下文定义信息(把上下文定义文件当成类路径资源)。

(2) FileSystemXmlApplicationContext。从文件系统中的 XML 配置文件载入上下文定义信息。

(3) XmlWebApplicationContext。从 Web 系统中(即 Web 应用目录 WEB-INF 中)的 XML 配置文件载入上下文定义信息。

其中，FileSystemXmlApplicationContext 只能在指定的路径中查找 XML 配置文件，而 ClassPathXmlApplicationContext 则可在整个类路径中查找 XML 配置文件。

例如：

```
ApplicationContext myApplicationContext=new ClassPathApplicationContext
("applicationContext.xml");
ApplicationContext myApplicationContext=new
FileSystemXmlApplicationContext ("c:/ applicationContext.xml");
ApplicationContext myApplicationContext=
    WebApplicationContextUtils.getWebApplicationContext
(request.getSession().getServletContext ());
```

对于 ApplicationContext 来说，最常用的方法为 getBean()，且该方法的功能与用法类似于 BeanFactory 的 getBean()方法。例如：

```
MyBean myBean = (MyBean)myApplicationContext.getBean("myBean");
```

除了 ApplicationContext 所提供的附加功能以外，ApplicationContext 与 BeanFactory 的另外一个重要区别在于单例 Bean 的加载方式。BeanFactory 延迟加载所有的 Bean，直至 getBean()方法被调用时相应的 Bean 实例才会被创建。而 ApplicationContext 则在上下文启动后便预先加载所有的单例 Bean，这样在需要时即可直接使用相应的 Bean 实例。

8.5 Spring AOP

8.5.1 AOP 简介

AOP 即面向切面编程(Aspect-Oriented Programming)，又称为面向方面编程。作为一种编程思想，AOP 是 OOP(Object-Oriented Programing，面向对象编程)的一种补充与完善。OOP 将程序分解为各个层次的对象，而 AOP 则将程序的运行过程分解为各个切面。

一般来说，AOP 专用于处理系统中分布于各个模块(或不同方法)中的交叉关注点问题。在 Java EE 应用中，则通常使用 AOP 来处理一些具有横切性质的系统级服务，如日志记录、性能统计、安全控制、权限检查、事务管理、异常处理等。具体地说，就是将系统级服务的行为代码从业务逻辑中分离出来，从而避免在改变这些行为时影响到业务逻辑代码。

其实，AOP 的目的就是保证程序员在不修改源代码的前提下为系统中业务组件的有关方法添加某种通用的功能。但从本质上看，要为业务方法添加通用功能，还是要修改其源代码的，只不过源代码的修改由 AOP 框架完成，无须程序员动手。

目前已实现 AOP 的框架主要有 Spring、AspectJ、Jboss 等。其中，Spring 框架为 AOP 提供了极为丰富的支持。基于 Spring AOP，在 Java EE 应用开发中，业务对象只需实现其应该完成的业务逻辑即可，而无须负责各种系统级服务。

8.5.2 AOP 的相关术语

与 AOP 相关的术语为数不少，主要包括关注点、连接点、增强、切入点、切面、引入、目标对象、AOP 代理、织入等。

1. 关注点(Concern)

关注点就是所关注的与业务无关的公共服务或系统级服务，如日志记录、权限检查、事务管理等。关注点通常又称为横切关注点(cross-cutting concern)，表示“要做什么”。

2. 连接点(Joinpoint)

连接点就是程序执行过程中明确的阶段点，如方法的调用或异常的抛出等。其实，关注点的功能就是添加到相应的连接点。因此，连接点表示“在哪里做”。在 Spring AOP 中，一个连接点总代表一个方法的执行。

3. 增强(Advice)

增强(在有些资料中又称之为通知)就是在特定的连接点处所执行的处理逻辑，也就是向连接点注入的代码。增强通常又称为增强处理，表示“具体怎么做”。

增强可分为不同的类型，Spring AOP 所支持的增强类型主要如下。

(1) Before Advice。前置增强，在连接点执行前被调用。

(2) AfterReturning Advice。后置增强，在连接点成功执行后被调用。

(3) AfterThrowing Advice。异常增强，在连接点抛出异常后被调用。

(4) Around Advice。环绕增强，在连接点执行前与成功执行后被调用。这种增强近似于前置增强(Before Advice)与后置增强(AfterReturning Advice)的组合，功能更加强大、灵活，可控制连接点的执行、改变连接点的参数值与返回值等。

4. 切入点(Pointcut)

切入点就是可以插入增强的连接点。换言之，当某个连接点满足指定的要求而被添加增强处理时，该连接点就转变为切入点。切入点用于确定增强的触发时机，如何使用表达式来定义切入点是 AOP 的核心。Spring 默认使用 AspectJ 切入点语法。

5. 切面(Aspect)

切面就是对系统中的横切关注点逻辑进行模块化封装的概念实体，可包含多个切入点(Pointcut)以及相应的增强(Advice)定义。切面可以横切多个业务对象。在 Spring 中，切面可通过增强器(Advisor)或拦截器(Intercepter)实现。

6. 引入(Introduction)

引入就是将方法或字段添加到被处理的类中。Spring 允许将新的接口引入任何被处理的对象中。

7. 目标对象(Target Object)

目标对象就是被 AOP 框架进行增强的对象，又称被增强对象或被代理对象。增强被应用到目标对象上，而目标对象则包含相应的连接点。

8. AOP 代理(AOP Proxy)

AOP 代理就是 AOP 框架所创建的对目标对象进行增强的对象。在 Spring 中，AOP 代理可以是 JDK 动态代理，也可以是 CGLib(Code Generation Library，代码生成库)动态代理。其中，JDK 动态代理为实现接口的目标对象的代理，而 CGLib 动态代理为不实现接口的目标对象的代理。

9. 织入(Weaving)

织入就是将增强添加到目标对象中，并创建一个 AOP 代理(即被增强的对象)的过程。织入的实现方式有两种，即编译时增强(如 AspectJ)与运行时增强(如 Spring AOP)。与其他纯 Java AOP 框架一样，Spring 在运行时完成织入。

8.5.3 AOP 的实现机制

AOP 是基于代理(Proxy)机制来实现的。代理分为两大类，即静态代理(static proxy)与动态代理(dynamic proxy)。静态代理是在编译阶段生成的，而动态代理则是在运行阶段生成的。下面通过一个简单实例对 AOP 的实现机制进行简要说明。

现假定有一个业务类 DoSomething，内含 3 个分别执行不同功能的方法，即

executeFunction1()、executeFunction2()与 executeFunction3()。该类的具体代码如下：

```
public class DoSomething {
    public void executeFunction1(){
        System.out.println("执行功能 1…");
    }
    public void executeFunction2(){
        System.out.println("执行功能 2…");
    }
    public void executeFunction3(){
        System.out.println("执行功能 3…");
    }
}
```

对于该业务类，若要求在执行各个方法前先进行权限检查，则最简单的做法就是直接在各个方法中添加相应的权限检查代码。这种直接修改源代码的方式虽然能达到目的，但不可避免地使业务类中出现一系列的重复代码。如果需求发生改变(如将权限检查改变为用户验证或日志记录等)，那么就必须逐一将有关的代码改掉。可见，这种方式工作量大，对有关功能的添加、修改或删除来说是极其不利的。

其实，对于业务类来说，权限检查、用户验证、日志记录等系统级服务功能并非其本身的业务逻辑，在业务类中直接添加服务功能代码，必然会额外增加业务类的执行逻辑，加重其负担，甚至会混淆其本身的职责。另外，当需求出现变化，或者不再需要有关的服务时，又无法简单地进行全面修改或从中移除。对于此类问题，若使用代理机制，则可妥善解决。

1. 静态代理

采用静态代理方式时，代理类与被代理的业务类必须实现同一个接口。在代理类中可以实现权限检查、用户验证、日志记录等相关服务，并在需要时再呼叫被代理类。而在被代理类中，只需实现其本身的业务逻辑或功能即可。

【实例 8-11】静态代理应用示例：执行方法前先进行权限检查，如图 8.13 所示。

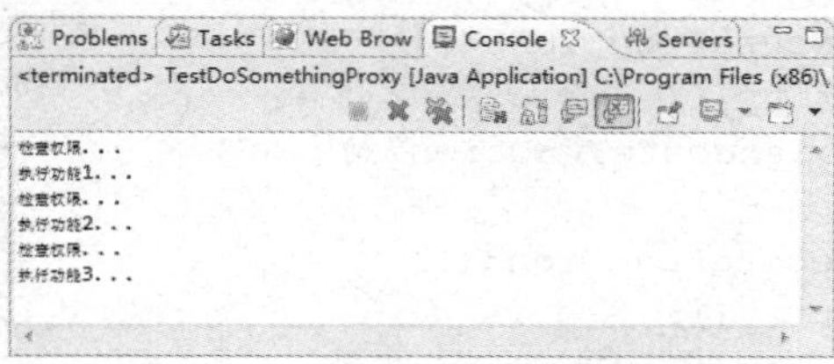

图 8.13　执行方法前先进行权限检查

主要步骤(在项目 web_08 中)如下。

(1) 在 org.etspace.abc.inte 包中新建一个接口 IDoSomething，其文件名为 IDoSomething.java，代码如下：

```
package org.etspace.abc.inte;
public interface IDoSomething {
    public void executeFunction1();
```

```
    public void executeFunction2();
    public void executeFunction3();
}
```

(2) 在 org.etspace.abc.inte.impl 包中新建一个 IDoSomething 接口的实现类 DoSomething。该类为业务类，其文件名为 DoSomething.java，代码如下：

```
package org.etspace.abc.inte.impl;
import org.etspace.abc.inte.IDoSomething;
public class DoSomething implements IDoSomething {
    public void executeFunction1(){
        System.out.println("执行功能 1…");
    }
    public void executeFunction2(){
        System.out.println("执行功能 2…");
    }
    public void executeFunction3(){
        System.out.println("执行功能 3…");
    }
}
```

(3) 在 org.etspace.abc.inte.impl 包中新建一个 IDoSomething 接口的实现类 DoSomethingProxy。该类为代理类，其文件名为 DoSomethingProxy.java，代码如下：

```
package org.etspace.abc.inte.impl;
import org.etspace.abc.inte.IDoSomething;
public class DoSomethingProxy implements IDoSomething {
    private IDoSomething doSomething;
    public DoSomethingProxy(IDoSomething doSomething){
        this.doSomething=doSomething;
    }
    public void executeFunction1(){
        checkAuthority();
        doSomething.executeFunction1();
    }
    public void executeFunction2(){
        checkAuthority();
        doSomething.executeFunction2();
    }
    public void executeFunction3(){
        checkAuthority();
        doSomething.executeFunction3();
    }
    public void checkAuthority(){
        System.out.println("检查权限…");
    }
}
```

(4) 在 org.etspace.abc.test 包中新建一个测试类 TestDoSomethingProxy，其文件名为 TestDoSomethingProxy.java，代码如下：

```
package org.etspace.abc.test;
import org.etspace.abc.inte.IDoSomething;
import org.etspace.abc.inte.impl.DoSomething;
import org.etspace.abc.inte.impl.DoSomethingProxy;
public class TestDoSomethingProxy {
    public static void main(String[] args) {
        IDoSomething myDoSomethingProxy=new DoSomethingProxy(new
DoSomething());
        myDoSomethingProxy.executeFunction1();
        myDoSomethingProxy.executeFunction2();
        myDoSomethingProxy.executeFunction3();
    }
}
```

运行结果如图 8.13 所示。TestDoSomethingProxy.java 为 Java 应用程序，因此可以直接运行。

解析：

(1) 被代理的业务类 DoSomething 与代理类 DoSomethingProxy 所实现的接口是一样的，均为 IDoSomething。

(2) 业务类 DoSomething 只需实现其本身的业务逻辑即可，无须添加任何涉及权限检查的代码。

(3) 在代理类 DoSomethingProxy 中，权限检查功能通过 checkAuthority()方法加以实现，并在业务逻辑开始前先行调用。

(4) 在测试类 TestDoSomethingProxy 中，创建接口对象时应用代理类对象，并将被代理类对象作为构造方法的参数。

由【实例 8-11】可见，静态代理要求为每种类型的被代理类都创建一个基于同样接口的代理类，因此对于规模很大的系统静态代理就难以胜任了。

2. 动态代理

与静态代理方式不同，动态代理方式所创建的代理类是一个泛类代理，可为各种不同的被代理业务类提供服务。动态代理只有在 JDK 1.3 或以上的 Java 运行环境下方可使用，可分为 JDK 动态代理(实现接口的目标对象的代理)与 CGLib 动态代理(不实现接口的目标对象的代理)两种，在此仅介绍 JDK 动态代理。

【实例 8-12】动态代理应用示例：执行方法前先进行权限检查，如图 8.13 所示。

主要步骤(在项目 web_08 中，在【实例 8-11】的基础上)如下。

(1) 在项目的 src 文件夹中新建一个包 org.etspace.abc.proxy(该包用于存放有关的动态代理类)。

(2) 在 org.etspace.abc.proxy 包中新建一个动态代理类 CheckAuthorityProxy，其文件名为 CheckAuthorityProxy.java，代码如下：

```
package org.etspace.abc.proxy;
import java.lang.reflect.InvocationHandler;
import java.lang.reflect.Method;
import java.lang.reflect.Proxy;
```

```
public class CheckAuthorityProxy implements InvocationHandler {
    private Object object;
   public CheckAuthorityProxy(Object object){
    this.object = object;
   }
   public Object bind(){
    return Proxy.newProxyInstance(object.getClass().getClassLoader(),
object.getClass().getInterfaces(),this);
   }
   public Object invoke(Object proxy, Method method, Object[] args)
        throws Throwable{
    Object result=null;
    try{
        checkAuthority();
        result=method.invoke(object, args);
    }catch(Exception e){
        e.printStackTrace();
    }
      return result;
    }
    public void checkAuthority(){
        System.out.println("检查权限…");
    }
}
```

动态代理类需要实现 InvocationHandler 接口，该接口的 invoke()回调方法负责处理被代理类方法的调用，其第 1 个参数为调用方法的代理实例，第 2 个参数为被代理类实例的方法名，第 3 个参数为相应方法的参数数组。可见，调用 invoke()方法会传入被代理对象的方法名与参数，然后通过 method.invoke(object, args)调用被代理类实例的方法，返回的结果即为被代理类实例中方法的返回值。

(3) 在 org.etspace.abc.test 包中新建一个测试类 TestCheckAuthorityProxy，其文件名为 TestCheckAuthorityProxy.java，代码如下：

```
package org.etspace.abc.test;
import org.etspace.abc.inte.IDoSomething;
import org.etspace.abc.inte.impl.DoSomething;
import org.etspace.abc.proxy.CheckAuthorityProxy;
public class TestCheckAuthorityProxy {
    public static void main(String[] args) {
        DoSomething myDoSomething=new DoSomething();
        CheckAuthorityProxy myCheckAuthorityProxy=new
CheckAuthorityProxy(myDoSomething);
        IDoSomething
myDoSomethingProxy=(IDoSomething)myCheckAuthorityProxy.bind();
        myDoSomethingProxy.executeFunction1();
        myDoSomethingProxy.executeFunction2();
        myDoSomethingProxy.executeFunction3();
    }
}
```

运行结果如图 8.13 所示。TestDoSomethingProxy.java 为 Java 应用程序，因此可直接运行之。

解析：

(1) 在动态代理类 CheckAuthorityProxy 的 bind()方法中，使用 Proxy 类的静态方法 newProxyInstance()创建一个代理对象。newProxyInstance()方法有 3 个参数，其中第 1 个参数用于指定被代理类的类加载器，第 2 个参数用于指定被代理类所实现的接口，第 3 个参数用于指定指派方法调用的处理程序(在此为当前类本身)。

(2) 在测试类 TestCheckAuthorityProxy 中，使用了【实例 8-11】中的接口 IDoSomething 以及实现该接口的业务类 DoSomething。

由【实例 8-12】可见，使用动态代理方式时，无须再为特定的业务类创建相应的代理类，因此可极大地减少编码的工作量。

综上所述，不管是静态代理方式还是动态代理方式，均可在不修改原有业务类的基础上，为其添加相应的系统级服务。其实，这正是 AOP 的基本思想。具体到此处的业务类 DoSomething，权限检查就是 AOP 中的关注点。将原始实现方式中直接添加到业务类中的与业务逻辑无关的权限检查代码片段提取出来，设计为独立的代理类，就成为 AOP 中的切面。在 AOP 编程中，切面通过一定的规则在应用程序需要时介入其中，为其提供服务；而在不需要的时候，因其独立性又可方便地从应用程序中分离出来。在此过程中，应用程序中的业务组件无须进行任何修改，因而可提高其可重用性。

8.5.4 Spring AOP 的基本应用

不同的 AOP 框架对于 AOP 的实现方式各有不同。Spring AOP 是基于动态代理机制实现的，其应用的关键在于增强(Advice)类的具体设计与相应代理的正确配置。

1. 增强(Advice)类的设计

Spring AOP 支持多种增强类型，在此仅以前置增强(Before Advice)为例进行介绍，其他增强类型请查阅有关的参考资料。

在 Spring AOP 中，前置增强类只需实现 MethodBeforeAdvice 接口，并覆盖其 before()方法即可。在 before()方法中，可根据需要编写相应的前置服务代码。

【实例 8-13】前置增强应用示例：执行方法前先进行权限检查，如图 8.13 所示。

主要步骤(在项目 web_08、【实例 8-11】的基础上)如下。

(1) 在项目的 src 文件夹中新建一个包 org.etspace.abc.advice(该包用于存放有关的增强实现类)。

(2) 在 org.etspace.abc.advice 包中新建一个前置增强类 AdviceBeforeDoSomething。其文件名为 AdviceBeforeDoSomething.java，代码如下：

```
package org.etspace.abc.advice;
import java.lang.reflect.Method;
import org.springframework.aop.MethodBeforeAdvice;
public class AdviceBeforeDoSomething implements MethodBeforeAdvice {
    public void before(Method arg0, Object[] arg1, Object arg2)
```

```
            throws Throwable {
        System.out.println("检查权限…");
    }
}
```

(3) 在配置文件 applicationContext.xml 添加相应的 bean 元素，注册业务类、前置增强类与代理类。具体代码如下：

```
<!-- 注册业务类 -->
<bean id="doSomething"
class="org.etspace.abc.inte.impl.DoSomething"></bean>
<!-- 注册前置增强类 -->
<bean id="adviceBeforeDoSomething"
class="org.etspace.abc.advice.AdviceBeforeDoSomething"></bean>
<!-- 注册代理类 -->
<bean id="doSomethingProxy"
class="org.springframework.aop.framework.ProxyFactoryBean">
    <property name="proxyInterfaces">
        <value>org.etspace.abc.inte.IDoSomething</value>
    </property>
    <property name="target">
        <ref bean="doSomething" />
    </property>
    <property name="interceptorNames">
        <list>
            <value>adviceBeforeDoSomething</value>
        </list>
    </property>
</bean>
```

在此，业务类为接口 IDoSomething 的实现类 DoSomething，其 bean 元素的 id 为“doSomething”。前置通知类为 AdviceBeforeDoSomething，其 bean 元素的 id 为“adviceBeforeDoSomething”。代理类为 ProxyFactoryBean，其 bean 元素的 id 为“doSomethingProxy”。在代理类所对应的 bean 元素中，通过 proxyInterfaces 属性告知代理可运行的接口(在此为 IDoSomething)，通过 target 属性指定目标对象(在此为 DoSomething 的实例)，通过 interceptorNames 属性指定要应用的增强实例(在此为 AdviceBeforeDoSomething 的实例)。由于未指定目标方法，因此相应的增强会被织入接口中所定义的各个方法处。

(4) 在 org.etspace.abc.test 包中新建一个测试类 TestAdviceOfDoSomething，其文件名为 TestAdviceOfDoSomething.java，代码如下：

```
package org.etspace.abc.test;
import org.etspace.abc.inte.IDoSomething;
import org.springframework.context.ApplicationContext;
import
org.springframework.context.support.ClassPathXmlApplicationContext;
public class TestAdviceOfDoSomething {
    public static void main(String[] args) {
```

```
        ApplicationContext myApplicationContext=new
ClassPathXmlApplicationContext("applicationContext.xml");
        IDoSomething
myDoSomethingProxy=(IDoSomething)myApplicationContext.getBean("doSomethi
ngProxy");
        myDoSomethingProxy.executeFunction1();
        myDoSomethingProxy.executeFunction2();
        myDoSomethingProxy.executeFunction3();
    }
}
```

运行结果如图 8.13 所示。TestAdviceOfDoSomething.java 为 Java 应用程序，因此可以直接运行。

解析：

(1) 在本实例中，DoSomething 与 AdviceBeforeDoSomething 是两个独立的类。对于 DoSomething 来说，无须知道 AdviceBeforeDoSomething 的存在；而 AdviceBeforeDoSomething 也可以运行到其他类之上。因此，DoSomething 与 AdviceBeforeDoSomething 都是可重用的。

(2) 代理类无须另外创建，而只需在配置文件 applicationContext.xml 中直接注册即可。

(3) 在测试类 TestAdviceOfDoSomething 中，代理实例是根据代理类所对应的 bean 元素的 id 创建的。

2. 增强器(Advisor)的使用

在 Spring AOP 中，增强器(Advisor)其实就是切入点(Pointcut)与增强(Advice)的适配器，用于将二者结合在一起，即决定在何处进行何种增强。

NameMatchMethodPointcutAdvisor 与 RegexpMethodPointcutAdvisor 是 Spring AOP 中常用的两种增强器，二者均可用于指定目标对象中要进行增强处理的方法，但前者可使用通配符，后者则使用正则表达式。

【实例 8-14】NameMatchMethodPointcutAdvisor 应用示例：执行 DoSomething 中的 executeFunction3()方法前先进行权限检查，如图 8.14 所示。

图 8.14　执行指定方法前先进行权限检查

主要步骤(在项目 web_08 中、【实例 8-13】的基础上)如下。

(1) 在配置文件 applicationContext.xml 添加一个 bean 元素，注册 NameMatchMethodPointcutAdvisor。具体代码如下：

```
<bean id="doSomethingAdvisor"
class="org.springframework.aop.support.NameMatchMethodPointcutAdvisor">
```

```
    <property name="mappedName">
        <value>*3</value>
    </property>
    <property name="advice">
        <ref bean="adviceBeforeDoSomething"></ref>
    </property>
</bean>
```

(2) 修改代理类 ProxyFactoryBean 所对应的 bean 元素(其 id 为“doSomethingProxy”)，将其中的 adviceBeforeDoSomething 替换为 doSomethingAdvisor。具体代码如下：

```
    <bean id="doSomethingProxy"
class="org.springframework.aop.framework.ProxyFactoryBean">
        <property name="proxyInterfaces">
            <value>org.etspace.abc.inte.IDoSomething</value>
        </property>
        <property name="target">
            <ref bean="doSomething" />
        </property>
        <property name="interceptorNames">
            <list>
                <value>doSomethingAdvisor</value>
            </list>
        </property>
    </bean>
```

至此，即可直接运行测试程序 TestAdviceOfDoSomething.java，结果如图 8.14 所示。

解析：在 NameMatchMethodPointcutAdvisor 所对应的 bean 元素中，为 mappedName 属性指定了“*3”，因此在调用目标对象中名称以“3”结尾的方法时，就会应用 AdviceBeforeDoSomething 的服务逻辑。

【实例 8-15】 RegexpMethodPointcutAdvisor 应用示例：执行 DoSomething 中的 executeFunction3()方法前先进行权限检查如图 8.14 所示。

主要步骤(在项目 web_08 中、【实例 8-14】的基础上)如下。

在配置文件 applicationContext.xml 中直接修改 id 为“doSomethingAdvisor”的 bean 元素，以注册 RegexpMethodPointcutAdvisor。具体代码为：

```
    <bean id="doSomethingAdvisor"
class="org.springframework.aop.support.RegexpMethodPointcutAdvisor">
        <property name="pattern">
            <value>.*3</value>
        </property>
        <property name="advice">
            <ref bean="adviceBeforeDoSomething"></ref>
        </property>
    </bean>
```

至此，即可直接运行测试程序 TestAdviceOfDoSomething.java，结果如图 8.14 所示。

解析：在 RegexpMethodPointcutAdvisor 所对应的 bean 元素中，为 pattern 属性指定了

“.*3”，因此在调用目标对象中名称以“3”结尾的方法时，就会应用 AdviceBeforeDoSomething 的服务逻辑。

8.6　Spring 事务支持

Spring 的事务处理是基于 Spring 的 AOP 实现的。Spring 事务的中心接口是 org.springframework.transaction.PlatformTransactionManager，其实现类在 Spring 中有许多。应用程序面向与平台无关的接口编程，当底层采用不同的持久层技术时，系统只需使用不同的接口实现类即可，而这种转换通常是由 Spring 容器负责管理的。

PlatformTransactionManager 接口的代码如下：

```
public interface PlatformTransactionManager {
    //获取当前事务
    TransactionStatus getTransaction(TransactionDefinition definition)
throws TransactionException;
    //提交事务
    void commit(TransactionStatus status) throws TransactionException;
    //回滚事务
    void rollback(TransactionStatus status) throws TransactionException;
}
```

其中，getTransaction()方法返回一个 TransactionStatus 对象，该对象可能是一个新的事务，也可能是一个已经存在的事务。若当前执行的线程正处于事务管理之下，则返回当前线程的事务对象，否则系统将创建一个新的事务对象并返回。

TransactionStatus 接口代表着当前的事务，其代码如下：

```
public interface TransactionStatus{
    //判断是否为一个新的事务
    boolean isNewTransaction();
    //判断是否存在保存点
    boolean hasSavepoint();
    //设置为只读事务
    void setRollbackOnly();
    //判断是否为只读事务
    boolean isRollbackOnly();
    //判断一个事务是否已经完成
    boolean isCompleted();
}
```

TransactionDefinition 接口代表着事务处理的一些属性定义(如事务的名称、传播行为、隔离层次等)，其代码如下：

```
public interface TransactionDefinition {
    //获取事务的传播行为
    int getPropagationBehavior();
    //获取事务的隔离层次
    int getIsolationLevel();
    //判断事务是否超时
```

```
    int getTimeout();
    //判断是否为只读事务
    boolean isReadOnly();
    //获取事务的名称
    String getName();
}
```

Spring 在 TransactionDefinition 接口中定义了 7 种类型的事务传播行为，分别规定了事务方法发生嵌套调用时事务如何进行传播，具体如下。

(1) PROPAGATION_REQUIRED。若当前没有事务，则新建一个事务；若已存在一个事务，则加入该事务中。这种类型最为常用。

(2) PROPAGATION_SUPPORTS。支持当前事务。若当前没有事务，则以非事务方式执行。

(3) PROPAGATION_MANDATORY。使用当前的事务。若当前没有事务，则抛出异常。

(4) PROPAGATION_REQUIRES_NEW。新建事务。若当前存在事务，则将当前事务挂起。

(5) PROPAGATION_NOT_SUPPORTED。以非事务方式执行。若当前存在事务，则将当前事务挂起。

(6) PROPAGATION_NEVER。以非事务方式执行。若当前存在事务，则抛出异常。

(7) PROPAGATION_NESTED。若当前存在事务，则在嵌套事务内执行；若当前没有事务，则执行与 PROPAGATION_REQUIRED 类似的操作。使用这种类型时，底层的数据源必须基于 JDBC 3.0 或以上，同时实现者需支持保存点事务机制。

Spring 同时支持编程式事务策略与声明式事务策略，但在大多数情况下均采用声明式事务策略。与编程式事务管理相比，声明式事务管理无须在代码中关注事务逻辑，其优势是极其明显的。由于声明式事务管理无须与具体的事务逻辑耦合，因此可方便地在不同的事务逻辑之间切换。

声明式事务管理需要进行相应的配置，其配置方式通常有 4 种，即使用 TransactionProxyFactoryBean 创建事务代理、使用 Bean 继承配置事务代理、使用 BeanNameAutoProxyCreator 自动创建事务代理与使用 DefaultAdvisorAutoProxyCreator 自动创建事务代理。

8.6.1 使用 TransactionProxyFactoryBean 创建事务代理

使用 TransactionProxyFactoryBean 为目标 Bean 创建事务代理是最为传统的声明式事务管理配置方式，其配置文件最为臃肿。采用这种方式时，对于每一个 Bean 来说，均需完成两个相关的配置，分别为目标 Bean 与使用 TransactionProxyFactoryBean 配置的代理 Bean，因此配置文件的增长是非常快的。如以下示例：

```
<?xml version="1.0" encoding="UTF-8"?>
<beans …>
    …
    <!--定义事务管理器，在此使用适用于 Hibernate 的事务管理器-->
```

```
    <bean id="transactionManager" class="org.springframework.orm.
hibernate3.HibernateTransactionManager">
        <!--为 HibernateTransactionManager bean 注入 SessionFactory 引用-->
        <property name="sessionFactory">
            <ref local="sessionFactory"/>
        </property>
    </bean>
    <!--为 oneDao 配置事务-->
    <bean id="oneDao" class="org.springframework.transaction.
interceptor.TransactionProxyFactoryBean">
        <!--为事务代理 Bean 注入事务管理器-->
        <property name="transactionManager">
            <ref bean="transactionManager"/>
        </property>
        <!--设置事务属性-->
        <property name="transactionAttributes">
            <props>
                <!--所有以 save 开头的方法均采用 REQUIRED 的事务策略，且只读-->
                <prop key="save*">PROPAGATION_REQUIRED,readOnly</prop>
                <!--其他方法均采用 REQUIRED 的事务策略-->
                <prop key="*">PROPAGATION_REQUIRED</prop>
            </props>
        </property>
        <!--为事务代理 Bean 设置目标 Bean-->
        <property name="target">
            <!--采用嵌套 Bean 配置目标 Bean-->
            <bean class="org.etspace.abc.dao.impl.OneDao">
                <!--为 DAO bean 注入 SessionFactory 引用-->
                <property name="sessionFactory">
                    <ref local="sessionFactory"/>
                </property>
            </bean>
        </property>
    </bean>
    …
</beans>
```

在以上配置文件中，oneDAO 的配置包括两部分，分别为 oneDAO 的目标 Bean 与事务代理 Bean。其中，目标 Bean 被配置为嵌套 Bean，以免因直接暴露在 Spring 容器中而被错误引用。

8.6.2　使用 Bean 继承配置事务代理

通常情况下，各事务代理的事务属性都是大同小异的，事务代理的实现类均为 TransactionProxyFactoryBean，且事务代理 Bean 必须注入事务管理器。因此，使用 Spring 提供的 Bean 继承机制，可有效简化事务代理的配置。具体地说，可先创建一个事务模板，并在其中完成大部分通用的配置，然后通过继承事务模板的方式创建各个实际的事务代理 Bean，从而减掉了原来一些重复的配置代码。如以下示例：

```
<?xml version="1.0" encoding="UTF-8"?>
<beans …>
    …
    <!--定义事务管理器，使用适用于 Hibernate 的事务管理器-->
    <bean id="transactionManager"
        class="org.springframework.orm.hibernate3.HibernateTransactionManager">
        <!--为 HibernateTransactionManager bean 注入 SessionFactory 引用-->
        <property name="sessionFactory">
            <ref local="sessionFactory"/>
        </property>
    </bean>
    <!--配置事务模板，模板 Bean 被设置为抽象 Bean-->
    <bean id="baseTransaction"
    class="org.springframework.transaction.interceptor.TransactionProxyF
actoryBean"
        lazy-init="true" abstract="true">
        <!--为事务模板注入事务管理器-->
        <property name="transactionManager">
            <ref bean="transactionManager"/>
        </property>
        <!--设置事务属性-->
        <property name="transactionAttributes">
            <props>
                <prop key="save*">PROPAGATION_REQUIRED,readOnly</prop>
                <prop key="*">PROPAGATION_REQUIRED</prop>
            </props>
        </property>
    </bean>
    <!-- 实际的事务代理 Bean，继承自事务模板-->
    <bean id="oneDao" parent="baseTransaction">
        <!--采用嵌套 Bean 配置目标 Bean -->
        <property name="target">
            <bean class="org.etspace.abc.dao.impl.OneDao">
                <property name="sessionFactory">
                    <ref local="sessionFactory"/>
                </property>
            </bean>
        </property>
    </bean>
    …
</beans>
```

由以上配置文件可见，与直接采用 TransactionProxyFactoryBean 的事务代理配置方式相比，通过 Bean 继承的方式极大减少了相关配置的代码量。各个事务代理 Bean 都继承事务模板 Bean，无须重复指定事务代理的实现类，也无须重复指定事务的传播属性。更重要的是，若事务代理需要额外的事务属性，可以自行指定之。此时，事务代理 Bean(子 Bean)的属性覆盖事务模板 Bean(父 Bean)的属性。当然，对于各个事务代理 Bean 来说，仍需配置自己的目标 Bean。因此，这种事务代理的配置依然是增量式的(虽然增量已经减少，但所有事务代理都必须单独进行配置)。

8.6.3　使用 BeanNameAutoProxyCreator 自动创建事务代理

使用 BeanNameAutoProxyCreator 自动创建事务代理是一种优秀的事务代理配置策略，可完全避免增量式配置。一方面，所有的事务代理均由系统自动创建；另一方面，容器中的目标 Bean 自动消失，因此无须使用嵌套 Bean 来保证目标 Bean 不可被访问。如以下示例：

```
<?xml version="1.0" encoding="UTF-8"?>
<beans …>
    …
    <!--定义事务管理器，使用适用于 Hibernate 的事务管理器-->
    <bean id="transactionManager"
        class="org.springframework.orm.hibernate3.HibernateTransactionManager">
        <!--为 HibernateTransactionManager bean 注入 SessionFactory 引用-->
        <property name="sessionFactory">
            <ref local="sessionFactory"/>
        </property>
    </bean>
    <!--配置事务拦截器-->
    <bean id="transactionInterceptor"
        class="org.springframework.transaction.interceptor.
TransactionInterceptor">
        <!--为事务拦截器注入事务管理器 -->
        <property name="transactionManager">
            <ref bean="transactionManager"/>
        </property>
        <!--设置事务属性-->
        <property name="transactionAttributes">
            <props>
                <prop key="save*">PROPAGATION_REQUIRED,readOnly</prop>
                <prop key="*">PROPAGATION_REQUIRED</prop>
            </props>
        </property>
    </bean>
    <!--定义 BeanNameAutoProxyCreator，该 Bean 根据事务拦截器为目标 Bean 自动创建
事务代理。由于无须被引用，因此该 Bean 无 id 属性-->
    <bean class="org.springframework.aop.framework.autoproxy.
BeanNameAutoProxyCreator">
        <!--指定需要自动创建事务代理的 Bean 的名称(Name) -->
        <property name="beanNames">
            <list>
                <value>oneDao</value>
                <!--可在此处增加其他需要自动创建事务代理的 Bean-->
            </list>
        </property>
        <!--指定 BeanNameAutoProxyCreator 所需的事务拦截器-->
        <property name="interceptorNames">
            <list>
```

```
                <value>transactionInterceptor</value>
                <!--可在此处可增加其他的事务拦截器-->
            </list>
        </property>
    </bean>
    <!--定义 DAO Bean(由 BeanNameAutoProxyCreator 自动为其生成事务代理)-->
    <bean id="oneDao" class="org.etspace.abc.dao.impl.OneDao">
        <property name="sessionFactory">
            <ref local="sessionFactory"/>
        </property>
    </bean>
    …
</beans>
```

其中，TranscationInterceptor 为事务拦截器，BeanNameAutoProxyCreator 为事务代理创建器(可根据 Bean 的名称自动生成相应的事务代理)。

在配置 TranscationInterceptor 所对应的 Bean 时，需为其设置相应的 transactionManager 属性与 transactionAttributes 属性。其中，transactionManager 属性用于指定所使用的事务管理器 Bean(在此为 transactionManager)，transactionAttributes 属性用于指定事务拦截器的事务属性(在此利用 props 子元素定义了两个事务传播规则，即以 save 开头的方法采用 PROPAGATION_REQUIRED 事务传播规则，并且只读；其他方法采用 PROPAGATION_REQUIRED 事务传播规则)。

在配置 BeanNameAutoProxyCreator 所对应的 Bean 时，需为其设置相应的 beanNames 属性与 interceptorNames 属性。其中，beanNames 属性用于指定需要为其自动生成代理的 Bean(在此为 oneDao)，interceptorNames 属性用于指定所使用事务拦截器 Bean(在此为 transactionInterceptor)。

8.6.4 使用 DefaultAdvisorAutoProxyCreator 自动创建事务代理

与使用 BeanNameAutoProxyCreator 自动创建事务代理的方式相比，使用 DefaultAdvisorAutoProxyCreator 自动创建事务代理是一种更加简洁的事务代理配置方式。作为一种事务代理生成器，DefaultAdvisorAutoProxyCreator 自动搜索 Spring 容器中的 Advisor，并为容器中所有的 Bean 创建代理。如以下示例：

```
<?xml version="1.0" encoding="UTF-8"?>
<beans …>
    …
    <!--定义事务管理器，使用适用于 Hibernate 的事务管理器-->
    <bean id="transactionManager"

    class="org.springframework.orm.hibernate3.HibernateTransactionManager">
        <!--为 HibernateTransactionManager bean 注入 SessionFactory 引用-->
        <property name="sessionFactory">
            <ref local="sessionFactory"/>
        </property>
    </bean>
```

```
    <!--配置事务拦截器-->
    <bean id="transactionInterceptor"
        class="org.springframework.transaction.interceptor.
TransactionInterceptor">
        <!--为事务拦截器注入事务管理器 -->
        <property name="transactionManager">
            <ref bean="transactionManager"/>
        </property>
        <!--设置事务属性-->
        <property name="transactionAttributes">
            <props>
                <prop key="save*">PROPAGATION_REQUIRED,readOnly</prop>
                <prop key="*">PROPAGATION_REQUIRED</prop>
            </props>
        </property>
    </bean>
    <!--定义事务 Advisor-->
    <bean class="org.springframework.transaction.interceptor.
TransactionAttributeSourceAdvisor">
        <!--指定 Advisor 所需的事务拦截器-->
        <property name="transactionInterceptor">
            <ref  local="transactionInterceptor"/>
        </property>
    </bean>
    <!--DefaultAdvisorAutoProxyCreator 搜索容器中的 Advisor，并为每个 Bean 创建
代理 -->
    <bean
class="org.springframework.aop.framework.autoproxy.DefaultAdvisorAutoPro
xyCreator"/>
    <!--定义 DAO Bean(由 DefaultAdvisorAutoProxyCreator 自动为其生成事务代理)-
->
    <bean id="oneDao" class="org.etspace.abc.dao.impl.OneDao">
        <property name="sessionFactory">
            <ref local="sessionFactory"/>
        </property>
    </bean>
    …
</beans>
```

使用 DefaultAdvisorAutoProxyCreator 自动创建事务代理的配置方式，容器可自动为目标 Bean 生成事务代理，因此在增加新的目标 Bean 时无须添加任何额外的代码，从而使配置文件变得更加简单。

8.7　Spring 与 Struts 2 的整合应用

在实际的 Java EE 应用开发中，Spring 通常作为容器使用，因为 Spring 可将系统中的各种组件置于其中，并统一以 Bean 的方式进行管理。基于此思想，Struts 2 的各个 Action 实现类均可作为系统的组件全部交由 Spring 容器管理。图 8.15 所示即为 Java EE 应用开发

的 JSP+JDBC+JavaBean+Struts 2+Spring 模式，其中也表明了 Spring 框架与 Struts 2 框架的整合原理。在此模式中，系统前端各个 Action 模块均由 Spring 容器统一管理与部署，从而实现了 Struts 2 与 Action 之间的完全解耦。其实，Spring 与 Struts 2 整合应用的根本动机也正在于此。

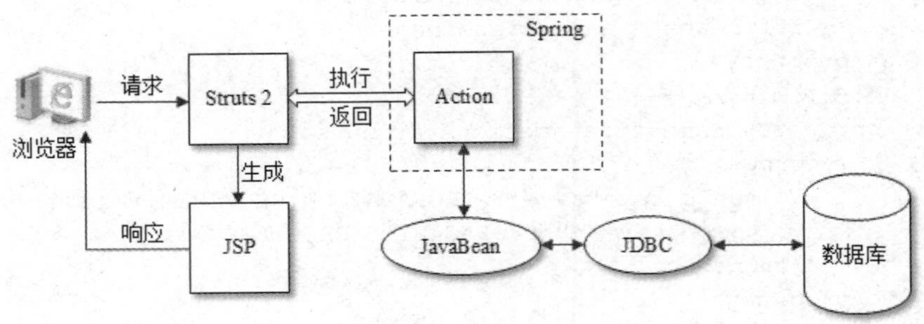

图 8.15　JSP+JDBC+JavaBean+Struts 2+Spring 模式

下面以 JSP+JDBC+JavaBean+Struts2+Spring 模式实现 Web 应用系统中的系统登录功能，其运行结果与第 6 章 6.9.1 小节的系统登录案例相同。

实现步骤(在项目 web_08 中)如下。

(1)　添加 Spring 开发能力，具体方法请参见本章的【实例 8-2】。

(2)　配置 Spring 框架的监听器。为此，只需对 Web 项目的配置文件 web.xml 进行相应的修改即可。具体代码如下：

```
<?xml version="1.0" encoding="UTF-8"?>
<web-app …>
   …
   <listener>
      <listener-class>
         org.springframework.web.context.ContextLoaderListener
      </listener-class>
   </listener>
   <context-param>
      <param-name>contextConfigLocation</param-name>
      <param-value>/WEB-INF/classes/applicationContext.xml</param-value>
   </context-param>
   …
</web-app>
```

在此，先通过 listener 元素指定 Spring 框架监听器的实现类，然后通过 context-param 元素指定配置文件及其具体位置。

(3)　加载 Struts 2 类库。为此，只需将 Struts 2 的 5 个基本类库、4 个附加类库与 1 个 Spring 支持包(struts2-spring-plugin.jar)复制到项目的 WebRoot\WEB-INF\lib 目录下即可。

注意： 在此一定要同时加载 Spring 支持包(struts2-spring-plugin.jar)。

(4)　配置 Struts 2 框架的核心过滤器。为此，只需对 Web 项目的配置文件 web.xml 进行相应的修改即可。具体代码如下：

```
<?xml version="1.0" encoding="UTF-8"?>
<web-app …>
    …
    <filter>
        <filter-name>struts2</filter-name>
        <filter-
class>org.apache.struts2.dispatcher.FilterDispatcher</filter-class>
    </filter>
    <filter-mapping>
        <filter-name>struts2</filter-name>
        <url-pattern>/*</url-pattern>
    </filter-mapping>
…
</web-app>
```

(5) 在项目的 src 文件夹下创建一个 properties 文件 struts.properties(该文件其实是一个文本文件)，并在其中将 Struts 2 的对象工厂指定为 Spring 容器，其代码如下：

```
struts.objectFactory=spring
```

(6) 将 SQL Server 2005/2008 的 JDBC 驱动程序 sqljdbc4.jar 添加到项目的 WebRoot\WEB-INF\lib 文件夹中。

(7) 在项目的 src 文件夹中新建一个包 org.etspace.abc.jdbc。

(8) 在 org.etspace.abc.jdbc 包中新建一个 JavaBean——DbBean，其文件名为 DbBean.java，代码与第 6 章 6.9.1 小节系统登录案例中的相同。

(9) 在项目的 WebRoot 文件夹中新建一个子文件夹 syslogin_struts_spring。

(10) 在子文件夹 syslogin_struts_spring 中添加一个新的 JSP 页面 login.jsp，其代码与第 6 章 6.9.1 小节系统登录案例中的相同。

(11) 在项目的 src 文件夹中新建一个包 org.etspace.abc.action。

(12) 在 org.etspace.abc.action 包中新建 Action 实现类 LoginAction，其文件名为 LoginAction.java，代码与第 6 章 6.9.1 小节系统登录案例中的相同。

(13) 在 Spring 的配置文件 applicationContext.xml 中添加一个 bean 元素，其实现类为 LoginAction，id 为“login”。有关代码如下：

```
<bean id="login" class="org.etspace.abc.action.LoginAction"></bean>
```

说明： Action 组件经过注册后，即可在运行时由 Spring 框架根据需要自动生成其实例。

(14) 在项目的 src 文件夹中创建配置文件 struts.xml，并在其中完成 LoginAction(Action)的配置，其代码如下：

```
<?xml version="1.0" encoding="UTF-8"?>
<!DOCTYPE struts PUBLIC
    "-//Apache Software Foundation//DTD Struts Configuration 2.0//EN"
    "http://struts.apache.org/dtds/struts-2.0.dtd">
<struts>
    <package name="login" extends="struts-default">
```

```
        <action name="login" class="login">
            <result
name="success">/syslogin_struts_spring/welcome.jsp</result>
            <result
name="error">/syslogin_struts_spring/error.jsp</result>
        </action>
    </package>
</struts>
```

在此，action 元素的 class 属性无须再指明相应 Action 实现类的全称类名，而只需设置为相应 Bean 的 id 即可。这样，即可将 Action 组件的管理统一交由 Spring 负责，从而实现了 Struts 2 与 Action 模块间的完全解耦。

(15) 在子文件夹 syslogin_struts 中添加一个新的 JSP 页面 welcome.jsp，其代码与第 6 章 6.9.1 小节系统登录案例中的相同。

(16) 在子文件夹 syslogin_struts 中添加一个新的 JSP 页面 error.jsp，其代码与第 6 章 6.9.1 小节系统登录案例中的相同。

Spring 与 Struts 2 整合应用的关键步骤主要如下。

(1) 添加 Struts 2 相关类库(包括 Spring 支持包 struts2-spring-plugin.jar)。
(2) 添加 Spring 开发能力。
(3) 创建 struts.properties 文件，并在其中指定 Spring 为容器。
(4) 修改 web.xml 文件，配置 Struts 2 框架的核心过滤器与 Spring 框架的监听器。
(5) 创建 Action 实现类，并在 Spring 的配置文件 applicationContext.xml 中进行注册。
(6) 创建 Struts 2 配置文件 struts.xml，并在其中完成相应 Action 的配置。
(7) 创建有关的 JSP 页面。

8.8 Spring 与 Hibernate 的整合应用

在 Java EE 应用开发中使用 Hibernate 框架时，一般是通过 DAO 接口访问数据库的。作为系统持久层的组件，各个 DAO 实现类以及 Hibernate 框架本身均可交由 Spring 容器管理。图 8.16 所示即为 Java EE 应用开发的 JSP+Spring+Hibernate+DAO 模式，其中也表明了 Spring 框架与 Hibernate 框架的整合原理。在此模式中，DAO 实现类与 Hibernate 框架一起置于 Spring 容器之中，而 Spring 容器就如同是一个工厂，向前台 JSP 程序提供了所需要的 DAO，同时管理着 Hibernate 对数据库的各种操作。其实，Spring 对 Hibernate 的管理正是通过其所提供的对数据源及 SessionFactory 的注入来实现的。

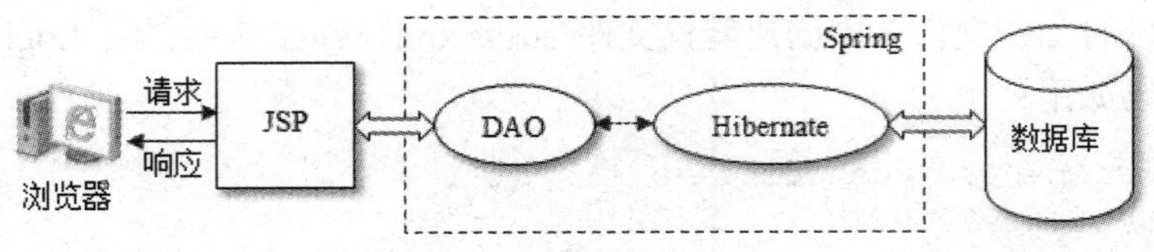

图 8.16 JSP+Spring+Hibernate+DAO 模式

下面以 JSP+Spring+Hibernate+DAO 模式实现 Web 应用系统中的系统登录功能，其运行结果与第 7 章 7.8.2 小节的系统登录案例相同。

实现步骤(在项目 web_08 中)如下。

(1)　添加 Spring 开发能力，具体方法请参见本章的【实例 8-2】。

(2)　配置 Spring 框架的监听器，具体方法请参见本章 8.7 节的系统登录案例。

(3)　在 MyEclipse 中创建对数据库(在此为 SQL Server 数据库 rsgl)的连接(在此将其命名为 rsgl)。其操作步骤与第 7 章【实例 7-1】中的相同。

(4)　添加 Hibernate 开发能力。为此，可按以下步骤进行操作。

①　右击项目名，并在其快捷菜单中选择 MyEclipse→Add Hibernate Capabilities...菜单项，打开图 8.17 所示的 Add Hibernate Capabilities 窗口。

②　在 Hibernate Specification 选项区选择相应的 Hibernate 框架版本(在此为 Hibernate 3.3)，并在其下面的列表框中选中需要的类库(在此为 Hibernate 3.3 Annotations & Entity Manager 与 Hibernate 3.3 Core Libraries)，然后单击 Next 按钮，界面切换至图 8.18 所示的定义 Hibernate 配置文件的 Hibernate Configuration 对话框(在此对话框，提示是使用 Hibernate 的配置文件还是使用 Spring 的配置文件进行 SessionFactory 的配置)。

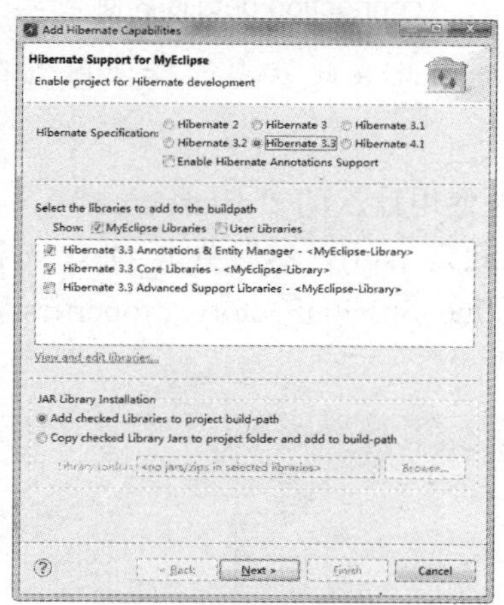
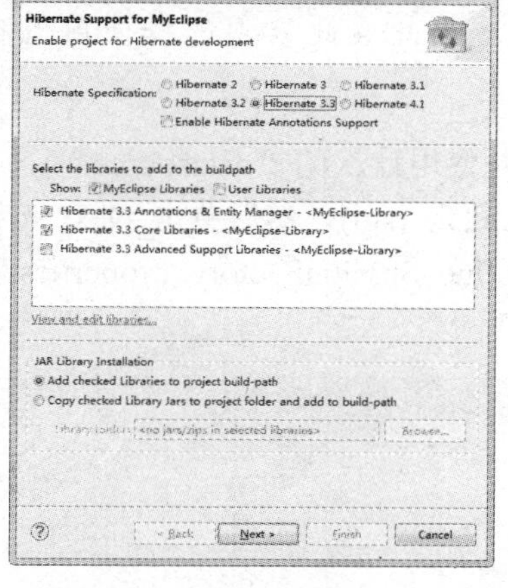

图 8.17　Add Hibernate Capabilities 对话框

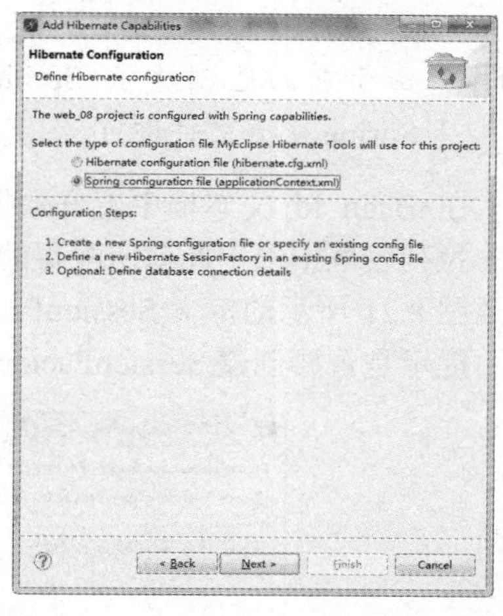

图 8.18　Hibernate Configuration 界面

③　选中 Spring configuration file (applicationContext.xml)单选按钮以使用 Spring 的配置文件，然后单击 Next 按钮，界面切换至图 8.19 所示的定义 Spring-Hibernate 配置文件的 Define Spring-Hibernate Configuration 对话框(在此对话框中，提示是创建一个新的配置文件还是使用已有的配置文件)。

说明： 在本项目中，使用 Spring 来对 Hibernate 进行管理。因此，无须再创建 Hibernate 的配置文件 hibernate.cfg.xml。

④　选中 Existing Spring configuration file 单选按钮以使用已存在的 Spring 配置文件，并在 Spring Config 下拉列表框中选择 src/applicationContext.xml 选项，在 SessionFactory Id 文本框中指定 Hibernate 的 SessionFactory 所对应的 Bean 的 Id(在此使用默认的 sessionFactory)，然后单击 Next 按钮，界面切换至图 8.20 所示的指定 Spring 数据源连接细

节的 Specify new Spring DataSource connection details(在此对话框中，要求设定具体的数据库连接信息)。

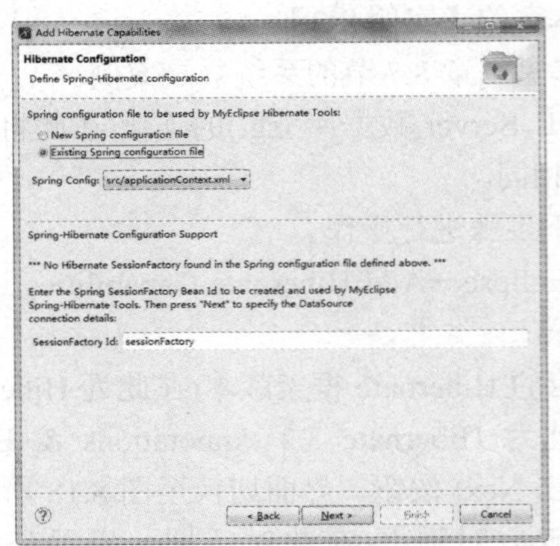

图 8.19　Define Spring-Hibernate configuration 界面

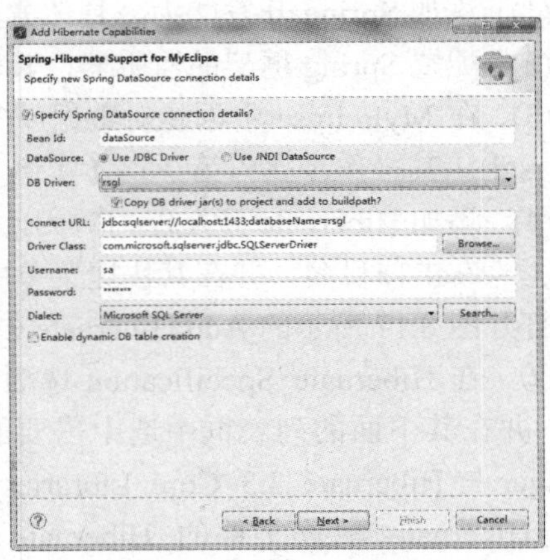

图 8.20　Specify new Spring DataSource connection details 界面

说明： 由于前面已创建了 Spring 的配置文件，因此在此只需选择使用已存在的 Spring 配置文件即可。

⑤ 在 Bean Id 文本框中置好数据源的名称(在此使用默认的“datasource”)，在 DB Driver 下拉列表框中选中此前所配置的数据库连接(在此为 rsgl)，然后单击 Next 按钮，界面切换至图 8.21 所示的定义 SessionFactory 属性的 Define SessionFactory properties(在此对话框中，提示是否要创建 SessionFactory 类)。

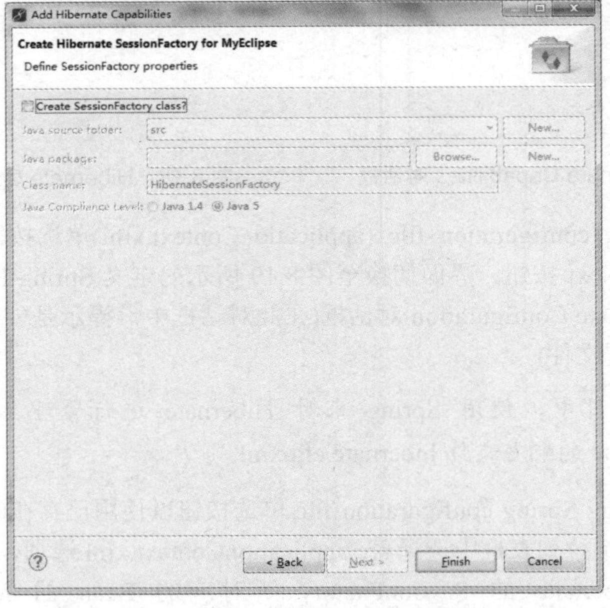

图 8.21　Define SessionFactory properties 界面

⑥ 取消选中 Create SessionFactory class? 复选框，然后单击 Finish 按钮关闭对话框。

说明： 在本项目中，由于 SessionFactory 实例的创建是通过 Spring 框架注入 id 为“sessionFactory”的 Bean 的实例来实现的，因此无须另外创建 SessionFactory 类。

至此，Hibernate 配置完毕。与此同时，项目中增加了一些 Hibernate 类库(Jar 包)，并在配置文件 applicationContext.xml 中添加了两个 id 分别为“dataSource”与“sessionFactory”的 bean 元素。此外，SQL Server 2005/2008 的 JDBC 驱动程序 sqljdbc4.jar 也会自动添加到项目的 WebRoot\WEB-INF\lib 文件夹中。

说明： 配置文件 applicationContext.xml 中 id 分别为“dataSource”与“sessionFactory”的 bean 元素的代码如下：

```
    <bean id="dataSource"
        class="org.apache.commons.dbcp.BasicDataSource">
        <property name="driverClassName"
            value="com.microsoft.sqlserver.jdbc.SQLServerDriver">
        </property>
        <property name="url"

    value="jdbc:sqlserver://localhost:1433;databaseName=rsgl">
        </property>
        <property name="username" value="sa"></property>
        <property name="password" value="abc123!"></property>
    </bean>
    <bean id="sessionFactory"

    class="org.springframework.orm.hibernate3.LocalSessionFactory
Bean">
        <property name="dataSource">
            <ref bean="dataSource" />
        </property>
        <property name="hibernateProperties">
            <props>
                <prop key="hibernate.dialect">
                    org.hibernate.dialect.SQLServerDialect
                </prop>
            </props>
        </property>
    </bean>
```

(5) 在项目的 src 文件夹中新建一个包 org.etspace.abc.vo(该包用来存放与数据库表相对应的 Java 类或 POJO)。

(6) 在 org.etspace.abc.vo 包中生成与数据库表 users 相对应的 Java 类 Users(其文件名为 Users.java)与映射文件 Users.hbm.xml。其操作步骤与第 7 章【实例 7-1】中的类似，操作完毕后会自动在配置文件 applicationContext.xml 的 id 为“sessionFactory”的 bean 元素中完成相应的映射文件配置。有关代码如下：

```
<bean id="sessionFactory"
    class="org.springframework.orm.hibernate3.LocalSessionFactoryBean">
    <property name="dataSource">
        <ref bean="dataSource" />
    </property>
    <property name="hibernateProperties">
        <props>
            <prop key="hibernate.dialect">
                org.hibernate.dialect.SQLServerDialect
            </prop>
        </props>
    </property>
    <property name="mappingResources">
        <list>
            <value>org/etspace/abc/vo/Users.hbm.xml</value></list>
    </property>
</bean>
```

(7) 在项目的 WebRoot 文件夹中新建一个子文件夹 syslogin_spring_hibernate_dao。

(8) 在子文件夹 syslogin_spring_hibernate_dao 中添加一个新的 JSP 页面 login.jsp，其代码与第 7 章 7.8.2 小节系统登录案例中的相同。

(9) 在项目的 src 文件夹中新建一个包 org.etspace.abc.dao(该包用于存放有关的 DAO 接口)。

(10) 在 org.etspace.abc.dao 包中新建一个接口 IUsersDAO，其文件名为 IUsersDAO.java，代码与第 7 章 7.8.2 小节系统登录案例中的相同。

(11) 在项目的 src 文件夹中新建一个包 org.etspace.abc.dao.impl(该包用于存放有关 DAO 接口的实现类及其父类)。

(12) 在 org.etspace.abc.dao.impl 包中新建一个接口实现类的父类 BaseDAO，其文件名为 BaseDAO.java，代码如下：

```
package org.etspace.abc.dao.impl;
import org.hibernate.SessionFactory;
import org.hibernate.Session;
public class BaseDAO {
    private SessionFactory sessionFactory;
    public SessionFactory getSessionFactory(){
        return sessionFactory;
    }
    public void setSessionFactory(SessionFactory sessionFactory){
        this.sessionFactory=sessionFactory;
    }
    public Session getSession(){
        Session session=sessionFactory.openSession();
        return session;
    }
}
```

(13) 在 org.etspace.abc.dao.impl 包中新建一个 IUsersDAO 接口的实现类 UsersDAO，其文件名为 UsersDAO.java，代码如下：

```
package org.etspace.abc.dao.impl;
import org.etspace.abc.dao.IUsersDAO;
import org.etspace.abc.vo.Users;
import java.util.*;
import org.hibernate.*;
public class UsersDAO extends BaseDAO implements IUsersDAO {
    public Users validateUsers(String username, String password) {
        Session mySession=getSession();
        Query query=mySession.createQuery("from Users where username=?
and password=?");
        query.setParameter(0, username);
        query.setParameter(1, password);
        List list=query.list();
        if(list.size()!=0)
        {
            Users users=(Users)list.get(0);
            return users;
        }
        mySession.close();
        return null;
    }
}
```

(14) 在 Spring 的配置文件 applicationContext.xml 中添加两个 bean 元素，其实现类分别为 BaseDAO 与 UsersDAO，id 分别为“baseDAO”与“usersDAO”。有关代码如下：

```
<bean id="baseDAO" class="org.etspace.abc.dao.impl.BaseDAO">
    <property name="sessionFactory">
        <ref bean="sessionFactory"></ref>
    </property>
</bean>
<bean id="usersDAO" class="org.etspace.abc.dao.impl.UsersDAO"
parent="baseDAO"></bean>
```

(15) 在子文件夹 syslogin_spring_hibernate_dao 中添加一个新的 JSP 页面 validate.jsp。其代码如下：

```
<%@ page language="java" pageEncoding="utf-8" %>
<%@ page import="org.etspace.abc.dao.*,org.etspace.abc.dao.impl.*" %>
<%@ page import="org.springframework.context.*,org.springframework.
context.support.*" %>
<%request.setCharacterEncoding("utf-8"); %>
<html>
    <head>
        <title>验证页面</title>
        <meta http-equiv="Content-Type" content="text/html;charset=utf-8">
    </head>
```

```
    <body>
        <%
        String username=request.getParameter("username");
        String password=request.getParameter("password");
        boolean validated=false;  //验证标识
        ApplicationContext myApplicationContext=new
ClassPathXmlApplicationContext("applicationContext.xml");
        IUsersDAO usersDAO=(IUsersDAO)myApplicationContext.getBean
("usersDAO");
        if(usersDAO.validateUsers(username, password)!=null)
        {
            validated=true;
        }

        if(validated)  //验证成功跳转到成功页面
        {
        %>
            <jsp:forward page="welcome.jsp"></jsp:forward>
        <%
        }
        else  //验证失败跳转到失败页面
        {
        %>
            <jsp:forward page="error.jsp"></jsp:forward>
        <%
        }
        %>
    </body>
</html>
```

(16) 在子文件夹 syslogin_hibernate_dao 中添加一个新的 JSP 页面 welcome.jsp，其代码与第 7 章 7.8.2 小节系统登录案例中的相同。

(17) 在子文件夹 syslogin_hibernate_dao 中添加一个新的 JSP 页面 error.jsp，其代码与第 7 章 7.8.2 小节系统登录案例中的相同。

Spring 与 Hibernate 整合应用的关键步骤如下。

(1) 添加 Spring 开发能力。

(2) 修改 web.xml 文件，配置 Spring 框架的监听器。

(3) 添加 Hibernate 开发能力。

(4) 创建表的 POJO 类及映射文件。

(5) 创建 DAO 接口，编写其实现类，并在 Spring 的配置文件 applicationContext.xml 中进行注册。

(6) 创建有关的 JSP 页面。

8.9　Spring 与 Struts 2、Hibernate 的整合应用

在实际的 Java EE 应用开发中，通常会将 Struts 2、Spring、Hibernate 这三个框架完全整合起来，以充分发挥它们各自的长处，从而以最优的方式实现系统的各项功能。其基本思想是将 Spring 作为一个统一的大容器使用，以容纳系统中的各种组件，包括 Action 实现类、DAO 实现类与 Hibernate 框架等。图 8.22 所示即为 Java EE 应用开发的 JSP+Struts2+Spring+Hibernate+DAO 模式(通常简称为 SSH2 模式)，其中也表明了 Struts 2、Spring、Hibernate 三个框架的整合原理。在此模式中，Struts 2 将原来包含在 JSP 中的控制分离出来，并通过调用 Spring 容器中的 Action 组件来完成控制逻辑的具体处理。在此过程中，若 Action 组件需要访问数据库，则会调用 DAO 组件所提供的有关方法。而 DAO 组件对数据库的各种操作，最终是通过 Hibernate 框架实现的。可见，整个系统是以 Spring 为核心的，所有的系统组件(包括 Hibernate 框架)均由 Spring 容器统一进行管理。

下面以 JSP+Struts2+Spring+Hibernate+DAO 模式实现 Web 应用系统中的系统登录功能，其运行结果与第 6 章 6.9.1 小节(或本章 8.7 节)系统登录案例中的相同。

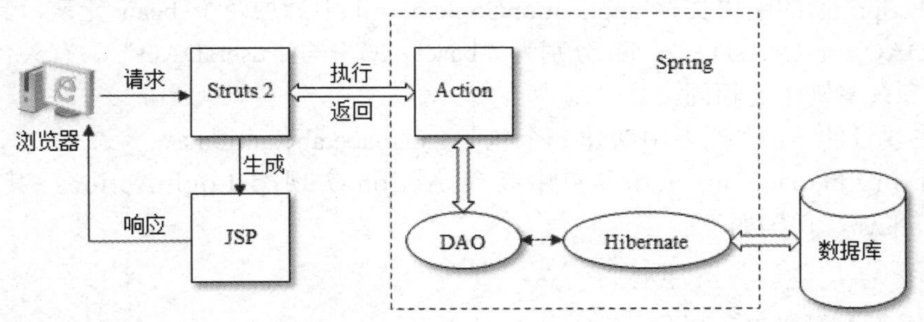

图 8.22　JSP+Struts 2+Spring+Hibernate+DAO 模式(SSH2 模式)

实现步骤(在项目 web_08 中)如下。

(1)　添加 Spring 开发能力，具体方法请参见本章的【实例 8-2】。

(2)　配置 Spring 框架的监听器，具体方法请参见本章 8.7 节的系统登录案例。

(3)　加载 Struts 2 类库，包括 Struts 2 的 5 个基本类库、4 个附加类库与 1 个 Spring 支持包(struts2-spring-plugin.jar)，具体方法请参见本章 8.7 节的系统登录案例。

(4)　配置 Struts 2 框架的核心过滤器，具体方法请参见本章 8.7 节的系统登录案例。

(5)　在项目的 src 文件夹下创建一个 properties 文件 struts.properties，并在其中将 Struts 2 的对象工厂指定为 Spring 容器，具体方法请参见本章 8.7 节的系统登录案例。

(6)　在 MyEclipse 中创建对数据库(在此为 SQL Server 数据库 rsgl)的连接(在此将其命名为 rsgl)，具体方法请参见本章 8.8 节的系统登录案例。

(7)　添加 Hibernate 开发能力，具体方法请参见本章 8.8 节的系统登录案例。

(8)　在项目的 src 文件夹中新建一个包 org.etspace.abc.vo(该包用来存放与数据库表相对应的 Java 类或 POJO)。

(9)　在 org.etspace.abc.vo 包中生成与数据库表 users 相对应的 Java 类 Users(其文件名

为 Users.java)与映射文件 Users.hbm.xml，具体方法请参见本章 8.8 节的系统登录案例。

(10) 在项目的 WebRoot 文件夹中新建一个子文件夹 syslogin_struts_spring_hibernate_dao。

(11) 在子文件夹 syslogin_struts_spring_hibernate_dao 中添加一个新的 JSP 页面 login.jsp，其代码与第 6 章 6.9.1 小节(或本章 8.7 节)系统登录案例中的相同。

(12) 在项目的 src 文件夹中新建一个包 org.etspace.abc.dao(该包用于存放有关的 DAO 接口)。

(13) 在 org.etspace.abc.dao 包中新建一个接口 IUsersDAO，其文件名为 IUsersDAO.java，代码与第 7 章 7.8.2 节(或本章 8.8 节)系统登录案例中的相同。

(14) 在项目的 src 文件夹中新建一个包 org.etspace.abc.dao.impl(该包用于存放有关 DAO 接口的实现类及其父类)。

(15) 在 org.etspace.abc.dao.impl 包中新建一个接口实现类的父类 BaseDAO，其文件名为 BaseDAO.java，代码与本章 8.8 节系统登录案例中的相同。

(16) 在 org.etspace.abc.dao.impl 包中新建一个 IUsersDAO 接口的实现类 UsersDAO，其文件名为 UsersDAO.java，代码与本章 8.8 节系统登录案例中的相同。

(17) 在 Spring 的配置文件 applicationContext.xml 中添加两个 bean 元素，其实现类分别为 BaseDAO 与 UsersDAO，id 分别为“baseDAO”与“usersDAO”。有关代码与本章 8.8 节系统登录案例中的相同。

(18) 在项目的 src 文件夹中新建一个包 org.etspace.abc.action。

(19) 在 org.etspace.abc.action 包中新建 Action 实现类 LoginAction，其文件名为 LoginAction.java，代码如下：

```
package org.etspace.abc.action;
import org.etspace.abc.dao.*;
import org.etspace.abc.dao.impl.*;
import org.springframework.context.*;
import org.springframework.context.support.*;
import com.opensymphony.xwork2.ActionSupport;
public class LoginAction extends ActionSupport{
    private String username;
    private String password;
    public String getUsername(){
        return username;
    }
    public void setUsername(String username){
        this.username=username;
    }
    public String getPassword(){
        return password;
    }
    public void setPassword(String password){
        this.password=password;
    }
```

```
    public String execute() throws Exception{
        boolean validated=false;  //验证标识
        ApplicationContext myApplicationContext=new
ClassPathXmlApplicationContext("applicationContext.xml");
        IUsersDAO usersDAO=(IUsersDAO)myApplicationContext.getBean
("usersDAO");
        if(usersDAO.validateUsers(username, password)!=null)
        {
            validated=true;
        }
        if(validated)
        {
            //验证成功返回"success"
            return "success";
        }
        else
        {
            //验证失败返回"error"
            return "error";
        }
    }
}
```

(20) 在 Spring 的配置文件 applicationContext.xml 中添加一个 bean 元素，其实现类为 LoginAction，id 为“login”。有关代码与本章 8.7 节系统登录案例中的相同。

(21) 在项目的 src 文件夹中创建配置文件 struts.xml，并在其中完成 LoginAction(Action)的配置，其代码如下：

```
<?xml version="1.0" encoding="UTF-8"?>
<!DOCTYPE struts PUBLIC
    "-//Apache Software Foundation//DTD Struts Configuration 2.0//EN"
    "http://struts.apache.org/dtds/struts-2.0.dtd">
<struts>
    <package name="login" extends="struts-default">
        <action name="login" class="login">
            <result
name="success">/syslogin_struts_spring_hibernate_dao/welcome.jsp</result
>
            <result
name="error">/syslogin_struts_spring_hibernate_dao/error.jsp</result>
        </action>
    </package>
</struts>
```

(22) 在子文件夹 syslogin_struts_spring_hibernate_dao 中添加一个新的 JSP 页面 welcome.jsp，其代码与第 6 章 6.9.1 小节(或本章 8.7 节)系统登录案例中的相同。

(23) 在子文件夹 syslogin_struts_spring_hibernate_dao 中添加一个新的 JSP 页面 error.jsp，其代码与第 6 章 6.9.1 小节(或本章 8.7 节)系统登录案例中的相同。

Spring 的 Hibernate ORM 框架提供了一个 HibernateDaoSupport 类，该类提供了一个 getHibernateTemplate()方法。若 DAO 接口的实现类继承自 HibernateDaoSupport 类，则可通过调用 getHibernateTemplate()方法返回的实例方便地完成各种持久化操作(或与数据库访问有关的操作)。另外，基于 HibernateDaoSupport 类的 DAO 组件必须获得一个 SessionFactory 的引用，方可完成相应的持久化操作。因此，在将其作为 Bean 进行配置时，应注意为其 sessionFactory 属性的注入设置好相应的 SessionFactory 引用。

例如，在本案例中，可弃用各接口实现类的父类 BaseDAO，并将 IUsersDAO 接口的实现类 UsersDAO 的代码改写为：

```
package org.etspace.abc.dao.impl;
import org.springframework.orm.hibernate3.support.HibernateDaoSupport;
import org.etspace.abc.dao.IUsersDAO;
import org.etspace.abc.vo.Users;
import java.util.*;
public class UsersDAO extends HibernateDaoSupport implements IUsersDAO {
    public Users validateUsers(String username, String password) {
        List list=getHibernateTemplate().find("from Users where
username=? and password=?", username, password);
        if(list.size()!=0)
        {
            Users users=(Users)list.get(0);
            return users;
        }
        return null;
    }
}
```

与此同时，应将配置文件 applicationContext.xml 中 BaseDAO 类所对应的 bean 元素删除，并将 UsersDAO 类所对应的 bean 元素的代码修改为：

```
<bean id="usersDAO" class="org.etspace.abc.dao.impl.UsersDAO">
    <property name="sessionFactory">
        <ref bean="sessionFactory"></ref>
    </property>
</bean>
```

Spring 与 Struts 2、Hibernate 整合应用的关键步骤主要如下。

(1) 添加 Struts 2 相关类库(包括 Spring 支持包 struts2-spring-plugin.jar)。

(2) 添加 Spring 开发能力。

(3) 创建 struts.properties 文件，并在其中指定 Spring 为容器。

(4) 修改 web.xml 文件，配置 Struts 2 框架的核心过滤器与 Spring 框架的监听器。

(5) 添加 Hibernate 开发能力。

(6) 创建表的 POJO 类及映射文件。

(7) 创建 DAO 接口，编写其实现类，并在 Spring 的配置文件 applicationContext.xml 进行注册。

(8) 创建 Action 实现类，并在 Spring 的配置文件 applicationContext.xml 中进行注册。

(9) 创建 Struts 2 配置文件 struts.xml，并在其中完成相应 Action 的配置。

(10) 创建有关的 JSP 页面。

本章小结

本章首先介绍了 Spring 框架的分层架构，然后通过具体实例讲解了 Spring 的基本应用技术、Spring 关键配置的设置方法、Spring 核心接口的基本用法、Spring AOP 的基本应用技术以及 Spring 事务支持的主要配置方法，最后再通过具体案例说明了 Spring 的综合应用技术。通过本章的学习，读者应熟练掌握 Spring 框架的相关应用技术，并能使用 JSP+JDBC+JavaBean+Struts 2+Spring 模式、JSP+Spring+Hibernate+DAO 模式与 JSP+Struts 2+Spring+Hibernate+DAO 模式(即 SSH2 模式)开发相应的 Web 应用系统。

思考题

1. 请简述 Spring 框架的分层架构。
2. 如何在 Web 项目中添加 Spring 开发能力？
3. Spring 的依赖注入方式主要有哪两种？其设计要点是什么？
4. Spring 默认的核心配置文件是什么？其主要作用是什么？
5. 如何完成 Bean 的基本定义？
6. 如何进行 Bean 的依赖配置？
7. 如何进行 Bean 的别名设置？
8. 如何进行 Bean 的作用域设置？
9. 如何进行 Bean 的生命周期方法设置？
10. Spring 的核心接口有哪几个？请简述之。
11. ApplicationContext 的常用实现有哪些？
12. 何为 AOP？其目的是什么？
13. AOP 的实现机制是什么？
14. Spring AOP 的增强(Advice)类型主要有哪些？
15. Spring 的声明式事务管理配置方式有哪几种？
16. 请简述 Spring 与 Struts 2 框架的整合方法。
17. 请简述 Spring 与 Hibernate 框架的整合方法。
18. 请简述 Spring 与 Struts 2、Hibernate 框架的整合方法。

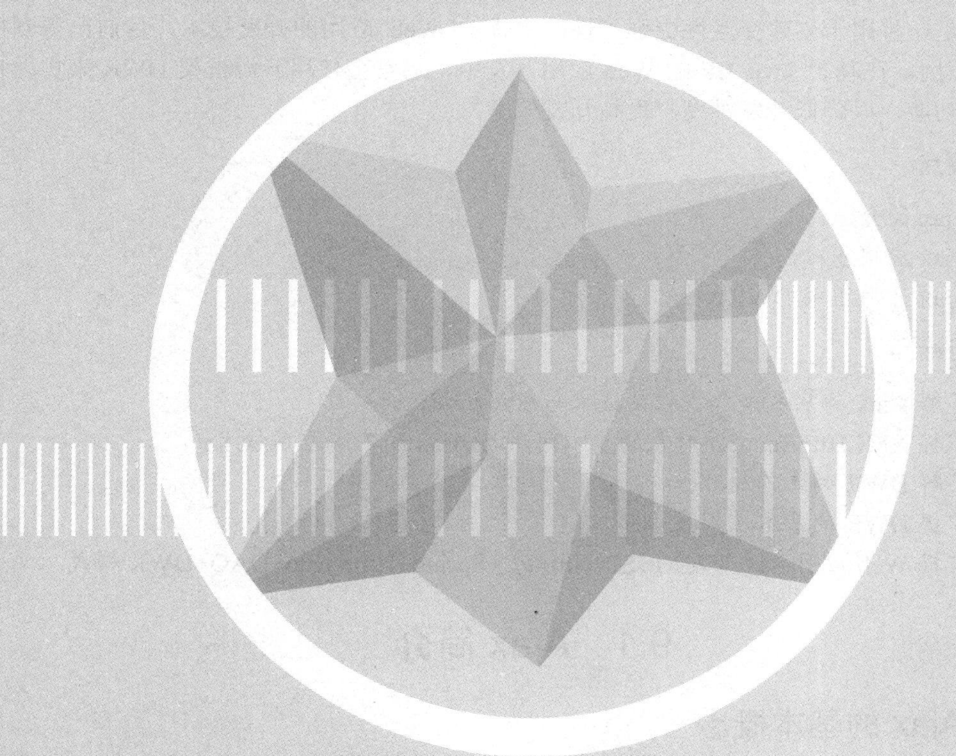

第 9 章

Ajax 应用

Ajax 是一种用于创建交互性更强、性能更优的 Web 应用的开发技术，目前已得到极其广泛的应用。在基于 Java EE 的 Web 应用开发中，通常会使用开源框架 DWR 来实现相应的 Ajax 功能，以降低开发难度，提高开发效率。

本章要点：

- Ajax 简介；
- Ajax 应用基础；
- Ajax 开源框架 DWR。

学习目标：

- 了解 Ajax 的基本概念、相关技术与应用场景；
- 掌握 XMLHttpRequest 对象的基本用法与 Ajax 的基本应用技术；
- 了解 DWR 的工作原理；
- 掌握 DWR 的基本应用技术；
- 掌握 Web 应用系统开发的 JSP+Struts 2+Spring+Hibernate+DAO+DWR 模式。

9.1 Ajax 简介

9.1.1 Ajax 的基本概念

Ajax(Asynchronous JavaScript and XML，异步 JavaScript 和 XML)，由 Jesse James Garrett 所创造，指的是一种创建交互式网页应用的开发技术。Ajax 经 Google 公司大力推广后已成为一种炙手可热的流行技术，而 Google 公司所发布的 Gmail、Google Suggest 等应用也最终让人们体验了 Ajax 的独特魅力。

Ajax 的核心理念是使用 XMLHttpRequest 对象发送异步请求，最初为 XMLHttpRequest 对象提供浏览器支持的是微软公司。1998 年，微软公司在开发 Web 版的 Outlook 时，即以 ActiveX 控件的方式为 XMLHttpRequest 对象提供了相应的支持。

实际上，Ajax 并非一种全新的技术，而是多种技术的相互融合。Ajax 所包含的各种技术均有其独到之处，相互融合在一起便成为一种功能强大的新技术。

Ajax 的相关技术主要包括：

(1) HTML/XHTML，实现页面内容的表示；

(2) CSS，格式化页面内容；

(3) DOM，对页面内容进行动态更新；

(4) XML，实现数据交换与格式转换；

(5) XMLHttpRequest 对象，实现与服务器的异步通信；

(6) JavaScript，实现各种技术的融合。

9.1.2 Ajax 的应用场景

众所周知，浏览器默认使用同步方式发送请求并等待响应。在 Web 应用中，请求的发送是通过浏览器进行的，在同步方式下，用户通过浏览器发出请求后，就只能等待服务器的响应；而在服务器返回响应之前，用户将无法执行任何进一步的操作，只能空等。反之，如果将请求与响应改为异步方式(即非同步方式)，那么在请求发送后，浏览器就无须空等服务器的响应，而是让用户继续对其中的 Web 应用程序进行其他操作。当服务器处理请求并返回响应时，再告知浏览器按程序所设定的方式进行相应的处理。可见，与同步方式相比，异步方式的运行效率更高，而且用户的体验也更佳。

Ajax 技术的出现为异步请求的发送带来了福音，并有效地降低了相关应用的开发难度。Ajax 具有异步交互的特点，可实现 Web 页面的动态更新，因此特别适用于交互较多、数据读取较为频繁的 Web 应用。下面仅列举几个典型的 Ajax 应用场景。

1. 验证表单数据

对于表单中所输入的数据，通常需要对其有效性进行验证。例如，在注册新用户时，对于所输入的用户名，往往要验证其唯一性。传统的验证方式是先提交表单，然后再对数据进行验证。这种验证方式需将整个表单页面提交到服务器端，不仅验证时间长，而且会给服务器造成不必要的负担。另外一种稍加改进的验证方式是让用户单击专门提供的验证按钮以启动验证过程，然后再通过相应的浏览器窗口查看验证结果。由于这种方式需要设计专门的验证页面，同时还需要另外打开浏览器窗口，不仅增加了工作量，而且会使系统显示更加臃肿，在运行时也会耗费更多的系统资源。

若应用 Ajax 技术，则可以有效地解决此类问题。此时，可由 XMLHttpRequest 对象发出验证请求，并根据返回的 HTTP 响应判断验证是否成功。在此期间，无须将整个页面提交到服务器，也不用打开新的窗口以显示验证结果，既可快速完成验证过程，也不会加重服务器的负担。

2. 按需获取数据

在 Web 应用系统中，分类树(或树形结构)的使用较为普遍，主要用于分类显示有关的数据。在传统模式下，对于分类树的每一次操作，若采用调用后台以获取相关数据的方式，则必然会引起整个页面的刷新，而用户也必须为此等待一段时间。为解决此方法所带来的响应速度慢、需刷新整个页面的问题，并避免频繁向服务器发送请求，可采取另外一种方式，即一次性获取分类结构中的所有数据，并将其存入数组，然后再根据用户的操作需求，使用 JavaScript 脚本来控制有关节点的呈现。不过，在这种情况下，如果用户不对分类树进行操作，或者只对分类树中的部分数据进行操作，那么所获取的所有数据或剩余数据就会成为垃圾资源。如果分类结构较为复杂，而且各类数据量均较为庞大，那么这种方式的弊端将会更加明显。

Ajax 技术的出现为分类树的实现提供了一种全新的机制。在初始化页面时，只需获取分类树的一级分类数据并显示，当用户单击分类树的某个一级分类节点时，则通过 Ajax 向服务器请求当前一级分类所属的二级分类数据并显示；若继续单击已呈现的某个二级分类节点，则再次通过 Ajax 向服务器请求当前二级分类所属的三级分类数据并显示，以此类推。这样，一方面，当前页面只需根据用户的操作向服务器请求所需要的数据，从而有效地减少了数据的加载量。另一方面，在更新当前页面时，也无须刷新整个页面，而只需刷新页面中需要更新其内容的那部分区域即可。其实，这就是所谓的页面的局部刷新。

3. 自动更新页面

在 Web 应用中，有些数据的变化是较为频繁的，如股市数据、天气预报等。在传统方式下，为及时了解有关数据的变化，用户必须不断地手动刷新页面，或者让页面本身具有定时刷新功能。这种做法虽然可以达到目的，但也具有明显缺陷。例如，若某段时间数据并无变化，但用户并不知道，而仍然不断地刷新页面，从而做了过多的无用操作。又如，若某段时间数据变化较为频繁，但用户并没有及时刷新页面，从而错失获取数据变化的机会。

对于此类问题，可应用 Ajax 技术妥善解决。页面加载以后，通过 Ajax 引擎在后台定时向服务器发送请求，查看是否有最新的消息。如果有，则加载新的数据，并且在页面上进行局部的动态更新，然后通过一定的方式通知用户。这样，既避免了用户不断手动刷新页面的不便，也不会在页面定时重复刷新时造成资源浪费。

9.2 Ajax 应用基础

Ajax 应用程序必须是由客户端与服务器一同合作的应用程序。JavaScript 是编写 Ajax 应用程序的客户端语言，而 XML 则是请求或响应时建议使用的信息交换的格式。

9.2.1 XMLHttpRequest 对象简介

Ajax 的核心为 XMLHttpRequest 组件，该组件在 Firefox、NetScape、Safari、Opera 中称为 XMLHttpRequest，在 IE(Internet Explorer)中则是称为 Microsoft XMLHTTP 或 Msxml2.XMLHTTP 的 ActiveX 组件(但在 IE 7 中已更名为 XMLHttpRequest)。

XMLHttpRequest 组件的对象(或实例)可通过 JavaScript 创建。XMLHttpRequest 对象提供客户端与 HTTP 服务器进行异步通信的协议，通过该协议，Ajax 可以使页面像桌面应用程序一样，只同服务器进行数据层的交换，而不用每次都刷新界面，也不用每次都将数据处理工作提交给服务器来完成。这样，既减轻了服务器的负担，又加快了响应的速度、缩短了用户等候的时间。

在 Ajax 应用程序中，若使用的浏览器为 Mozilla、Firefox 或 Safari，则可通过 XMLHttpRequest 对象来发送非同步请求；若使用的浏览器为 IE6 或之前的版本，则应使用 ActiveXObject 对象来发送非同步请求。因此，为兼容各种不同的浏览器，必须先进行测试，以正确创建 XMLHttpRequest 对象(即获取 XMLHttpRequest 或 ActiveXObject 对象)。例如：

```
var xmlHttp;
if(window.ActiveXObject){
    xmlHttp = new ActiveXObject("Microsoft.XMLHTTP");
}
else if(window.XMLHttpRequest){
    xmlHttp = new XMLHttpRequest();
}
```

创建了 XMLHttpRequest 对象后，为实现相应的 Ajax 的功能，可在 JavaScript 脚本中调用 XMLHttpRequest 对象的有关方法(见表 9.1)，或访问 XMLHttpRequest 对象的有关属性(见表 9.2)。

表 9.1　XMLHttpRequest 对象的方法

方　法	说　明
void open("method", "url" [,asyncFlag [,"userName" [, "password"]]])	建立与服务器的连接。method 参数可以是 GET、POST 或 PUT，url 参数可以是相对或绝对 URL，可选参数 asyncFlag、Username、password 分别为是否非同步标记、用户名、密码
void send(content)	向服务器发送请求
void setRequestHeader("header","value")	设置指定标头的值(在调用该方法之前必须先调用 open 方法)
void abort()	停止当前请求
string getAllResponseHeaders()	获取响应的所有标头(键/值对)
string getResponseHeader("header")	获取响应中指定的标头

表 9.2　XMLHttpRequest 对象的属性

属　性	说　明
onreadystatechange	状态改变事件触发器(每个状态的改变都会触发该事件触发器)
readyState	对象状态，包括：0＝ 未初始化；1＝ 正在加载；2＝ 已加载；3＝ 交互中；4＝ 已完成
responseText	服务器的响应(字符串)
responseXML	服务器的响应(XML)。该对象可以解析为一个 DOM 对象
status	服务器返回的 HTTP 状态码
statusText	HTTP 状态码的相应文本

9.2.2　Ajax 的请求与响应过程

Ajax 利用浏览器中网页的 JavaScript 脚本程序来完成数据的提交或请求，并将 Web 服务器响应后返回的数据由 JavaScript 脚本程序处理后呈现到页面上。Ajax 的请求与响应过程如图 9.1 所示，大致可分为 5 个基本步骤。

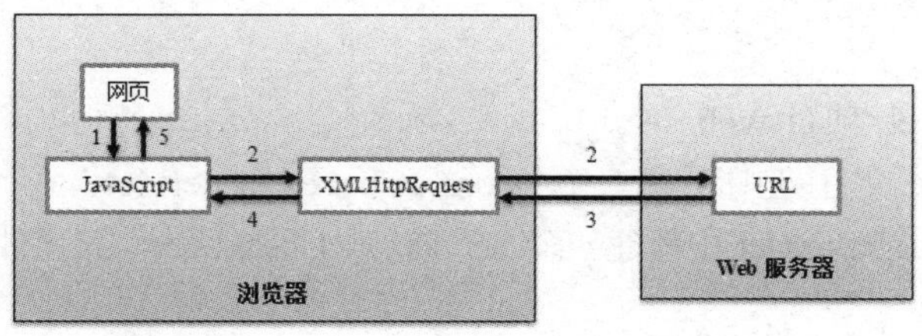

图 9.1　Ajax 的请求与响应过程

(1) 网页调用 JavaScript 脚本程序。

(2) JavaScript 利用浏览器提供的 XMLHttpRequest 对象向 Web 服务器发送请求。

(3) Web 服务器接受请求并由指定的 URL 处理后返回相应的结果给浏览器的 XMLHttpRequest 对象。

(4) XMLHttpRequest 对象调用指定的 JavaScript 处理方法。

(5) 被调用的 JavaScript 处理方法解析返回的数据并更新当前页面。

9.2.3 Ajax 的基本应用

下面通过一个简单的应用实例说明 Ajax 的基本应用技术。

【实例 9-1】 Ajax 应用实例：用户名验证。图 9.2 所示为“用户注册”页面。在其中的“用户名”文本框中输入用户名后再让其失去输入焦点，即可立即验证该用户名的唯一性，并显示相应的结果。若所输入的用户名已经被注册了，则显示图 9.3(a)所示的“对不起，该用户名已存在！”的信息，否则显示图 9.3(b)所示“恭喜你，该用户名可使用！”的信息。

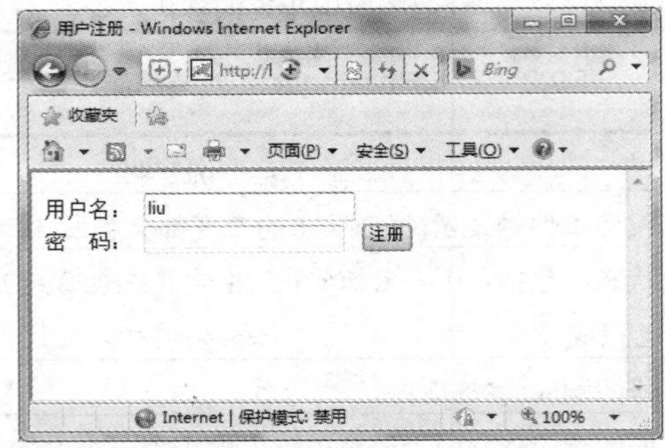

图 9.2　“用户注册”页面

(a)　　　　(b)

图 9.3　验证结果页面

主要步骤如下。

(1) 新建一个 Web 项目 web_09。

(2) 在项目的 src 文件夹中新建一个包 org.etspace.abc.servlet。

(3) 在 org.etspace.abc.servlet 中新建一个 Servlet——CheckUser，其文件名为 CheckUser.java，代码如下：

```
package org.etspace.abc.servlet;
import javax.servlet.http.HttpServlet;
```

```
import javax.servlet.http.HttpServletRequest;
import javax.servlet.http.HttpServletResponse;
import javax.servlet.ServletException;
import java.io.IOException;
import java.io.PrintWriter;
public class CheckUser extends HttpServlet {
    public void doGet(HttpServletRequest request,HttpServletResponse
response)
        throws ServletException,IOException{
        response.setContentType("text/html");
        PrintWriter out = response.getWriter();
        //为方便起见，在此假设数据库中已存在一些用户名，
        //在真正的应用中应该是从数据库中查询出来的。
        String [] usernames={"zhang","li","wang","zhao","liu"};
        //获取用户输入的用户名
        String username=request.getParameter("username");
        //设置响应内容
        String responseContext="true";
        for(int i=0;i<usernames.length;i++){
            //若存在该用户名，则修改响应内容
            if(username.equals(usernames[i]))
                responseContext="false";
        }
        //将处理结果返回给客户端
        out.println(responseContext);
        out.flush();
        out.close();
    }
    public void doPost(HttpServletRequest request,HttpServletResponse
response)
            throws ServletException,IOException{
        doGet(request,response);
    }
}
```

(4) 在项目的配置文件 web.xml 中添加 CheckUser(Servlet)的配置代码，具体代码如下：

```
<?xml version="1.0" encoding="UTF-8"?>
<web-app …>
  …
  <servlet>
    <servlet-name>CheckUser</servlet-name>
    <servlet-class>org.etspace.abc.servlet.CheckUser</servlet-class>
  </servlet>
  <servlet-mapping>
    <servlet-name>CheckUser</servlet-name>
    <url-pattern>/CheckUser</url-pattern>
  </servlet-mapping>
  …
</web-app>
```

(5) 在项目的 WebRoot 文件夹中添加一个新的 JSP 页面 yhzc.jsp，其代码如下：

```
<%@ page language="java" pageEncoding="UTF-8"%>
<html>
    <head>
        <title>用户注册</title>
    </head>
    <script type="text/javascript">
        var xmlHttp;
        //创建 XMLHttpRequest 对象
        function createHttpRequest(){
            if(window.ActiveXObject){  //若能获取 ActiveXObject
                xmlHttp = new ActiveXObject("Microsoft.XMLHTTP");
            }
            else if(window.XMLHttpRequest){  //若能获取 XMLHttpRequest
                xmlHttp = new XMLHttpRequest();
            }
        }
        function check(){
            //获取用户输入的用户名
            var username=document.all.username.value;
            //若为空
            if(username.length < 1){
                alert("用户名不能为空!");
                return;
            }
            createHttpRequest();
            //将状态触发器绑定到一个函数
            xmlHttp.onreadystatechange = processor;
            //通过 Get 方法对指定的 URL 建立服务器的调用
            xmlHttp.open("get","CheckUser?username="+username);
            //发送请求
            xmlHttp.send(null);
        }
        //状态改变处理函数
        function processor(){
            var responseContext;
            //如果响应完成
            if(xmlHttp.readyState == 4){
                //如果返回成功
                if(xmlHttp.status == 200){
                    //获取响应内容
                    responseContext = xmlHttp.responseText;
                    //如果用户名有效
                    if(responseContext.indexOf("true")!=-1){
                        alert("恭喜你，该用户名可使用！");
                    }else{  //否则，用户名无效
                        alert("对不起，该用户名已存在！");
                    }
                }
```

```
            }
        }
    </script>
    <body>
        <form action="">
        用户名:
        <!-- 当文本框内容改变时执行 check()函数 -->
        <input type="text" name="username" onchange="check()"/><br>
        密    码:
        <input type="password" name="password"/>  
        <input type="submit" value="注册"/>
        </form>
    </body>
</html>
```

本实例基于 Ajax 技术，在无须刷新页面的情况下，实现了用户名唯一性的验证功能。

9.3　Ajax 开源框架 DWR

9.3.1　DWR 简介

DWR 是目前常用的一种用于创建基于 Java EE 平台的 Ajax Web 站点的开源框架，可极大地减少 Ajax 的编程工作量。借助于 DWR，开发人员无须具备专业的 JavaScript 知识即可轻松实现所需要的 Ajax 功能。

DWR 的开发包 dwr.jar 可从其官方网站(http://directwebremoting.org)或其他有关网站下载，在此下载的开发包的版本号为 2.0.11。

为使 Web 项目支持 DWR 框架，只需将 DWR 的开发包 dwr.jar 与 Struts 2 框架的 jar 包 commons-logging-Xxx.jar(如 commons-logging-1.1.3.jar)复制到项目的 WebRoot\WEB-INF\lib 目录下即可。

9.3.2　DWR 的工作原理

DWR 可在浏览器中使用 JavaScript 代码调用 Web 服务器上的 Java 代码，就如同 Java 代码在浏览器中一样。其工作原理如图 9.4 所示，主要有两点：①动态地将 Java 类生成 JavaScript 类，从而使客户端能够通过脚本远程调用服务器端的方法；②当数据从服务器端返回时，自动调用相应的回调(callback)方法进行处理。

其实，DWR 包含有两个主要部分：①运行在服务器端的 Java Servlet，用于处理请求并向浏览器发送响应；②运行在浏览器端的 JavaScript，用于发送请求并动态更新网页。

Java 从根本上说使用的是同步机制，然而 Ajax 却是异步的。因此，在调用远程方法后，一旦数据从服务器返回时，需通过回调方法进行相应的处理。DWR 就是一种支持回调功能的 Ajax 应用框架。

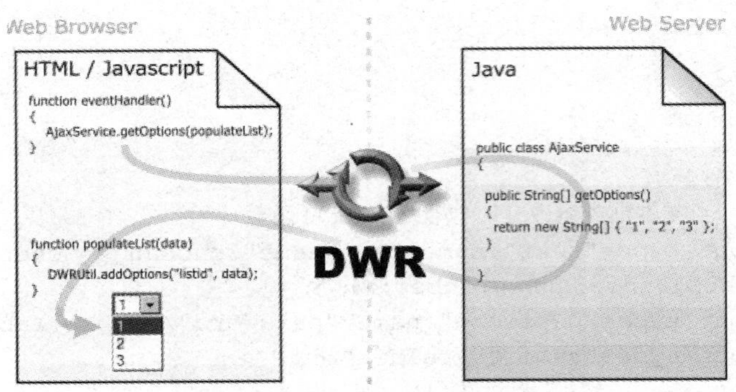

图 9.4　DWR 的工作原理

9.3.3　DWR 的基本应用

下面通过一个简单的应用实例，简要说明 DWR 的基本应用技术。

【实例 9-2】DWR 应用实例：时间查询。图 9.5 所示为“当前时间”页面，单击其中的“当前时间”按钮后，即可获取并显示出当前的系统时间，如图 9.6 所示。

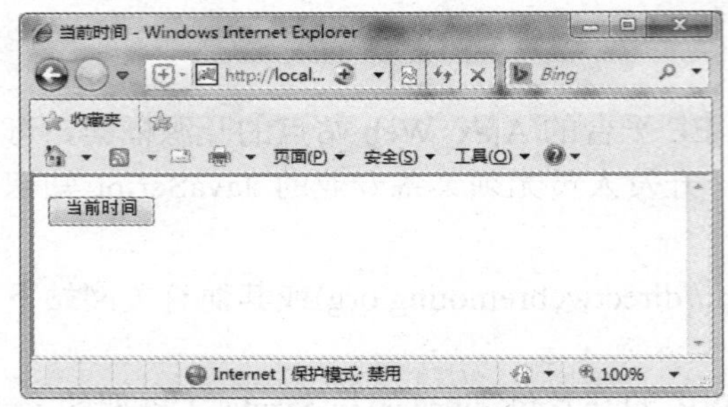

图 9.5　“当前时间”页面

图 9.6　“当前时间”对话框

主要步骤(在项目 web_09 中)如下。

(1) 将 DWR 的开发包 dwr.jar 与 Struts 2 框架的 jar 包 commons-logging-Xxx.jar(如 commons-logging-1.1.3.jar)复制到项目的 WebRoot\WEB-INF\lib 目录下。

(2) 修改配置文件 web.xml，为 DwrServlet(Servlet)添加相应的配置代码。具体代码如下。

```
<?xml version="1.0" encoding="UTF-8"?>
<web-app …>
  …
  <servlet>
    <servlet-name>dwr</servlet-name>
    <servlet-class>org.directwebremoting.servlet.DwrServlet</servlet-
class>
    <init-param>
    <param-name>debug</param-name>
        <param-value>true</param-value>
    </init-param>
```

```
        <init-param>
            <param-name>crossDomainSessionSecurity</param-name>
            <param-value>false</param-value>
        </init-param>
    </servlet>
    <servlet-mapping>
        <servlet-name>dwr</servlet-name>
        <url-pattern>/dwr/*</url-pattern>
    </servlet-mapping>
    …
</web-app>
```

通过以上配置，即可让 Web 应用程序将全部以/dwr 开始的 URL 所指向的请求都交给 org.directwebremoting.servlet.DwrServlet 这个 Java Servle 来进行处理。

(3) 在项目的 WebRoot\WEB-INF 文件夹中创建 DWR 部署描述文件 dwr.xml，其代码如下：

```
<?xml version="1.0" encoding="UTF-8"?>
<!DOCTYPE dwr PUBLIC
    "-//GetAhead Limited//DTD Direct Web Remoting 1.0//EN"
    "http://www.getahead.ltd.uk/dwr/dwr10.dtd">
<dwr>
    <allow>
    <create creator="new" javascript="AjaxDate">
            <param name="class" value="java.util.Date"/>
    </create>
    </allow>
</dwr>
```

部署描述文件 dwr.xml 用于指定哪些 Java 类可以被 DWR 框架创建并通过 JavaScript 远程调用。在此，指定了可以被 DWR 创建的 Java 类为 java.util.Date，并给该类指定一个 JavaScript 名称 AjaxDate。必要时，也可将自定义的 Java 类公开给 JavaScript 远程调用。

在部署描述文件 dwr.xml 中，create 元素的 creator 属性是必需的，用于指定使用哪种创建器。该属性的常用取值有三种，即 new、scripted 与 spring。其中，new 值表示使用 Java 类默认的无参数构造函数(方法)来创建类的对象(实例)，scripted 值表示使用脚本语言来创建 Java 类对象，spring 值表示通过 Spring 框架的 Bean 来创建 Java 类对象。

必要时，可通过 create 元素的子元素 include 指明要公开给 JavaScript 的方法。例如，若只需公开 Date 类的 toString 方法，则应将 create 元素修改为：

```
<create creator="new" javascript="AjaxDate">
    <param name="class" value="java.util.Date"/>
    <include method="toString" />
</create>
```

(4) 在项目的 WebRoot 文件夹中添加一个新的 JSP 页面 dqsj.jsp，其代码如下：

```
<%@ page language="java" pageEncoding="UTF-8"%>
<html>
    <head>
```

```
        <title>当前时间</title>
        <script type="text/javascript" src="dwr/engine.js"></script>
        <script type="text/javascript" src="dwr/util.js"></script>
        <script type="text/javascript"
src="dwr/interface/AjaxDate.js"></script>
        <script type="text/javascript">
           function dqsj() {
             AjaxDate.toString(show);
           }
            //显示当前时间
           function show(data) {
              window.alert("当前时间为: "+data);
           }
        </script>
    </head>
    <body>
        <input type="button" value="当前时间" onClick="dqsj()">
    </body>
</html>
```

解析：

(1) 在 dqsj.jsp 页面中，必须添加以下几行代码：

```
<script type="text/javascript" src="dwr/engine.js"></script>
<script type="text/javascript" src="dwr/util.js"></script>
<script type="text/javascript" src="dwr/interface/AjaxDate.js"></script>
```

其中，AjaxDate.js 是由 DWR 根据部署描述文件 dwr.xml 的具体配置自动生成的，内含 AjaxDate.toString()等方法的具体定义。

(2) 在 dqsj.jsp 页面中，通过“当前时间”按钮的单击事件触发 JavaScript 函数 dqsj()的执行。在 dqsj()函数中，通过调用 AjaxDate.toString()方法实现对 Date.toString()方法的远程调用，从而获取到当前的系统时间，然后再通过回调方法 show()显示。在 show()方法中，参数 data 为调用 AjaxDate.toString()方法返回的数据(即当前的系统时间)。

说明： DWR 提供了许多功能，允许通过部署描述文件 dwr.xml 迅速而简单地创建服务器端 Java 类的 Ajax 接口，而无须编写任何 Servlet 代码、对象序列化代码或客户端 XMLHttpRequest 代码。

DWR 能动态地将 Java 类生成 JavaScript 类，其代码就像 Ajax 一样，调用好像发生在浏览器端，但实际上是发生在服务器端，DWR 则负责数据的传送与转换。这种从 Java 到 JavaScript 的远程调用方式使 DWR 用起来有点像 RMI 或 SOAP 的常规 RPC 机制。此外，DWR 还有一个明显的优点，也就是不需要任何浏览器插件就能在网页上运行。

9.3.4 DWR 与 Struts 2、Spring、Hibernate 的整合应用

在目前的各类 Web 应用系统中，Ajax 技术的使用是相当广泛的。在开发 Java EE 应用系统时，通过整合 DWR 框架，即可轻松地在有关的 JSP 页面中添加相应的 Ajax 功能。

图 9.7 所示即为 Java EE 应用开发的 JSP+Struts 2+Spring+Hibernate+DAO+DWR 模式，其中也表明了 DWR 与 Struts 2、Spring、Hibernate 框架的整合原理。在此模式中，所有的系统组件(包括 Action 实现类、DAO 实现类与 Hibernate 框架等)均由 Spring 容器统一进行管理，而 DWR 框架的作用就是根据具体配置自动创建相应 DAO 组件中类的 Ajax 接口，供相关的 JSP 页面通过 JavaScript 进行调用，从而实现所需要的 Ajax 功能。

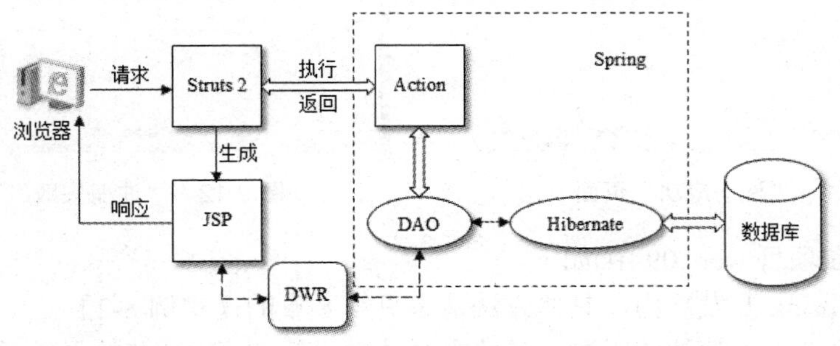

图 9.7　JSP+Struts 2+Spring+Hibernate+DAO+DWR 模式

下面以 JSP+Struts 2+Spring+Hibernate+DAO+DWR 模式实现 Web 应用系统中的用户注册功能，并从中说明 DWR 与 Struts 2、Spring、Hibernate 的整合应用技术。

图 9.8 所示为“用户注册”页面，在“用户名”文本框中输入用户名并让其失去焦点，即可验证所输入的用户名的唯一性。若该用户名已被注册过，则显示“该用户名已被使用！”的信息(见图 9.9)，否则显示“该用户名可以使用！”的信息(见图 9.10)。在表单中单击“注册”按钮后，若能成功注册，则显示图 9.11 的“注册成功”页面，否则显示图 9.12 所示的“注册失败”页面。

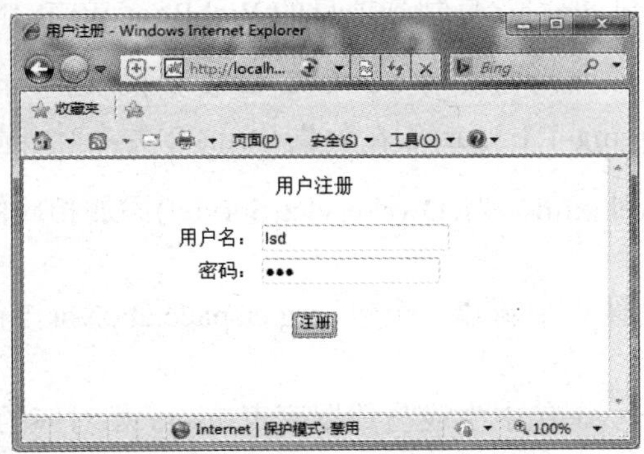

图 9.8　“用户注册”页面

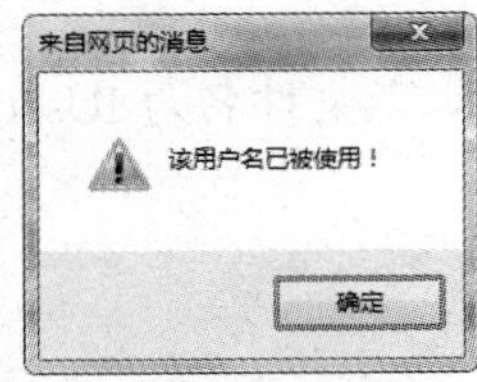

图 9.9　“该用户名已被使用”提示框

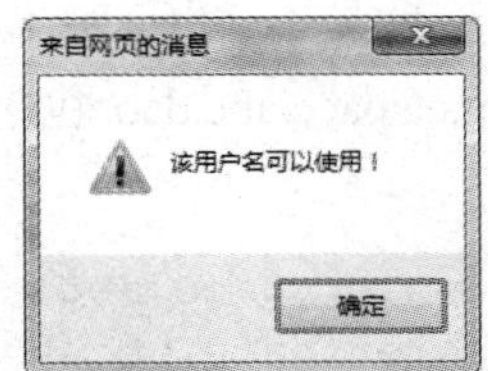

图 9.10　“该用户名可以使用”提示框

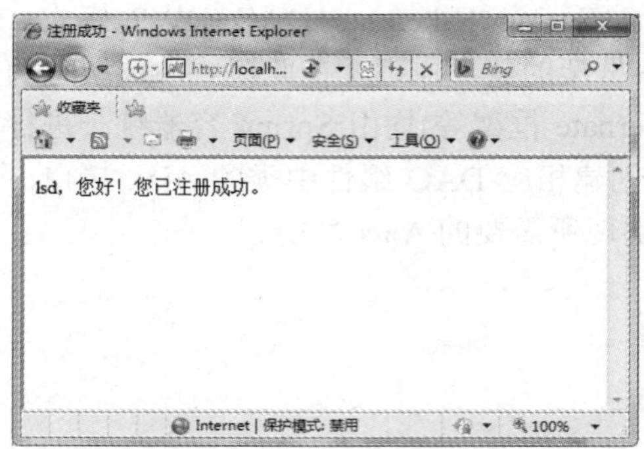

图 9.11 “注册成功”页面

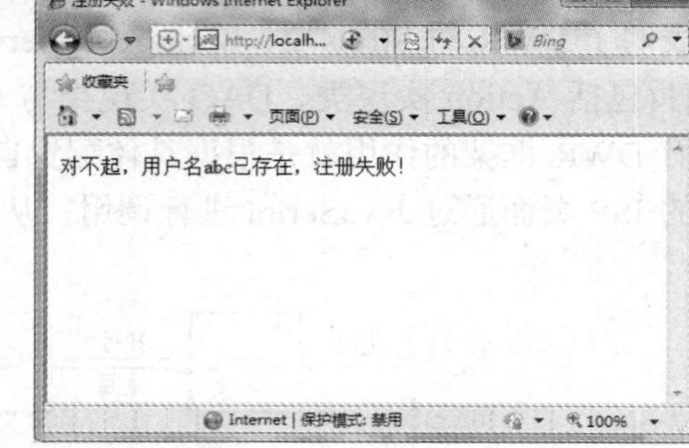

图 9.12 “注册失败”页面

主要步骤(在项目 web_09 中)如下。

(1) 添加 Spring 开发能力，具体方法请参见第 8 章的【实例 8-2】。

(2) 配置 Spring 框架的监听器，具体方法请参见第 8 章 8.7 节的系统登录案例。

(3) 加载 Struts 2 类库，包括 Struts 2 的 5 个基本类库、4 个附加类库与 1 个 Spring 支持包(struts2-spring-plugin.jar)。

(4) 配置 Struts 2 框架的核心过滤器，具体方法请参见第 8 章 8.7 节的系统登录案例。

(5) 在项目的 src 文件夹下创建一个 properties 文件 struts.properties，并在其中将 Struts 2 的对象工厂指定为 Spring 容器，具体方法请参见第 8 章 8.7 节的系统登录案例。

(6) 在 MyEclipse 中创建对数据库(在此为 SQL Server 数据库 rsgl)的连接(在此将其命名为 rsgl)。

(7) 添加 Hibernate 开发能力，具体方法请参见第 8 章 8.8 节的系统登录案例。

(8) 将 DWR 的开发包 dwr.jar 复制到项目的 WebRoot\WEB-INF\lib 目录下。

说明： 在此，DWR 应用所需要的 Struts 2 框架的 jar 包 commons-logging-Xxx.jar(如 commons-logging-1.1.3.jar)已在加载 Struts 2 类库时一起添加到项目中。

(9) 修改配置文件 web.xml，为 DwrServlet(Servlet)添加相应的配置代码，具体方法请参见本章的【实例 9-2】。

(10) 在项目的 src 文件夹中新建一个包 org.etspace.abc.vo(该包用来存放与数据库表相对应的 Java 类或 POJO)。

(11) 在 org.etspace.abc.vo 包中生成与数据库表 users 相对应的 Java 类 Users(其文件名为 Users.java)与映射文件 Users.hbm.xml。

(12) 在项目的 src 文件夹中新建一个包 org.etspace.abc.dao(该包用于存放有关的 DAO 接口)。

(13) 在 org.etspace.abc.dao 包中新建一个接口 IUsersDAO，其文件名为 IUsersDAO.java，代码如下：

```
package org.etspace.abc.dao;
import org.etspace.abc.vo.Users;
public interface IUsersDAO {
    public void saveUsers(Users users);
```

```
    public boolean existUsers(String username);
}
```

(14) 在项目的 src 文件夹中新建一个包 org.etspace.abc.dao.impl(该包用于存放有关的DAO 接口的实现类)。

(15) 在 org.etspace.abc.dao.impl 包中新建一个 IUsersDAO 接口的实现类 UsersDAO。其文件名为 UsersDAO.java，代码如下：

```
package org.etspace.abc.dao.impl;
import org.springframework.orm.hibernate3.support.HibernateDaoSupport;
import org.etspace.abc.dao.IUsersDAO;
import org.etspace.abc.vo.Users;
import java.util.*;
public class UsersDAO extends HibernateDaoSupport implements IUsersDAO {
    public void saveUsers(Users users) {
        getHibernateTemplate().save(users);
    }
    public boolean existUsers(String username) {
        List list=getHibernateTemplate().find("from Users where
username=?", username);
        if(list.size()>0)
        {
            return true;
        }
        return false;
    }
}
```

(16) 在 Spring 的配置文件 applicationContext.xml 中添加一个 bean 元素，其实现类分别为 UsersDAO，id 为 “usersDAO”。有关代码如下：

```
<bean id="usersDAO" class="org.etspace.abc.dao.impl.UsersDAO">
    <property name="sessionFactory">
        <ref bean="sessionFactory"></ref>
    </property>
</bean>
```

(17) 在项目的 WebRoot\WEB-INF 文件夹中创建 DWR 的部署描述文件 dwr.xml，并在其中完成相应的配置。其有关代码如下：

```
<?xml version="1.0" encoding="UTF-8"?>
<!DOCTYPE dwr PUBLIC
   "-//GetAhead Limited//DTD Direct Web Remoting 1.0//EN"
   "http://www.getahead.ltd.uk/dwr/dwr10.dtd">
<dwr>
    <allow>
    …
        <create creator="spring" javascript="AjaxUsersDAO">
            <param name="beanName" value="usersDAO"/>
            <include method="existUsers"/>
    </create>
```

```
    …
    </allow>
</dwr>
```

(18) 在项目的 src 文件夹中新建一个包 org.etspace.abc.action。

(19) 在 org.etspace.abc.action 包中新建 Action 实现类 UsersAction，其文件名为 UsersAction.java，代码如下：

```
package org.etspace.abc.action;
import org.etspace.abc.vo.Users;
import org.etspace.abc.dao.IUsersDAO;
import com.opensymphony.xwork2.ActionSupport;
public class UsersAction extends ActionSupport{
    private Users users;
    private IUsersDAO usersDAO;
    public Users getUsers() {
        return users;
    }
    public void setUsers(Users users) {
        this.users = users;
    }
    public void setUsersDAO(IUsersDAO usersDAO) {
        this.usersDAO = usersDAO;
    }
    public String execute() throws Exception{
            return SUCCESS;
    }
    public String register() throws Exception{
        if (usersDAO.existUsers(users.getUsername()))
        {
            return ERROR;
        }
        users.setUsertype("普通用户");
        usersDAO.saveUsers(users);
        return SUCCESS;
    }
}
```

(20) 在 Spring 的配置文件 applicationContext.xml 中添加一个 bean 元素，其实现类为 UsersAction，id 为“usersAction”。有关代码如下：

```
<bean id="usersAction" class="org.etspace.abc.action.UsersAction">
    <property name="usersDAO">
        <ref bean="usersDAO"></ref>
    </property>
</bean>
```

(21) 在项目的 src 文件夹创建配置文件 struts.xml，并在其中完成 UsersAction(Action) 的配置，其代码如下：

```
<?xml version="1.0" encoding="UTF-8"?>
```

```
<!DOCTYPE struts PUBLIC
    "-//Apache Software Foundation//DTD Struts Configuration 2.0//EN"
    "http://struts.apache.org/dtds/struts-2.0.dtd">
<struts>
    <package name="default" extends="struts-default">
        <action name="register" class="usersAction" method="register">
            <result name="success">/usersregister/success.jsp</result>
            <result name="error">/usersregister/error.jsp</result>
        </action>
    </package>
</struts>
```

(22) 在项目的 WebRoot 文件夹中新建一个子文件夹 usersregister。

(23) 在子文件夹 usersregister 中添加一个新的 JSP 页面 register.jsp，其代码如下：

```
<%@ page language="java" pageEncoding="utf-8" %>
<%@ taglib prefix="s" uri="/struts-tags" %>
<html>
    <head>
        <title>用户注册</title>
        <script type="text/javascript"
src="<%=request.getContextPath()%>/dwr/engine.js"></script>
        <script type="text/javascript"
src="<%=request.getContextPath()%>/dwr/util.js"></script>
        <script type="text/javascript"
src="<%=request.getContextPath()%>/dwr/interface/AjaxUsersDAO.js"></scri
pt>
        <script type="text/javascript">
         function check() {
           var username=document.all.username.value;
           if (username=="") {
             window.alert("用户名不能为空！");
             return;
           }
           AjaxUsersDAO.existUsers(username,show);
         }
         function show(data) {
           if (data) {
             window.alert("该用户名已被使用！");
             return;
           }
           window.alert("该用户名可以使用！");
         }
        </script>
    </head>
    <body>
        <div align="center">
        <form action="register.action" method="post">
            用户注册<br><br>
            <table>
```

```
            <tr><td align="right">用户名：</td><td><input type="text"
name="users.username" id="username" onblur="check()"></td></tr>
            <tr><td align="right">密码：</td><td><input type="password"
name="users.password" id="password"></td></tr>
            </table>
            <br>
            <input type="submit" value="注册">
        </form>
        </div>
    </body>
</html>
```

(24) 在子文件夹 usersregister 中添加一个新的 JSP 页面 success.jsp，其代码如下：

```
<%@ page language="java" pageEncoding="utf-8" %>
<%@ taglib prefix="s" uri="/struts-tags" %>
<html>
    <head>
        <title>注册成功</title>
    </head>
    <body>
        <s:property value="users.username" />，您好！您已注册成功。
    </body>
</html>
```

(25) 在子文件夹 usersregister 中添加一个新的 JSP 页面 error.jsp，其代码如下：

```
<%@ page language="java" pageEncoding="utf-8" %>
<%@ taglib prefix="s" uri="/struts-tags" %>
<html>
    <head>
        <title>注册失败</title>
    </head>
    <body>
        对不起，用户名<s:property value="users.username" />已存在，注册失败！
    </body>
</html>
```

本 章 小 结

本章首先介绍了 Ajax 的基本概念、相关技术与应用场景，然后通过具体实例讲解了 XMLHttpRequest 对象的基本用法与 Ajax 的基本应用技术，最后在简要介绍 DWR 框架工作原理的基础上，通过具体实例说明了 DWR 的基本应用技术与综合应用技术。通过本章的学习，读者应熟练掌握 Ajax 与 DWR 框架的相关应用技术，并能使用 JSP+Struts 2+Spring+Hibernate+DAO+DWR 模式开发相应的 Web 应用系统。

思 考 题

1. Ajax 是什么？其相关技术主要包括哪些？
2. 请列举几个典型的 Ajax 应用场景。

3. 如何创建 XMLHttpRequest 对象？

4. XMLHttpRequest 对象的常用方法与属性有哪些？

5. 请简述 Ajax 的请求与响应过程。

6. 请简述 DWR 框架的工作原理。

7. 如何在 Web 项目中应用 DWR 框架？

8. DWR 部署描述文件是什么？有何主要作用？

9. 在 JSP 页面中如何调用 DWR 框架所创建的类与方法？

10. 请简述 DWR 与 Struts 2、Spring、Hibernate 框架的整合方法。

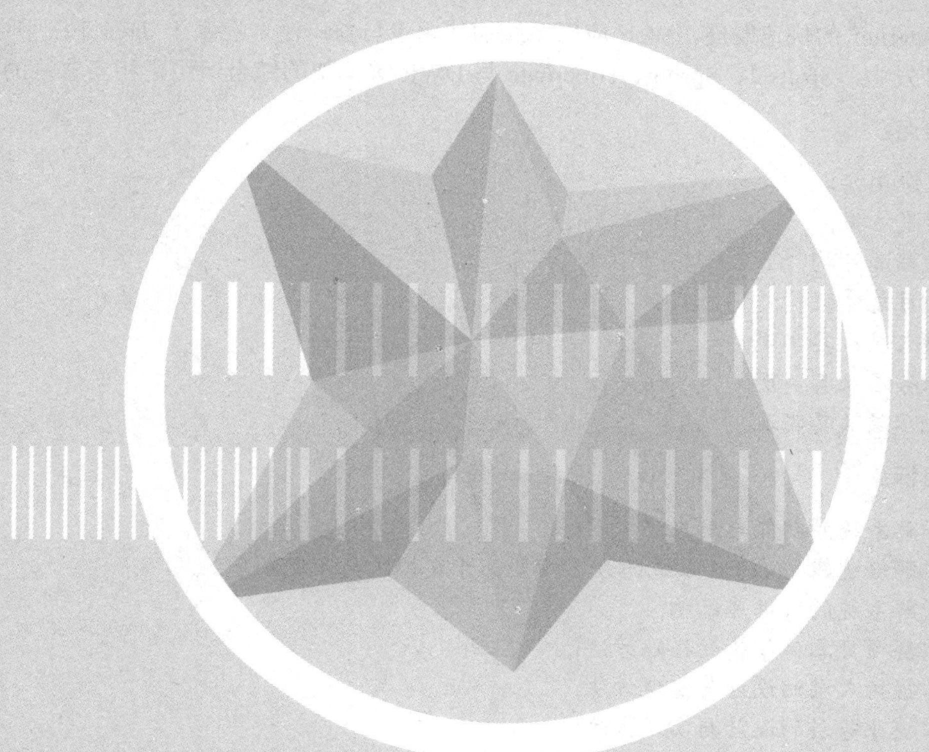

第 10 章

Web 应用案例

随着 Internet 的快速发展，Web 应用系统的使用也日益广泛。在基于 Java EE 平台的 Web 应用开发中，Struts 2、Spring、Hibernate 与 DWR 等框架的综合运用是相当普遍的。

本章要点：

- 系统简介；
- 开发方案；
- 数据库结构；
- 项目总体架构；
- 持久层及其实现；
- 业务层及其实现；
- 表示层及其实现。

学习目标：

- 了解系统的功能与用户；
- 了解轻量级 Java EE 系统的分层模型、开发模式与开发顺序；
- 了解系统的数据库结构；
- 掌握项目总体架构的搭建方法；
- 掌握持久层的设计与实现方法；
- 掌握业务层的设计与实现方法；
- 掌握表示层的设计与实现方法；
- 掌握 Web 应用系统开发的 JSP+Struts 2+Spring+Hibernate+DWR+DAO+Service 模式。

10.1 系统简介

通过分析典型的 Web 应用案例，可更好地理解并掌握有关的开发技术。在此，将以一个简单的人事管理系统为例，说明采用 Struts 2、Spring、Hibernate 与 DWR 框架开发 Web 应用系统的基本方案、主要过程与相关技术。

10.1.1 系统功能

本人事管理系统较为简单，仅用于对单位职工的基本信息进行相应的管理，其主要功能具体如下。

1. 部门管理

部门管理功能包括部门的查询、增加、修改与删除。每个部门的信息包括部门的编号与名称，其中，部门的编号是唯一的。

2. 职工管理

职工管理功能包括职工的查询、增加、修改与删除。每个职工的信息包括职工的编号、姓名、性别、出生日期、基本工资、岗位津贴与所在部门(编号)，其中，职工的编号是唯一的。

3. 用户管理

用户管理功能包括用户的查询、增加、修改与删除以及用户密码的重置与设置。每个用户的信息包括用户名、密码与用户类型，其中，用户名是唯一的。

10.1.2　系统用户

本系统的用户分为两种类型，即系统管理员与普通用户，各用户需登录成功后方可使用有关的功能，使用完毕后则可通过安全退出(或注销)功能退出系统。在使用过程中，各用户均可通过密码设置功能修改自己的登录密码，以提高安全性。

用户的操作权限是根据其类型确定的。在本系统中，系统管理员可执行系统的所有功能，但密码设置功能仅限于修改自己的密码。至于普通用户，则只能执行职工管理功能以及密码设置与安全退出功能(其中，密码设置功能也仅限于修改自己的密码)。

本系统规定，默认系统管理员的用户名为 admin，其初始密码为 12345。以默认系统管理员身份登录系统后，即可创建其他系统管理员以及所需要的普通用户(新建系统管理员与普通用户的初始密码与其用户名一致)。

10.2　开 发 方 案

本系统的开发遵循轻量级 Java EE 系统的分层模型与面向接口编程的基本思想，并采用多框架(包括 Struts 2、Spring、Hibernate 与 DWR 框架)整合的典型模式。

10.2.1　分层模型

轻量级 Java EE 系统的分层模型是一个通用的三层架构模型，由表示层、业务层与持久层构成，如图 10.1 所示。

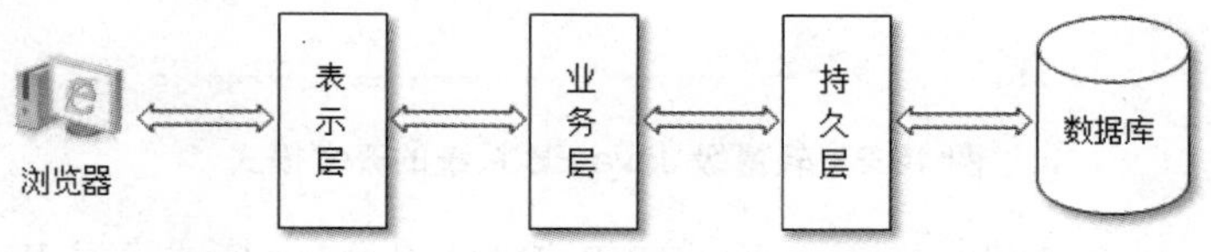

图 10.1　轻量级 Java EE 系统的分层模型

1. 表示层

表示层通常又称为 Web 层，用于实现 Web 前端界面与业务流程控制功能，是用户与 Java EE 系统进行直接交互的层面。在具体实现时，表示层使用业务层所提供的 Service(服务)来满足用户的各种需求。

2. 业务层

业务层通常又称为业务逻辑层，由一系列相互独立的 Service 构成。一个 Service 其实就是一个程序模块或组件，用于完成特定的应用功能。在具体实现时，Service 通过调用 DAO 接口所公开的方法，经由持久层间接地访问后台数据库。

3. 持久层

持久层通常又称为数据持久层，主要由持久化 POJO 类、DAO 接口及其实现类等构成。持久层屏蔽了底层 JDBC 连接与数据库操作的细节，为业务层的 Service 提供了简洁、统一、面向对象的数据访问接口。

10.2.2 开发模式

在基于 Java EE 的 Web 应用开发中，通常采用多框架整合的方式来实现系统的三层模型，其中最为典型的开发模式之一就是 JSP+Struts 2+Spring+Hibernate+DWR+DAO+Service(见图 10.2)，即以 JSP 为基础，以 Spring 框架为核心，同时整合 Struts 2，Hibernate 与 DWR 框架，并遵循面向接口编程的基本思想。

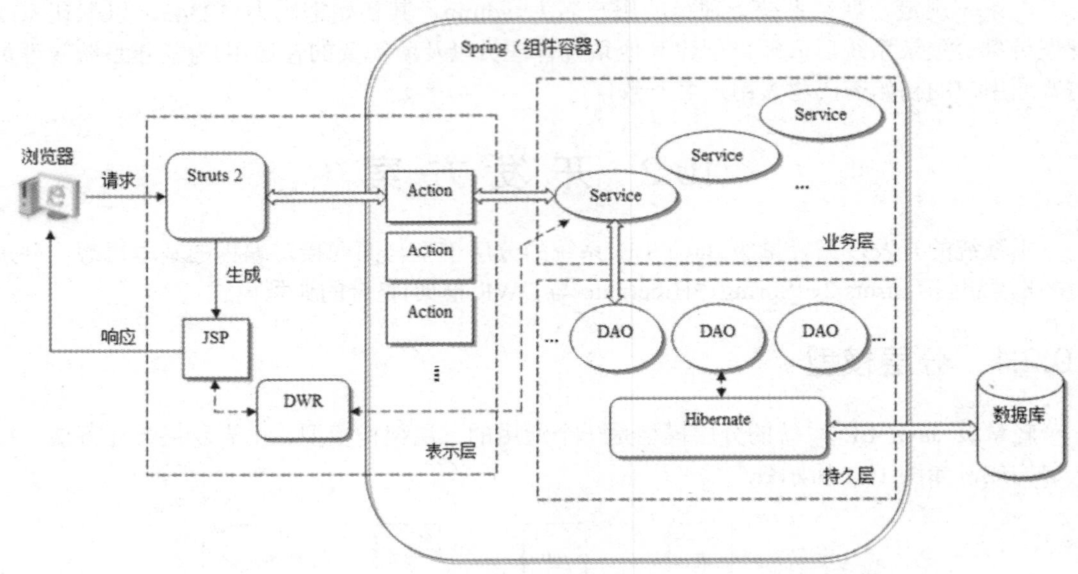

图 10.2　轻量级 Java EE 系统的开发模式

使用 JSP+Struts 2+Spring+Hibernate+DWR+DAO+Service 模式开发 Web 应用系统时，持久层包含一系列由 Spring 容器管理的 DAO 组件与 Hibernate 框架(包括由其生成的 POJO 类与映射文件)，业务层包含一系列由 Spring 容器管理的 Service 组件，而表示层则主要包括 JSP 页面(可包含 CSS 层叠样式表)、DWR 框架、Struts 2 控制器以及位于 Spring 容器之中的 Action 组件。可见，整个系统的所有组件(包括 DAO、Service 与 Action 等)与 Hibernate 均交由 Spring 统一管理。当用户需要扩充系统的功能时，只需将新增功能做成组件并置于 Spring 容器中，然后再适当修改前端的 JSP 页面即可。在此过程中，系统已有的结构与功能不会受到任何影响。

在具体开发过程中，对于持久层，持久化 POJO 对象及其映射文件可利用 Hibernate 的逆向工程能力自动生成(必要时需在此基础上再进行相应的修改)，程序员只需编写 DAO 接口及其实现类，并在 applicationContext.xml 文件中进行注册即可；对于业务层，程序员只需编写 Service 接口及其实现类，并在 applicationContext.xml 文件中进行注册即可；对于表示层，程序员只需完成有关 JSP 页面的设计(必要时需使用 DWR 框架实现一定的 Ajax 功能)与 Action 组件的编写，并在 struts.xml 文件中进行相应的配置以实现有关的控制逻辑即可。

基于多框架整合的系统开发方案的优点是显而易见的。一方面，可有效减少编程工作量，缩短开发周期并降低成本。另一方面，可使系统的架构更加清晰、合理，运行更加稳定、可靠。正因为如此，目前企业级的 Java EE 应用开发都会根据具体情况选用某些合适的框架，以达到快捷、高效的目的。

注意： MVC 是所有 Web 应用系统的通用开发模式。在 Java EE 系统的三层架构中，表示层包括了 MVC 的 V(视图)与 C(控制)，而业务层与持久层的各个组件则相当于 MVC 的广义的 M(模型)。对于 MVC 来说，其核心是控制器(通常由 Struts 2 担任)。而对于 Java EE 系统的三层架构来说，其核心是组件容器(通常由 Spring 担任)，至于其中的控制器(通常由 Struts 2 担任)，则只承担表示层的控制功能。

10.2.3 开发顺序

一个大型项目的开发往往需要一个团队而不是单个程序员来完成，团队开发的关键在于团队成员的分工与协作，而面向接口的编程对项目的团队开发来说是极其有利了。只要提供了相应的接口，各个程序员就可以直接调用其中所定义的有关方法，而无须顾及其具体实现。

在开发一个大型的 Java EE 项目时，通常会先运用已有的成熟软件来搭建出系统的主体框架，然后再编写特定功能的有关组件并将其集成到主体框架内。在此过程中，一般应先实现持久层的 DAO 接口，然后实现业务层的业务逻辑，最后再实现表示层的页面与控制逻辑。

10.3 数据库结构

本人事管理系统所使用的数据库为在 SQL Server 2005/2008 中创建的人事管理数据库 rsgl，其中有 3 个表，即部门表 bmb、职工表 zgb 与用户表 users。各表的结构见表 3.7～表 3.9。

为便于项目的开发及有关功能的调试，可将表 3.10~表 3.12 所示的记录输入相应的表中。

说明： 本系统的用户类型只有两种，即系统管理员与普通用户。相应地，用户表 users 中用户类型字段 usertype 的取值也只有两种，即“系统管理员”与“普通用户”。

注意： 在用户表 users 中，至少要包含一个默认系统管理员用户，其用户名为 admin，初始密码为 12345。

10.4 项目总体架构

在具体实现系统的有关功能前，应先创建相应的 Web 项目，并在其中完成有关框架的整合，同时根据需要确定项目的基本结构(即创建好相应的包与文件夹)。对于本人事管理系统来说，可在 MyEclipse 中按以下步骤完成项目的前期搭建工作。

(1) 新建一个 Web 项目 rsgl。

(2) 添加 Spring 开发能力，具体方法请参见第 8 章的【实例 8-2】。

(3) 配置 Spring 框架的监听器，具体方法请参见第 8.7 节的系统登录案例。

(4) 加载 Struts 2 类库，包括 Struts 2 的 5 个基本类库、4 个附加类库与 1 个 Spring 支持包(struts2-spring-plugin.jar)。

(5) 配置 Struts 2 框架的核心过滤器。具体方法请参见第 8.7 节的系统登录案例。

(6) 在项目的 src 文件夹下创建一个 properties 文件 struts.properties，并在其中将 Struts 2 的对象工厂指定为 Spring 容器。具体方法请参见第 8.7 节的系统登录案例。

(7) 在 MyEclipse 中创建对人事管理数据库 rsgl 的连接(在此将其命名为 rsgl)。

(8) 添加 Hibernate 开发能力，具体方法请参见第 8.8 节的系统登录案例。

(9) 集成 DWR 框架。具体方法请参见第 9 章【实例 9-2】的步骤(1)、(2)。

说明： 在此，DWR 应用所需要的 Struts 2 框架的 jar 包 commons-logging-Xxx.jar(如 commons-logging-1.1.3.jar)已在加载 Struts 2 类库时一起添加到项目中。

(10) 在项目的 WebRoot 文件夹中新建一个子文件夹 images，用于存放项目所需要的有关图片。

(11) 在项目的 src 文件夹中新建一个包 org.etspace.rsgl.vo，用来存放与数据库表相对应的 POJO 类及映射文件(*.hbm.xml)。

(12) 在项目的 src 文件夹中新建一个包 org.etspace.rsgl.action，用来存放有关的 Action 实现类。Action 实现类通过调用业务逻辑来处理用户请求，然后对流程进行相应的控制。

(13) 在项目的 src 文件夹中新建一个包 org.etspace.rsgl.dao，用来存放有关的 DAO 接口。DAO 接口中的方法由接口的实现类来具体实现，并用于与数据库进行交互。

(14) 在项目的 src 文件夹中新建一个包 org.etspace.rsgl. dao.impl，用来存放 DAO 接口的实现类。

(15) 在项目的 src 文件夹中新建一个包 org.etspace.rsgl.service，用来存放有关的业务逻辑接口。业务逻辑接口中的方法由接口的实现类来具体实现，并用于处理用户的请求。

(16) 在项目的 src 文件夹中新建一个包 org.etspace.rsgl.service.impl，用来存放业务逻辑接口的实现类。

(17) 在项目的 src 文件夹中新建一个包 org.etspace. rsgl.tool，用来存放公用的工具类，如分页类、拦截器类等。

至此，项目的前期搭建工作即告完成，其总体架构如图 10.3 所示。此后，即可逐步推进项目的后续开发工作。

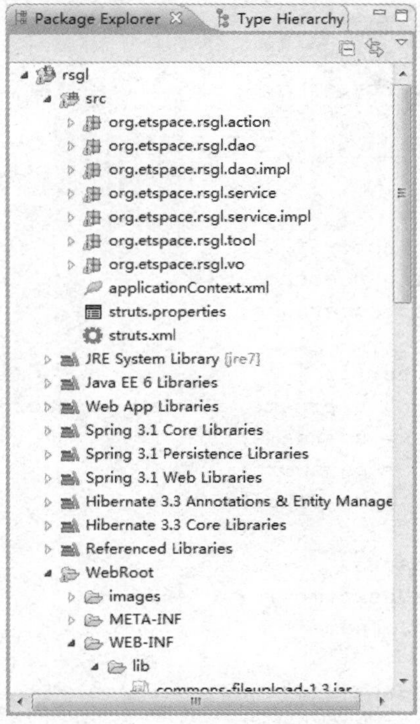

图 10.3　Web 项目的总体架构

10.5　持久层及其实现

持久层的开发主要包括 POJO 类与映射文件的生成以及各 DAO 组件的设计与实现。其中，DAO 组件用于实现与数据库的交互，即进行相应的 CRUD 操作(也就是记录的增加、查询、修改与删除操作)，从而完成对底层数据库的持久化访问。

10.5.1　POJO 类与映射文件

利用 Hibernate 的逆向工程向导，按照第 7 章【实例 7-1】中所述的方法与步骤，在 org.etspace.rsgl.vo 包中分别生成与数据库表相对应的 POJO 类及映射文件。其中，与用户表 users 相对应的 POJO 类为 Users(其文件名为 Users.java)，映射文件为 Users.hbm.xml；与部门表 bmb 相对应的 POJO 类为 Bmb(其文件名为 Bmb.java)，映射文件为 Bmb.hbm.xml；与职工表 zgb 相对应的 POJO 类为 Zgb(其文件名为 Zgb.java)，映射文件为 Zgb.hbm.xml。

Users.java 的代码如下：

```
package org.etspace.rsgl.vo;
/**
 * Users entity. @author MyEclipse Persistence Tools
 */
```

```
public class Users implements java.io.Serializable {
    // Fields
    private String username;
    private String password;
    private String usertype;
    // Constructors
    /** default constructor */
    public Users() {
    }
    /** minimal constructor */
    public Users(String username) {
        this.username = username;
    }
    /** full constructor */
    public Users(String username, String password, String usertype) {
        this.username = username;
        this.password = password;
        this.usertype = usertype;
    }
    // Property accessors
    public String getUsername() {
        return this.username;
    }
    public void setUsername(String username) {
        this.username = username;
    }
    public String getPassword() {
        return this.password;
    }
    public void setPassword(String password) {
        this.password = password;
    }
    public String getUsertype() {
        return this.usertype;
    }
    public void setUsertype(String usertype) {
        this.usertype = usertype;
    }
}
```

Users.hbm.xml 的代码如下：

```
<?xml version="1.0" encoding="utf-8"?>
<!DOCTYPE hibernate-mapping PUBLIC "-//Hibernate/Hibernate Mapping DTD
3.0//EN"
"http://hibernate.sourceforge.net/hibernate-mapping-3.0.dtd">
<!--
   Mapping file autogenerated by MyEclipse Persistence Tools
-->
<hibernate-mapping>
```

```
    <class name="org.etspace.rsgl.vo.Users" table="users" schema="dbo" 
catalog="rsgl">
        <id name="username" type="java.lang.String">
            <column name="username" length="10" />
            <generator class="assigned" />
        </id>
        <property name="password" type="java.lang.String">
            <column name="password" length="20" />
        </property>
        <property name="usertype" type="java.lang.String">
            <column name="usertype" length="10" />
        </property>
    </class>
</hibernate-mapping>
```

Bmb.java、Zgb.java 和 Bmb.hbm.xml、Zgb.hbm.xml 的代码分别与 Users.java 和 Users.hbm.xml 类似，在此不再详述。

在 Hibernate 框架的支持下，借助于与数据库表相对应的 POJO 类与映射文件，各 DAO 组件即可以面向对象的方式实现对数据库的有关操作。

说明： 利用 Hibernate 的逆向工程向导生成各表所对应的 POJO 类及其映射文件，会同时在 Spring 的配置文件 applicationContext.xml 中完成相应的映射文件注册操作。

注意： 对于 Hibernate 所生成 POJO 类与映射文件，有时候还需要根据具体情况进行相应的修改。

10.5.2 用户管理 DAO 组件及其实现

用户管理 DAO 组件主要用于实现对用户表 users 的有关操作，其实现步骤如下。

(1) 在 org.etspace.rsgl.dao 包中创建用户管理 DAO 接口 IUsersDAO。IUsersDAO 接口的文件名为 IUsersDAO.java，其代码如下：

```
package org.etspace.rsgl.dao;
import org.etspace.rsgl.vo.Users;
import java.util.List;
public interface IUsersDAO {
    //增加用户
    public void save(Users users);
    //删除用户
    public void delete(String username);
    //修改用户
    public void update(Users users);
    //查询用户
    public Users find(String username);
    //分页查询用户
    public List findAll(int pageNow,int pageSize,Users users);
    //查询用户总数
```

```
    public int findAllSize(Users users);
    //检查用户
    public Users check(String username,String password);
    //是否存在用户
    public boolean exist(String username);
}
```

(2) 在 org.etspace.rsgl.dao.impl 包中创建 IUsersDAO 接口的实现类 UsersDAO。UsersDAO 实现类的文件名为 UsersDAO.java，其代码如下：

```
package org.etspace.rsgl.dao.impl;
import org.springframework.orm.hibernate3.support.HibernateDaoSupport;
import org.etspace.rsgl.dao.IUsersDAO;
import org.etspace.rsgl.vo.Users;
import org.hibernate.Query;
import org.hibernate.Session;
import org.hibernate.Transaction;
import java.util.List;
public class UsersDAO extends HibernateDaoSupport implements IUsersDAO {
    public void save(Users users) {
        getHibernateTemplate().save(users);
    }
    public void delete(String username){
        getHibernateTemplate().delete(find(username));
    }
    public void update(Users users){
        getHibernateTemplate().update(users);
    }
    public Users find(String username){
        List list=getHibernateTemplate().find("from Users where username=?",
username);
        if(list.size()>0)
            return (Users)list.get(0);
        else
            return null;
    }
    public List findAll(int pageNow,int pageSize,Users users){
        Session session=getHibernateTemplate().getSessionFactory().
openSession();
        Transaction ts=session.beginTransaction();
        String hql="";
        if (users==null)
            hql="from Users order by username";
        else{
            hql="from Users where username like '%"+users.getUsername().
trim()+"%' order by username";
        }
        Query query=session.createQuery(hql);
        int firstResult=(pageNow-1)*pageSize;
        query.setFirstResult(firstResult);
        query.setMaxResults(pageSize);
```

```
        List list=query.list();
        ts.commit();
        session.close();
        session=null;
        return list;

    }
    public int findAllSize(Users users){
        String hql="";
        if (users==null)
            hql="from Users";
        else
            hql="from Users where username like
'%"+users.getUsername().trim()+"%'";
        return getHibernateTemplate().find(hql).size();
    }
    public Users check(String username, String password) {
        List list=getHibernateTemplate().find("from Users where username=?
and password=?",username,password);
        if(list.size()>0)
            return (Users)list.get(0);
        else
            return null;
    }
    public boolean exist(String username) {
        List list=getHibernateTemplate().find("from Users where username=?",
username);
        if(list.size()>0)
            return true;
        else
            return false;
    }
}
```

说明： 本项目在实现 DAO 接口时，使用了 Spring 整合 Hibernate 后提供的 HibernateDaoSupport 类，从而简化了有关功能代码的编写。

(3) 将 UsersDAO 作为一个组件交由 Spring 容器管理。为此，只需在 Spring 的配置文件 applicationContext.xml 中为 UsersDAO 添加一个相应的 bean 元素即可。有关代码如下：

```
<bean id="usersDAO" class="org.etspace.rsgl.dao.impl.UsersDAO">
    <property name="sessionFactory">
        <ref bean="sessionFactory"></ref>
    </property>
</bean>
```

该 bean 元素的 id 为 usersDAO，实现类为 UsersDAO，并通过 property 子元素为属性 sessionFactory 依赖注入一个 SessionFactory 实例(SessionFactory 所对应 Bean 的 id 为 sessionFactory)。

说明：继承 HibernateDaoSupport 类的 DAO 组件必须获得一个 SessionFactory 实例，方可实现对数据库的持久化访问。

10.5.3 部门管理 DAO 组件及其实现

部门管理 DAO 组件主要用于实现对部门表 bmb 的有关操作。其实现步骤如下。

(1) 在 org.etspace.rsgl.dao 包中创建部门管理 DAO 接口 IBmbDAO。IBmbDAO 接口的文件名为 IBmbDAO.java，其代码如下：

```
package org.etspace.rsgl.dao;
import org.etspace.rsgl.vo.Bmb;
import java.util.List;
public interface IBmbDAO {
    //增加部门
    public void save(Bmb bmb);
    //删除部门
    public void delete(String bmbh);
    //修改部门
    public void update(Bmb bmb);
    //查询部门
    public Bmb find(String bmbh);
    //分页查询部门
    public List findAll(int pageNow,int pageSize,Bmb bmb);
    //查询部门总数
    public int findAllSize(Bmb bmb);
    //获取所有部门
    public List getAll();
    //是否存在部门
    public boolean exist(String bmbh);
}
```

(2) 在 org.etspace.rsgl.dao.impl 包中创建 IBmbDAO 接口的实现类 BmbDAO。BmbDAO 实现类的文件名为 BmbDAO.java，其代码如下：

```
package org.etspace.rsgl.dao.impl;
import org.springframework.orm.hibernate3.support.HibernateDaoSupport;
import org.etspace.rsgl.dao.IBmbDAO;
import org.etspace.rsgl.vo.Bmb;
import java.util.List;
import org.hibernate.*;
public class BmbDAO extends HibernateDaoSupport implements IBmbDAO {
    public void save(Bmb bmb) {
        getHibernateTemplate().save(bmb);
    }
    public void delete(String bmbh) {
        getHibernateTemplate().delete(find(bmbh));
    }
    public void update(Bmb bmb) {
        getHibernateTemplate().update(bmb);
```

```
    }
    public Bmb find(String bmbh) {
        List list=getHibernateTemplate().find("from Bmb where bmbh=?",bmbh);
        if(list.size()>0)
            return (Bmb)list.get(0);
        else
            return null;
    }
    public List findAll(int pageNow, int pageSize, Bmb bmb) {
        Session session=getHibernateTemplate().getSessionFactory().
        openSession();
        Transaction ts=session.beginTransaction();
        String hql="";
        if (bmb==null)
            hql="from Bmb order by bmbh";
        else{
            hql="from Bmb where bmbh like '%"+bmb.getBmbh()+"%' order by
            bmbh";
        }
        Query query=session.createQuery(hql);
        int firstResult=(pageNow-1)*pageSize;
        query.setFirstResult(firstResult);
        query.setMaxResults(pageSize);
        List list=query.list();
        ts.commit();
        session.close();
        session=null;
        return list;
    }
    public int findAllSize(Bmb bmb) {
        String hql="";
        if (bmb==null)
            hql="from Bmb";
        else
            hql="from Bmb where bmbh like '%"+bmb.getBmbh()+"%'";
        return getHibernateTemplate().find(hql).size();
    }
    public List getAll() {
        return getHibernateTemplate().find("from Bmb order by bmbh");
    }
    public boolean exist(String bmbh) {
        List list=getHibernateTemplate().find("from Bmb where bmbh=?",bmbh);
        if(list.size()>0)
            return true;
        else
            return false;
    }
}
```

(3) 将 BmbDAO 作为一个组件交由 Spring 容器管理。为此，只需在 Spring 的配置文件 applicationContext.xml 中为 BmbDAO 添加一个相应的 bean 元素即可。有关代码如下：

```
<bean id="bmbDAO" class="org.etspace.rsgl.dao.impl.BmbDAO">
    <property name="sessionFactory">
        <ref bean="sessionFactory" />
    </property>
</bean>
```

该 bean 元素的 id 为 bmbDAO，实现类为 BmbDAO，并通过 property 子元素为属性 sessionFactory 依赖注入一个 SessionFactory 实例(SessionFactory 所对应 Bean 的 id 为 sessionFactory)。

10.5.4 职工管理 DAO 组件及其实现

职工管理 DAO 组件主要用于实现对职工表 zgb 的有关操作。其实现步骤如下。

(1) 在 org.etspace.rsgl.dao 包中创建职工管理 DAO 接口 IZgbDAO。IZgbDAO 接口的文件名为 IZgbDAO.java，其代码如下：

```
package org.etspace.rsgl.dao;
import org.etspace.rsgl.vo.Zgb;
import java.util.List;
public interface IZgbDAO {
    //增加职工
    public void save(Zgb zgb);
    //删除职工
    public void delete(String bh);
    //修改职工
    public void update(Zgb zgb);
    //查询职工
    public Zgb find(String bh);
    //分页查询职工
    public List findAll(int pageNow,int pageSize,Zgb zgb);
    //查询职工总数
    public int findAllSize(Zgb zgb);
    //获取所有职工
    public List getAll();
    //获取指定部门的职工
    public List getAllByBmbh(String bmbh);
    //是否存在职工
    public boolean exist(String bh);
}
```

(2) 在 org.etspace.rsgl.dao.impl 包中创建 IZgbDAO 接口的实现类 ZgbDAO。ZgbDAO 实现类的文件名为 ZgbDAO.java，其代码如下：

```
package org.etspace.rsgl.dao.impl;
import org.springframework.orm.hibernate3.support.HibernateDaoSupport;
import org.etspace.rsgl.dao.IZgbDAO;
import org.etspace.rsgl.vo.Zgb;
import java.util.List;
```

```
import org.hibernate.*;
public class ZgbDAO extends HibernateDaoSupport implements IZgbDAO {
    public void save(Zgb zgb) {
        getHibernateTemplate().save(zgb);
    }
    public void delete(String bh) {
        getHibernateTemplate().delete(find(bh));
    }
    public void update(Zgb zgb) {
        getHibernateTemplate().update(zgb);
    }
    public Zgb find(String bh) {
        List list=getHibernateTemplate().find("from Zgb where bh=?",bh);
        if(list.size()>0)
            return (Zgb)list.get(0);
        else
            return null;
    }
    public List findAll(int pageNow, int pageSize, Zgb zgb) {
        Session session=getHibernateTemplate().getSessionFactory().
        openSession();
        Transaction ts=session.beginTransaction();
        String hql="";
        if (zgb==null)
            hql="from Zgb order by bh";
        else{
            hql="from Zgb where bh like '%"+zgb.getBh()+"%' and bm like
            '%"+zgb.getBm()+"%' order by bh";
        }
        Query query=session.createQuery(hql);
        int firstResult=(pageNow-1)*pageSize;
        query.setFirstResult(firstResult);
        query.setMaxResults(pageSize);
        List list=query.list();
        ts.commit();
        session.close();
        session=null;
        return list;
    }
    public int findAllSize(Zgb zgb) {
        String hql="";
        if (zgb==null)
            hql="from Zgb";
        else
            hql="from Zgb where bh like '%"+zgb.getBh()+"%' and bm like
            '%"+zgb.getBm()+"%'";
        return getHibernateTemplate().find(hql).size();
    }

    public List getAll() {
```

```
        return getHibernateTemplate().find("from Zgb order by bh");
    }
    public List getAllByBmbh(String bmbh) {
        return getHibernateTemplate().find("from Zgb where bm='"+bmbh+"'
        order by bh");
    }
    public boolean exist(String bh) {
        List list=getHibernateTemplate().find("from Zgb where bh=?",bh);
        if(list.size()>0)
            return true;
        else
            return false;
    }
}
```

(3) 将 ZgbDAO 作为一个组件交由 Spring 容器管理。为此，只需在 Spring 的配置文件 applicationContext.xml 中为 ZgbDAO 添加一个相应的 bean 元素即可。有关代码如下：

```
<bean id="zgbDAO" class="org.etspace.rsgl.dao.impl.ZgbDAO">
    <property name="sessionFactory">
        <ref bean="sessionFactory" />
    </property>
</bean>
```

该 bean 元素的 id 为 zgbDAO，实现类为 ZgbDAO，并通过 property 子元素为属性 sessionFactory 依赖注入一个 SessionFactory 实例(SessionFactory 所对应 Bean 的 id 为 sessionFactory)。

10.6 业务层及其实现

业务层的开发主要是实现一系列的 Service 组件，而各个 Service 组件其实就是相应业务逻辑接口的实现类。Service 组件用于为表示层的处理程序(Action)提供相应的服务，其具体实现依赖于持久层的 DAO 组件，是对 DAO 的进一步封装。

业务层 Service 组件的存在使得上层(表示层)的 Action 无须直接访问下层(持久层)DAO 组件的方法，而是调用面向应用的业务逻辑方法，从而彻底地屏蔽了下层的数据库操作，可让上层开发人员将精力主要集中在业务流程的具体实现上。

10.6.1 用户管理 Service 组件及其实现

用户管理 Service 组件主要用于提供与用户管理功能相关的各种操作。其实现步骤如下。

(1) 在 org.etspace.rsgl.service 包中创建用户管理 Service 接口 IUsersService。IUsersService 接口的文件名为 IUsersService.java，其代码如下：

```
package org.etspace.rsgl.service;
import org.etspace.rsgl.vo.Users;
import java.util.List;
```

```
public interface IUsersService {
    //增加用户
    public void save(Users users);
    //删除用户
    public void delete(String username);
    //修改用户
    public void update(Users users);
    //查询用户
    public Users find(String username);
    //分页查询用户
    public List findAll(int pageNow,int pageSize,Users users);
    //查询用户总数
    public int findAllSize(Users users);
    //检查用户
    public Users check(String username,String password);
    //是否存在用户
    public boolean exist(String username);
}
```

(2) 在 org.etspace.rsgl.service.impl 包中创建 IUsersService 接口的实现类 UsersService。UsersService 实现类的文件名为 UsersService.java，其代码如下：

```
package org.etspace.rsgl.service.impl;
import org.etspace.rsgl.dao.IUsersDAO;
import org.etspace.rsgl.vo.Users;
import org.etspace.rsgl.service.IUsersService;
import java.util.List;
public class UsersService implements IUsersService {
    private IUsersDAO usersDAO;
    public void setUsersDAO(IUsersDAO usersDAO) {
        this.usersDAO = usersDAO;
    }
    public void save(Users users) {
        usersDAO.save(users);
    }
    public void delete(String username) {
        usersDAO.delete(username);
    }
    public void update(Users users) {
        usersDAO.update(users);
    }
    public Users find(String username) {
        return usersDAO.find(username);
    }
    public List findAll(int pageNow, int pageSize, Users users) {
        return usersDAO.findAll(pageNow, pageSize, users);
    }
    public int findAllSize(Users users) {
        return usersDAO.findAllSize(users);
    }
    public Users check(String username, String password) {
```

```
        return usersDAO.check(username, password);
    }
    public boolean exist(String username) {
        return usersDAO.exist(username);
    }
}
```

(3) 将 UsersService 作为一个组件交由 Spring 容器管理。为此，只需在 Spring 的配置文件 applicationContext.xml 中为 UsersService 添加一个相应的 bean 元素即可。有关代码如下：

```
<bean id="usersService"
class="org.etspace.rsgl.service.impl.UsersService">
    <property name="usersDAO">
        <ref bean="usersDAO"></ref>
    </property>
</bean>
```

该 bean 元素的 id 为 usersService，实现类为 UsersService，并通过 property 子元素为属性 usersDAO 依赖注入一个 UsersDAO 实例(UsersDAO 所对应的 Bean 的 id 为 usersDAO)。

10.6.2 部门管理 Service 组件及其实现

部门管理 Service 组件主要用于提供与部门管理功能相关的各种操作。其实现步骤如下。

(1) 在 org.etspace.rsgl.service 包中创建部门管理 Service 接口 IBmbService。IBmbService 接口的文件名为 IBmbService.java，其代码如下：

```
package org.etspace.rsgl.service;
import org.etspace.rsgl.vo.Bmb;
import java.util.List;
public interface IBmbService {
    //增加部门
    public void save(Bmb bmb);
    //删除部门
    public void delete(String bmbh);
    //修改部门
    public void update(Bmb bmb);
    //查询部门
    public Bmb find(String bmbh);
    //分页查询部门
    public List findAll(int pageNow,int pageSize,Bmb bmb);
    //查询部门总数
    public int findAllSize(Bmb bmb);
    //获取所有部门
    public List getAll();
    //是否存在部门
    public boolean exist(String bmbh);
}
```

(2) 在 org.etspace.rsgl.service.impl 包中创建 IBmbService 接口的实现类 BmbService。BmbService 实现类的文件名为 BmbService.java，其代码如下：

```
package org.etspace.rsgl.service.impl;
import org.etspace.rsgl.service.IBmbService;
import org.etspace.rsgl.dao.IBmbDAO;
import org.etspace.rsgl.vo.Bmb;
import java.util.List;
public class BmbService implements IBmbService {
    private IBmbDAO bmbDAO;
    public void setBmbDAO(IBmbDAO bmbDAO) {
        this.bmbDAO = bmbDAO;
    }
    public void save(Bmb bmb) {
        bmbDAO.save(bmb);
    }
    public void delete(String bmbh) {
        bmbDAO.delete(bmbh);
    }
    public void update(Bmb bmb) {
        bmbDAO.update(bmb);
    }
    public Bmb find(String bmbh) {
        return bmbDAO.find(bmbh);
    }
    public List findAll(int pageNow, int pageSize, Bmb bmb) {
        return bmbDAO.findAll(pageNow, pageSize, bmb);
    }
    public int findAllSize(Bmb bmb) {
        return bmbDAO.findAllSize(bmb);
    }
    public List getAll() {
        return bmbDAO.getAll();
    }
    public boolean exist(String bmbh) {
        return bmbDAO.exist(bmbh);
    }
}
```

(3) 将 BmbService 作为一个组件交由 Spring 容器管理。为此，只需在 Spring 的配置文件 applicationContext.xml 中为 BmbService 添加一个相应的 bean 元素即可。有关代码如下：

```
<bean id="bmbService" class="org.etspace.rsgl.service.impl.BmbService">
    <property name="bmbDAO">
        <ref bean="bmbDAO"/>
    </property>
</bean>
```

该 bean 元素的 id 为 bmbService，实现类为 BmbService，并通过 property 子元素为属性 bmbDAO 依赖注入一个 BmbDAO 实例(BmbDAO 所对应的 Bean 的 id 为 bmbDAO)。

10.6.3 职工管理 Service 组件及其实现

职工管理 Service 组件主要用于提供与职工管理功能相关的各种操作。其实现步骤如下。

(1) 在 org.etspace.rsgl.service 包中创建职工管理 Service 接口 IZgbService。IZgbService 接口的文件名为 IZgbService.java，其代码如下：

```
package org.etspace.rsgl.service;
import org.etspace.rsgl.vo.Zgb;
import java.util.List;
public interface IZgbService {
    //增加职工
    public void save(Zgb zgb);
    //删除职工
    public void delete(String bh);
    //修改职工
    public void update(Zgb zgb);
    //查询职工
    public Zgb find(String bh);
    //分页查询职工
    public List findAll(int pageNow,int pageSize,Zgb zgb);
    //查询职工总数
    public int findAllSize(Zgb zgb);
    //获取所有职工
    public List getAll();
    //获取指定部门的职工
    public List getAllByBmbh(String bmbh);
    //是否存在职工
    public boolean exist(String bh);
}
```

(2) 在 org.etspace.rsgl.service.impl 包中创建 IZgbService 接口的实现类 ZgbService。ZgbService 实现类的文件名为 ZgbService.java，其代码如下：

```
package org.etspace.rsgl.service.impl;
import org.etspace.rsgl.service.IZgbService;
import org.etspace.rsgl.dao.IZgbDAO;
import org.etspace.rsgl.vo.Zgb;
import java.util.List;
public class ZgbService implements IZgbService {
    private IZgbDAO zgbDAO;
    public void setZgbDAO(IZgbDAO zgbDAO) {
        this.zgbDAO = zgbDAO;
    }
    public void save(Zgb zgb) {
        zgbDAO.save(zgb);
```

```
    }
    public void delete(String bh) {
        zgbDAO.delete(bh);
    }
    public void update(Zgb zgb) {
        zgbDAO.update(zgb);
    }
    public Zgb find(String bh) {
        return zgbDAO.find(bh);
    }
    public List findAll(int pageNow, int pageSize, Zgb zgb) {
        return zgbDAO.findAll(pageNow, pageSize, zgb);
    }
    public int findAllSize(Zgb zgb) {
        return zgbDAO.findAllSize(zgb);
    }
    public List getAll() {
        return zgbDAO.getAll();
    }
    public List getAllByBmbh(String bmbh) {
        return zgbDAO.getAllByBmbh(bmbh);
    }
    public boolean exist(String bh) {
        return zgbDAO.exist(bh);
    }
}
```

(3) 将 ZgbService 作为一个组件交由 Spring 容器管理。为此，只需在 Spring 的配置文件 applicationContext.xml 中为 ZgbService 添加一个相应的 bean 元素即可。有关代码如下：

```
<bean id="zgbService" class="org.etspace.rsgl.service.impl.ZgbService">
    <property name="zgbDAO">
        <ref bean="zgbDAO"/>
    </property>
</bean>
```

该 bean 元素的 id 为 zgbService，实现类为 ZgbService，并通过 property 子元素为属性 zgbDAO 依赖注入一个 ZgbDAO 实例(ZgbDAO 所对应的 Bean 的 id 为 zgbDAO)。

10.7　表示层及其实现

作为 Java EE 应用系统的最上层，表示层与用户的实际使用体验密切相关，其开发任务就是实现系统的各项具体功能，主要包括有关页面的设计、相应 Action 的编写与配置，以及相关拦截器的编写与配置等。总体来说，表示层开发人员只需调用业务层所提供的各项服务，合理实现面向特定应用的各项系统功能即可。

10.7.1 素材文件的准备

对于一个 Web 应用系统的开发者来说，通常要先准备好相应的素材文件，如页面设计所需要的图片文件与层叠样式表文件等。

1. 图片文件

本人事管理系统所需要的图片文件如图 10.4 所示，只需将其复制到项目的 images 子文件夹中即可。

图 10.4　项目所需要的图片文件

2. 层叠样式表文件

在项目的 WebRoot 文件夹中创建一个层叠样式表文件 stylesheet.css，其代码如下：

```
body
{
    font-size: 9pt;
    font-family: 宋体;
    color: #3366FF;
}
A
{
    FONT-SIZE: 9pt;
    TEXT-DECORATION: underline;
    color: #3366FF;
}
A:link {
    FONT-SIZE: 9pt;
    TEXT-DECORATION: none;
    color: #3366FF;
}
A:visited {
    FONT-SIZE: 9pt;
    TEXT-DECORATION: none;
    color: #3366FF;
}
A:active {
    FONT-SIZE: 9pt;
    TEXT-DECORATION: none;
    color: #3366FF;
}
A:hover {
    COLOR: red;
    TEXT-DECORATION: underline
}
TABLE {
```

```
    FONT-SIZE: 9pt
}
TR {
    FONT-SIZE: 9pt
}
TD {
    FONT-SIZE: 9pt;
}
```

10.7.2　公用模块的实现

本系统有关功能的实现依赖于一些公用的模块，主要包括一个分页类、一个登录检查拦截器类与一个权限检查拦截器类等。

1. 分页类 Pager

本项目在实现用户、部门与职工的查询功能时，均以分页方式显示相应的查询结果。在查询结果显示页面中，通常要显示首页、上一页、下一页与尾页的链接，以便于用户的浏览操作。为此，可先创建一个分页类，以利于分页功能的具体实现。

在本项目中，分页类为 Pager，创建于 org.etspace.rsgl.tool 包中，其文件名为 Pager.java，代码如下：

```
package org.etspace.rsgl.tool;
public class Pager {
    private int pageNow;          //当前页码
    private int pageSize;         //每页显示的记录数
    private int totalPage;        //总页数
    private int totalSize;        //记录总数
    private boolean hasFirst;     //是否有首页
    private boolean hasPre;       //是否有前一页
    private boolean hasNext;      //是否有下一页
    private boolean hasLast;      //是否有最后一页
    //构造方法(函数)
    public Pager(int pageNow,int pageSize,int totalSize){
        this.pageNow=pageNow;
        this.pageSize=pageSize;
        this.totalSize=totalSize;
    }
    public int getPageNow() {
        return pageNow;
    }
    public void setPageNow(int pageNow) {
        this.pageNow = pageNow;
    }
    public int getPageSize() {
        return pageSize;
    }
    public void setPageSize(int pageSize) {
        this.pageSize = pageSize;
```

```
    }
    public int getTotalPage() {
        totalPage=totalSize/pageSize;
        if(totalSize%pageSize!=0)
            totalPage++;
        return totalPage;
    }
    public void setTotalPage(int totalPage) {
        this.totalPage = totalPage;
    }
    public int getTotalSize() {
        return totalSize;
    }
    public void setTotalSize(int totalSize) {
        this.totalSize = totalSize;
    }
    public boolean isHasFirst() {
        //若当前页为第一页，则无首页
        if(pageNow==1)
            return false;
        else
            return true;
    }
    public void setHasFirst(boolean hasFirst) {
        this.hasFirst = hasFirst;
    }
    public boolean isHasPre() {
        //若有首页，则当前页不是第一页，因此有前一页
        if(this.isHasFirst())
            return true;
        else
            return false;
    }
    public void setHasPre(boolean hasPre) {
        this.hasPre = hasPre;
    }
    public boolean isHasNext() {
        //若有尾页，则当前页不是最后一页，因此有下一页
        if(isHasLast())
            return true;
        else
            return false;
    }
    public void setHasNext(boolean hasNext) {
        this.hasNext = hasNext;
    }
    public boolean isHasLast() {
        //若当前页为最后一页，或总页数为 0，则无尾页
        if(pageNow==this.getTotalPage() || this.getTotalPage()==0)
            return false;
```

```
        else
            return true;
    }
    public void setHasLast(boolean hasLast) {
        this.hasLast = hasLast;
    }
}
```

2. 登录检查拦截器类 LoginFilter

本系统规定，用户必须登录成功后，方可使用系统的有关功能。为检验用户是否已经登录，可设计一个专门的登录检查拦截器。其主要实现步骤如下。

(1) 在 org.etspace.rsgl.tool 包中创建一个登录检查拦截器类 LoginFilter，其文件名为 LoginFilter.java，代码如下：

```
package org.etspace.rsgl.tool;
import java.util.Map;
import org.etspace.rsgl.vo.Users;
import com.opensymphony.xwork2.Action;
import com.opensymphony.xwork2.ActionInvocation;
import com.opensymphony.xwork2.interceptor.AbstractInterceptor;
public class LoginFilter extends AbstractInterceptor{
    public String intercept(ActionInvocation arg0) throws Exception {
        Map session=arg0.getInvocationContext().getSession();
        Users user=(Users) session.get("user");
        if(user==null){
            return Action.LOGIN;
        }
        return arg0.invoke();
    }
}
```

(2) 在项目的 src 文件夹中创建配置文件 struts.xml，并在其中完成 LoginFilter 拦截器的配置，其代码如下：

```
<?xml version="1.0" encoding="UTF-8"?>
<!DOCTYPE struts PUBLIC
    "-//Apache Software Foundation//DTD Struts Configuration 2.0//EN"
    "http://struts.apache.org/dtds/struts-2.0.dtd">
<struts>
    <package name="rsgl" extends="struts-default">
        <interceptors>
            <!-- 定义了一个名为 loginFilter 的拦截器 -->
            <interceptor name="loginFilter"
             class="org.etspace.rsgl.tool.LoginFilter"/>
        </interceptors>
    </package>
</struts>
```

(3) 在项目的 struts.xml 文件中添加一个名为 login 的全局结果配置。其有关代码如下：

```
<package name="rsgl" extends="struts-default">
    …
    <!-- 全局跳转，凡名为 login 结果的均跳转到 login.jsp -->
    <global-results>
        <result name="login">/login.jsp</result>
    </global-results>
</package>
```

其中，login.jsp 为本系统的系统登录页面。

3. 权限检查拦截器类 AuthorityFilter

在本系统中，有些功能只能由系统管理员执行，因此，在开始执行某项功能前，应对用户的类型进行判断，以确定其是否具有相应的操作权限。为此，可设计一个专门的权限检查拦截器类。其主要实现步骤如下。

(1) 在 org.etspace.rsgl.tool 包中创建一个权限检查拦截器类 AuthorityFilter，其文件名为 AuthorityFilter.java，代码如下：

```
package org.etspace.rsgl.tool;
import java.util.Map;
import org.etspace.rsgl.vo.Users;
import com.opensymphony.xwork2.Action;
import com.opensymphony.xwork2.ActionInvocation;
import com.opensymphony.xwork2.interceptor.AbstractInterceptor;
public class AuthorityFilter extends AbstractInterceptor{
    public String intercept(ActionInvocation arg0) throws Exception {
        Map session=arg0.getInvocationContext().getSession();
        Users user=(Users) session.get("user");
        if(user.getUsertype().compareTo("系统管理员")!=0){
            return Action.NONE;
        }
        return arg0.invoke();
    }
}
```

(2) 在项目的 struts.xml 文件中完成 AuthorityFilter 拦截器的配置。其有关代码如下：

```
<interceptors>
    …
    <!-- 定义了一个名为 authorityFilter 的拦截器 -->
    <interceptor name="authorityFilter"
class="org.etspace.rsgl.tool.AuthorityFilter"/>
</interceptors>
```

10.7.3 登录功能的实现

本系统的系统登录页面如图 10.5 所示。在此页面中，输入正确的用户名与密码，再单击“确定”按钮，即可打开图 10.6 所示的系统主界面。反之，若输入的用户名或密码不正确，则系统将重新打开系统登录页面，等待用户再次登录。

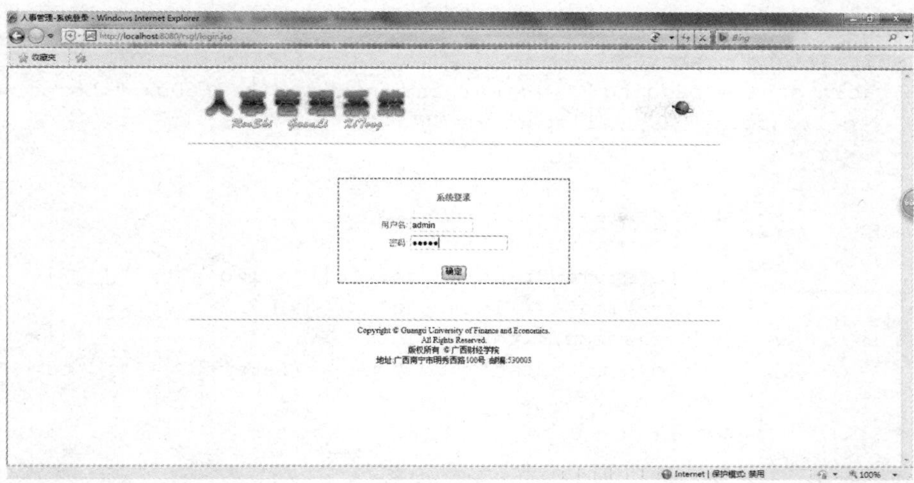

图 10.5　系统登录页面

图 10.6　系统主界面

系统登录功能的实现过程如下所述。

(1) 在项目的 WebRoot 文件夹中添加一个新的 JSP 页面 login.jsp，其代码如下：

```
<%@ page language="java" import="java.util.*" pageEncoding="UTF-8"%>
<%@ taglib prefix="s" uri="/struts-tags" %>
<!DOCTYPE HTML PUBLIC "-//W3C//DTD HTML 4.01 Transitional//EN">
<html>
  <head>
    <title>人事管理-系统登录</title>
    <meta http-equiv="pragma" content="no-cache">
    <meta http-equiv="cache-control" content="no-cache">
    <meta http-equiv="expires" content="0">
    <meta http-equiv="keywords" content="人事管理">
    <link rel="stylesheet" href="stylesheet.css" type="text/css"></link>
  </head>
```

```
<body>
  <div align="center">
  <table style="padding: 0px; margin: 0px; width: 800px;" border="0"
     cellpadding="0" cellspacing="0">
     <tr>
      <td>
           <table style="width:100%;">
              <tr>
                 <td style="text-align: left"><img alt=""
                 src="images/Title.png"></td>
                 <td> </td>
                 <td><img alt="" src="images/LuEarth.GIF"></td>
              </tr>
           </table>
      </td>
     </tr>
     <tr>
        <td><hr /></td>
     </tr>
     <tr>
        <td> </td>
     </tr>
     <tr>
        <td> </td>
     </tr>
     <tr>
        <td> </td>
     </tr>
     <tr>
        <td align="center">
              <s:form action="login.action" method="post"
               theme="simple">
              <table style="border: thin dashed #008080;" width="350"
               align="center">
              <tr>
              <td style="width: 30%"> </td>
              <td style="width: 70%"> </td>
              </tr>
              <tr>
              <td align="center" colspan="2">
              <b>系统登录</b>
              </td>
              </tr>
              <tr>
              <td> </td>
              <td> </td>
              </tr>
              <tr>
              <td align="right">用户名: </td>
              <td>
```

```
                    <s:textfield name="users.username" label="用户名"
                     size="10"/>
                </td>
                </tr>
                <tr>
                <td align="right">密码: </td>
                <td>
                    <s:password name="users.password" label="密码"
                     size="20"/>
                </td>
                </tr>
                <tr>
                <td> </td>
                <td> </td>
                </tr>
                <tr>
                <td align="center" colspan="2">
                <s:submit value="确定"/>
                </td>
                </tr>
                </table>
                </s:form>
          </td>
      </tr>
      <tr>
          <td> </td>
      </tr>
      <tr>
          <td> </td>
      </tr>
      <tr>
          <td><hr /></td>
      </tr>
      <tr>
          <td style="text-align: center">
                <font color="#330033">Copyright &copy; Guangxi University
                 of Finance and Economics.<br />
                All Rights Reserved.</font><br />
                <font color="#330033">版权所有 &copy; 广西财经学院</font><br />
                <font color="#330033">地址:广西南宁市明秀西路 100 号
                邮编:530003</font><br />
          </td>
      </tr>
  </table>
  </div>
 </body>
</html>
```

(2) 在 org.etspace.rsgl.action 包中新建一个 Action 实现类 UsersAction，其文件名为 UsersAction.java，代码如下：

```
package org.etspace.rsgl.action;
import com.opensymphony.xwork2.ActionSupport;
import com.opensymphony.xwork2.ActionContext;
import org.etspace.rsgl.vo.Users;
import org.etspace.rsgl.service.IUsersService;
import java.util.Map;
public class UsersAction extends ActionSupport{
    private Users users;
    private IUsersService usersService;
    public Users getUsers() {
        return users;
    }
    public void setUsers(Users users) {
        this.users = users;
    }
    public IUsersService getUsersService() {
        return usersService;
    }
    public void setUsersService(IUsersService usersService) {
        this.usersService = usersService;
    }
    public String execute()throws Exception{
        return SUCCESS;
    }
    public String login()throws Exception{
        Users user=usersService.check(users.getUsername(),
users.getPassword());
        if(user!=null){
            Map session=(Map)ActionContext.getContext().getSession();
            session.put("user", user);
            this.users=null;
            return SUCCESS;
        }else{
            return ERROR;
        }
    }
}
```

(3) 在 Spring 的配置文件 applicationContext.xml 中添加一个 bean 元素，其实现类为 UsersAction，id 为 usersAction。有关代码如下：

```
<bean id="usersAction" class="org.etspace.rsgl.action.UsersAction"
scope="prototype">
    <property name="usersService">
        <ref bean="usersService"></ref>
    </property>
</bean>
```

(4) 在项目的 struts.xml 文件中完成名为 login 的 Action 配置，其有关代码如下：

```
<action name="login" class="usersAction" method="login">
```

```
    <result name="success">/indexAdmin.jsp</result>
    <result name="error">/login.jsp</result>
</action>
```

(5) 在项目的 WebRoot 文件夹中添加一个新的 JSP 页面 indexAdmin.jsp，其代码如下：

```
<%@ page language="java" import="java.util.*" pageEncoding="UTF-8"%>
<!DOCTYPE HTML PUBLIC "-//W3C//DTD HTML 4.01 Transitional//EN">
<html>
  <head>
    <title>人事管理</title>
    <link rel="stylesheet" href="stylesheet.css" type="text/css"></link>
  </head>
  <frameset rows="100,*" cols="*">
      <frame src="./main.jsp" name="topFrame" scrolling="yes" id="topFrame">
      <frameset rows="*" cols="200,*">
          <frame src="./menu.jsp" name="leftFrame" scrolling="auto"
id="leftFrame">
          <frame src="./home.jsp" name="rightFrame" scrolling="yes"
id="rightFrame">
      </frameset>
  </frameset>
  <noframes>
  <body>
      <div>
      此网页使用了框架，但您的浏览器不支持框架。
      </div>
  </body>
  </noframes>
</html>
```

10.7.4　系统主界面的实现

系统的主界面由 indexAdmin.jsp 页面生成。该页面其实是一个框架页面，用于将主界面分为 3 个部分。其中，上方用于打开 main.jsp 页面，下方左侧用于打开 menu.jsp 页面，下方右侧用于打开 home.jsp 页面。

1. main.jsp 页面

main.jsp 页面主要用于显示系统的标题图片与当前用户的基本信息，并提供一个“注销”链接以安全退出系统。该页面的代码如下：

```
<%@ page language="java" import="java.util.*" pageEncoding="UTF-8"%>
<%@ taglib prefix="s" uri="/struts-tags" %>
<!DOCTYPE HTML PUBLIC "-//W3C//DTD HTML 4.01 Transitional//EN">
<html>
  <head>
    <title>人事管理</title>
```

```
    <link rel="stylesheet" href="./stylesheet.css"
type="text/css"></link>
  </head>
<body>
    <div>
          <table style="width: 100%;">
             <tr>
                 <td style="width: 80%">
                     <img alt="" src="./images/Title.png">
                 </td>
                 <td style="width: 20%" align="center">
                     <img alt="" src="./images/LuEarth.GIF">
                     <br />
                     [<s:property value="#session.user.username" />|<s:property
value="#session.user.usertype" />|<a href="./logout.action" target=
"_top">注销</a>]</td>
             </tr>
             <tr>
                 <td>  </td>
                 <td>  </td>
                 <td>  </td>
             </tr>
             </table>
    </div>
</body>
</html>
```

2. menu.jsp 页面

menu.jsp 页面主要用于显示系统的功能菜单，内含一系列用于执行相应功能的超链接。该页面的代码如下：

```
<%@ page language="java" import="java.util.*" pageEncoding="UTF-8"%>
<!DOCTYPE HTML PUBLIC "-//W3C//DTD HTML 4.01 Transitional//EN">
<html>
  <head>
    <title>人事管理</title>
    <link rel="stylesheet" href="./stylesheet.css"
type="text/css"></link>
  </head>
<body>
    <div>
        <table border="0" width="150px">
            <tr>
                <td align="center" bgcolor="#66CCFF">
                     </td>
            </tr>
            <tr>
                <td>
                     </td>
            </tr>
```

```
            <tr>
                <td>
                    <img alt="" src="./images/LuVred.png">部门管理</td>
            </tr>
            <tr>
                <td>
                     <img alt="" src="./images/LuArrow.gif">
                    <a href="bmbAddView.action" target="rightFrame">部门增加
</a></td>
            </tr>
            <tr>
                <td>
                     <img alt="" src="./images/LuArrow.gif">
                    <a href="bmbList.action" target="rightFrame">部门维护
</a></td>
            </tr>
            <tr>
                <td>
                     </td>
            </tr>
            <tr>
                <td>
                    <img alt="" src="./images/LuVblue.png">职工管理</td>
            </tr>
            <tr>
                <td>
                     <img alt="" src="./images/LuArrow.gif">
                    <a href="zgbAddView.action" target="rightFrame">职工增加
<br></a></td>
            </tr>
            <tr>
                <td>
                     <img alt="" src="./images/LuArrow.gif">
                    <a href="zgbList.action" target="rightFrame">职工维护
</a></td>
            </tr>
            <tr>
                <td>
                     </td>
            </tr>
            <tr>
                <td>
                    <img alt="" src="./images/LuVred.png">用户管理</td>
            </tr>
            <tr>
                <td>
                     <img alt="" src="./images/LuArrow.gif">
                    <a href="usersAddView.action" target="rightFrame">用户增
加</a></td>
            </tr>
```

```
            <tr>
                <td>
                     <img alt="" src="./images/LuArrow.gif">
                    <a href="usersList.action" target="rightFrame">用户维护
</a></td>
            </tr>
            <tr>
                <td>
                     </td>
            </tr>
            <tr>
                <td>
                    <img alt="" src="./images/LuVblue.png"> 当前用户</td>
            </tr>
            <tr>
                <td>
                     <img alt="" src="./images/LuArrow.gif">
                    <a href="usersSetPwdView.action" target="rightFrame">密
码设置</a></td>
            </tr>
            <tr>
                <td>
                     <img alt="" src="./images/LuArrow.gif">
                    <a href="logout.action" target="_top">安全退出</a></td>
            </tr>
            <tr>
                <td>
                     </td>
            </tr>
            <tr>
                <td bgcolor="#66CCFF">
                     </td>
            </tr>
        </table>
    </div>
</body>
</html>
```

3. home.jsp 页面

home.jsp 页面用于显示一张欢迎图片，其实是系统工作区的初始界面。该页面的代码如下：

```
<%@ page language="java" import="java.util.*" pageEncoding="UTF-8"%>
<!DOCTYPE HTML PUBLIC "-//W3C//DTD HTML 4.01 Transitional//EN">
<html>
  <head>
    <title>人事管理</title>
    <link rel="stylesheet" href="./stylesheet.css"
type="text/css"></link>
```

```
  </head>
<body>
    <div>
        <table style="width:500px;" align="center">
            <tr>
                <td align="center" height="180px">
                     </td>
            </tr>
            <tr>
                <td align="center">
                    <img alt="" src="./images/welcome.png">
                </td>
            </tr>
            <tr>
                <td align="center" height="180px">
                     </td>
            </tr>
            <tr>
                <td align="center">
                    <hr /></td>
            </tr>
            <tr>
                <td align="center">
                      <font color="#330033">Copyright &copy;All Rights
Reserved.</font></td>
            </tr>
        </table>
    </div>
</body>
</html>
```

10.7.5 当前用户功能的实现

当前用户功能仅针对当前用户(系统管理员或普通用户)自身，可由当前用户根据需要随时执行。在本系统中，当前用户功能共有两项，即密码设置与安全退出。其中，密码设置功能用于设置或更改当前用户的登录密码，安全退出功能用于清除当前用户的有关信息并退出系统。

1. 密码设置

在系统主界面中单击“密码设置”链接，将打开图 10.7 所示的密码设置页面，在其中输入欲设置的密码后，再单击“确定”按钮，即可完成密码的设置或更改，并显示图 10.8 所示的操作成功页面。若未输入密码便直接单击“确定”按钮，则会显示图 10.9 所示的操作失败页面。

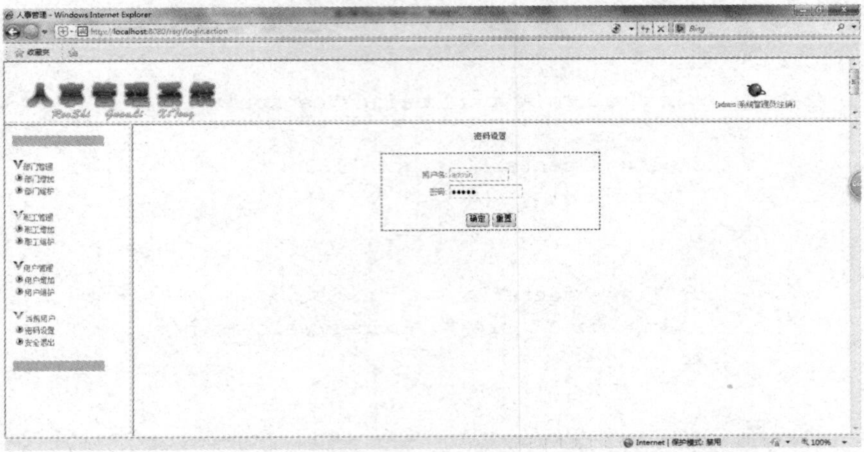

图 10.7　密码设置页面

图 10.8　操作成功页面

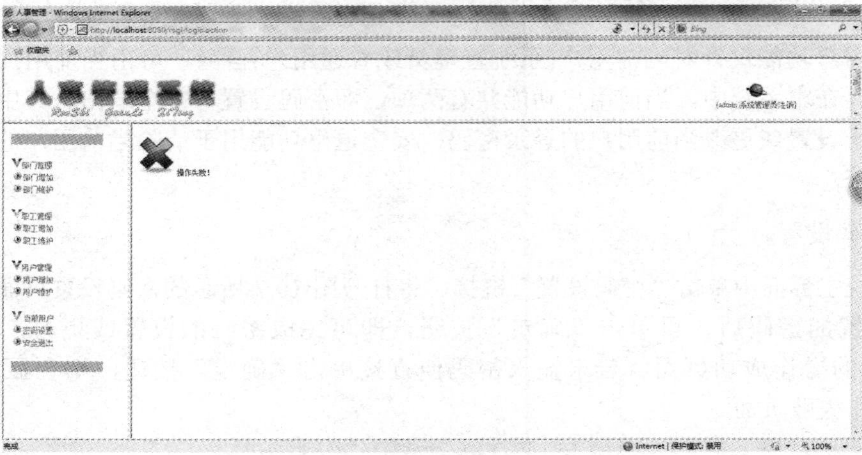

图 10.9　操作失败页面

密码设置功能的实现过程如下所述。

(1) 在 UsersAction 实现类中添加 usersSetPwdView()与 usersSetPwd()方法，其有关代码如下：

```
    public String usersSetPwdView()throws Exception{
        Map session=(Map)ActionContext.getContext().getSession();

    users=usersService.find(((Users)session.get("user")).getUsername());
        return SUCCESS;
    }
    public String usersSetPwd()throws Exception{
        if (users.getPassword().length()!=0){
            usersService.update(users);
            Users user=usersService.check(users.getUsername(),
users.getPassword());
            Map session=(Map)ActionContext.getContext().getSession();
            session.put("user", user);
            this.users=null;
            return SUCCESS;
        }else{
            return ERROR;
        }
    }
```

(2) 在项目的 struts.xml 文件中完成名为 usersSetPwdView 与 usersSetPwd 的 Action 配置，其有关代码如下：

```
<action name="usersSetPwdView" class="usersAction"
method="usersSetPwdView">
    <result name="success">/usersSetPwdView.jsp</result>
    <interceptor-ref name="defaultStack"></interceptor-ref>
    <interceptor-ref name="loginFilter"/>
</action>
<action name="usersSetPwd" class="usersAction" method="usersSetPwd">
    <result name="success">/success.jsp</result>
    <result name="error">/error.jsp</result>
    <interceptor-ref name="defaultStack"></interceptor-ref>
    <interceptor-ref name="loginFilter"/>
</action>
```

(3) 在项目的 WebRoot 文件夹中添加一个新的 JSP 页面 usersSetPwdView.jsp，其代码如下：

```
<%@ page language="java" import="java.util.*" pageEncoding="UTF-8"%>
<%@ taglib prefix="s" uri="/struts-tags" %>
<%  //request.setCharacterEncoding("UTF-8"); %>
<html>
  <head>
    <title>系统用户一密码设置</title>
    <link rel="stylesheet" href="./stylesheet.css"
type="text/css"></link>
```

```
  </head>
  <body>
  <div align="center">
  <b>密码设置</b><br>
  <br>
  <s:form action="usersSetPwd.action" method="post" theme="simple">
  <table align="center" width="350" style="border: thin dashed rgb(0, 128,
128);">
  <tr>
  <td style="width: 30%;"> </td>
  <td style="width: 70%;"> </td>
  </tr>
  <tr>
  <td align="right">用户名:</td>
  <td>
    <s:textfield name="users.username" label="用户名" size="10"
maxlength="10" disabled="true"></s:textfield>
  </td>
  </tr>
  <tr>
  <td align="right">密码:</td>
  <td>
    <s:password name="users.password" label="密码" size="15"
maxlength="20"></s:password>
  </td>
  </tr>
  <tr>
  <td> </td>
  <td>
    <s:hidden name="users.username"></s:hidden>
    <s:hidden name="users.usertype"></s:hidden>
  </td>
  </tr>
  <tr>
  <td align="center" colspan="2">
  <s:submit value="确定"></s:submit>
  <s:reset value="重置"></s:reset>
  </td>
  </tr>
  </table>
  </s:form>
  </div>
  </body>
</html>
```

(4) 在项目的 WebRoot 文件夹中添加一个新的 JSP 页面 success.jsp，其代码如下：

```
<%@ page language="java" import="java.util.*" pageEncoding="UTF-8"%>
<%@ taglib prefix="s" uri="/struts-tags" %>
<html>
  <head>
```

```
    <title>操作信息</title>
    <link rel="stylesheet" href="./stylesheet.css"
type="text/css"></link>
  </head>
  <body>
    <img src="./images/LuRight.jpg">
    <FONT color="green">操作成功！</FONT>
  </body>
</html>
```

(5)　在项目的 WebRoot 文件夹中添加一个新的 JSP 页面 error.jsp，其代码如下：

```
<%@ page language="java" import="java.util.*" pageEncoding="UTF-8"%>
<%@ taglib prefix="s" uri="/struts-tags" %>
<html>
  <head>
    <title>操作信息</title>
    <link rel="stylesheet" href="./stylesheet.css"
type="text/css"></link>
  </head>
  <body>
    <img src="./images/LuWrong.jpg">
    <FONT color="red">操作失败！</FONT>
    <s:fielderror/>
  </body>
</html>
```

2. 安全退出

在系统主界面中单击“安全退出”链接(或“注销”链接)，将直接关闭系统的主界面，并打开图 10.10 所示的系统登录页面。

图 10.10　系统登录页面

安全退出功能的实现过程如下所述。

(1)　在 UsersAction 实现类中添加 logout()方法，其有关代码如下：

```
public String logout()throws Exception{
    Map session=(Map)ActionContext.getContext().getSession();
    session.remove("user");
    this.users=null;
    return SUCCESS;
}
```

(2) 在项目的 struts.xml 文件中完成名为 logout 的 Action 配置。其有关代码如下：

```
<action name="logout" class="usersAction" method="logout">
    <result name="success">/login.jsp</result>
</action>
```

10.7.6 用户管理功能的实现

用户管理功能包括用户的增加与维护，而用户的维护又包括用户的查询、修改、删除与密码重置。本系统规定，用户管理功能只能由系统管理员使用。

1. 用户增加

在系统主界面中单击“用户增加”链接，若当前用户为普通用户，将打开图 10.11 所示的“您无此操作权限！”页面；反之，若当前用户为系统管理员，将打开图 10.12 所示的“用户增加”页面。在其中输入用户名并选定相应的用户类型后，再单击“确定”按钮，若能成功添加用户，将显示图 10.13 所示的“操作成功”页面；否则，将显示图 10.14 所示的“操作失败”页面。

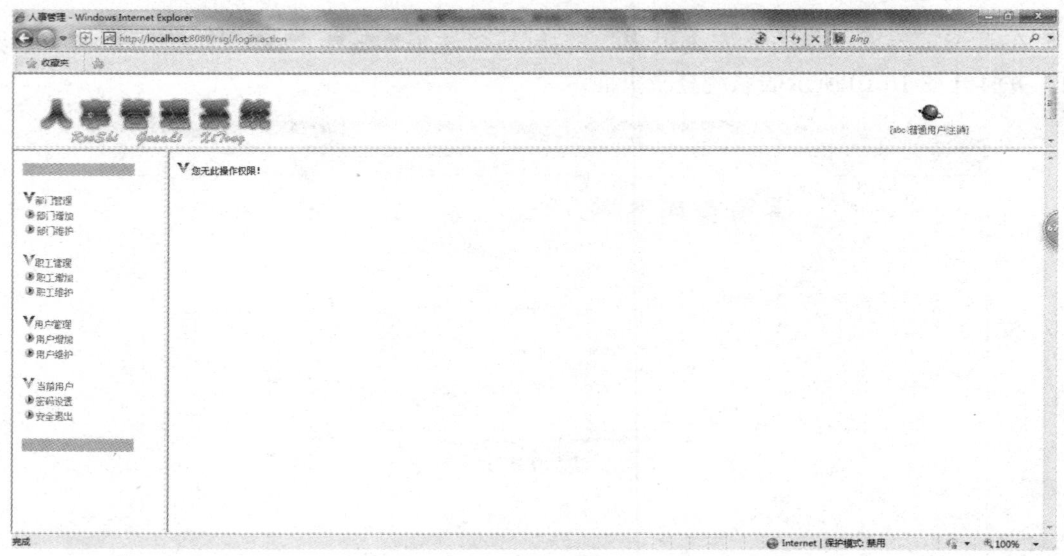

图 10.11 “您无此操作权限！”页面

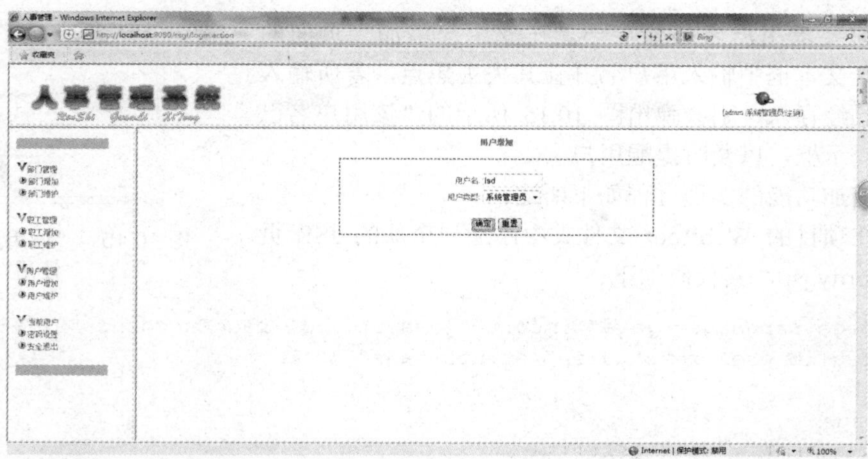

图 10.12　“用户增加”页面

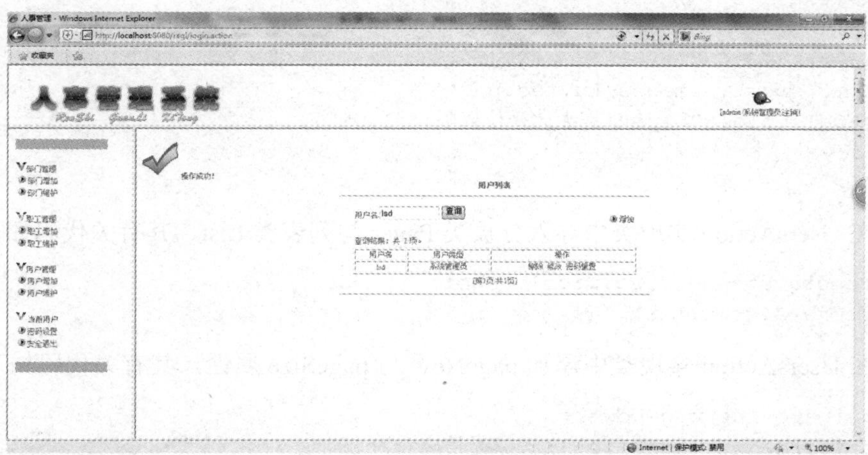

图 10.13　“操作成功！”页面

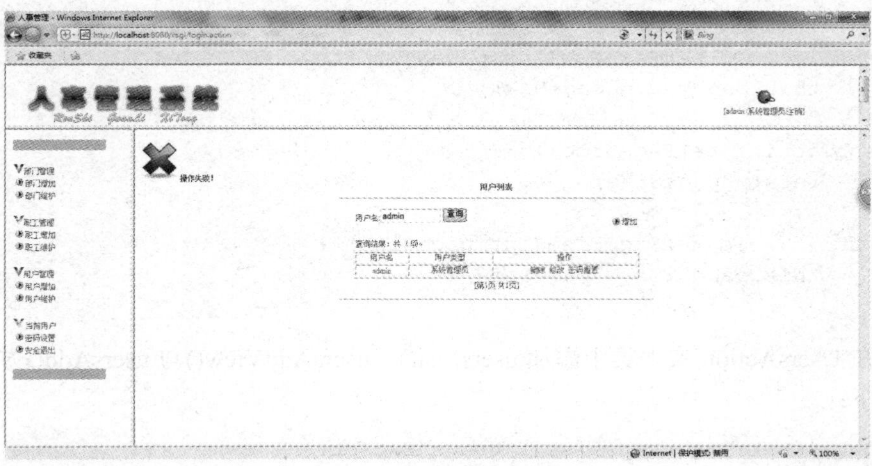

图 10.14　“操作失败！”页面

本系统要求用户名必须唯一。在“用户增加”页面的“用户名”文本框中输入用户名并让其失去焦点，若所输入的用户名已经存在，则会弹出图 10.15 所示的“该用户名已被使用”提示框，以及时提醒用户。

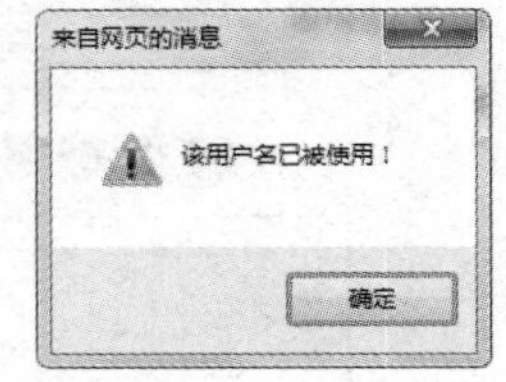

图 10.15　“该用户名已被使用”提示框

用户增加功能的实现过程如下所述。

(1) 在项目的 WebRoot 文件夹中添加一个新的 JSP 页面 noAuthority.jsp，其代码如下：

```
<%@ page language="java" import="java.util.*" pageEncoding="UTF-8"%>
<%@ taglib prefix="s" uri="/struts-tags" %>
<html>
  <head>
    <title>操作信息</title>
    <link rel="stylesheet" href="./stylesheet.css"
type="text/css"></link>
  </head>
  <body>
    <img src="./images/LuVred.png">
    <FONT color="blue">您无此操作权限！</FONT>
  </body>
</html>
```

(2) 在 UsersAction 实现类中导入分页类 Pager 与列表类 List，其有关代码如下：

```
import org.etspace.rsgl.tool.Pager;
import java.util.List;
```

(3) 在 UsersAction 实现类中添加 pageNow 与 pageSize 属性，其有关代码如下：

```
    private int pageNow=1;
    private int pageSize=2;  //通常应设为10
    public int getPageNow() {
        return pageNow;
    }
    public void setPageNow(int pageNow) {
        this.pageNow = pageNow;
    }
    public int getPageSize() {
        return pageSize;
    }
    public void setPageSize(int pageSize) {
        this.pageSize = pageSize;
    }
```

(4) 在 UsersAction 实现类中添加 usersList()、usersAddView()与 usersAdd()方法，其有关代码如下：

```
    public String usersList()throws Exception{
        List list=usersService.findAll(pageNow, pageSize, users);
        Map request=(Map)ActionContext.getContext().get("request");
```

```
        Pager page=new Pager(pageNow,pageSize,
        usersService.findAllSize(users));
        request.put("list", list);
        request.put("page", page);
        return SUCCESS;
    }
    public String usersAddView()throws Exception{
        return SUCCESS;
    }
    public String usersAdd()throws Exception{
        if (users.getUsername().length()!=0 && usersService.find
        (users.getUsername())==null){
            users.setPassword(users.getUsername());
            usersService.save(users);
            return SUCCESS;
        }else{
            return ERROR;
        }
    }
```

(5) 在项目的 struts.xml 文件中完成名为 usersList、usersAddView 与 usersAdd 的 Action 配置，其有关代码如下：

```
<action name="usersList" class="usersAction" method="usersList">
    <result name="success">/usersList.jsp</result>
    <result name="none">/noAuthority.jsp</result>
    <interceptor-ref name="defaultStack"></interceptor-ref>
    <interceptor-ref name="loginFilter"/>
    <interceptor-ref name="authorityFilter"/>
</action>
<action name="usersAddView" class="usersAction"
 method="usersAddView">
    <result name="success">/usersAddView.jsp</result>
    <result name="none">/noAuthority.jsp</result>
    <interceptor-ref name="defaultStack"></interceptor-ref>
    <interceptor-ref name="loginFilter"/>
    <interceptor-ref name="authorityFilter"/>
</action>
<action name="usersAdd" class="usersAction" method="usersAdd">
    <result name="success">/usersSuccess.jsp</result>
    <result name="error">/usersError.jsp</result>
    <result name="none">/noAuthority.jsp</result>
    <interceptor-ref name="defaultStack"></interceptor-ref>
    <interceptor-ref name="loginFilter"/>
    <interceptor-ref name="authorityFilter"/>
</action>
```

(6) 在项目的 WebRoot\WEB-INF 文件夹中创建 DWR 部署描述文件 dwr.xml，其代码如下：

```
<?xml version="1.0" encoding="UTF-8"?>
```

```
<!DOCTYPE dwr PUBLIC
   "-//GetAhead Limited//DTD Direct Web Remoting 1.0//EN"
   "http://www.getahead.ltd.uk/dwr/dwr10.dtd">
<dwr>
    <allow>
    <create creator="spring" javascript="AjaxUsersService">
            <param name="beanName" value="usersService"/>
            <include method="exist"/>
    </create>
    </allow>
</dwr>
```

(7) 在项目的 WebRoot 文件夹中添加一个新的 JSP 页面 usersAddView.jsp，其代码如下：

```
<%@ page language="java" import="java.util.*" pageEncoding="UTF-8"%>
<%@ taglib prefix="s" uri="/struts-tags" %>
<html>
  <head>
    <title>系统用户</title>
    <link rel="stylesheet" href="./stylesheet.css"
type="text/css"></link>
        <script type="text/javascript"
src="<%=request.getContextPath()%>/dwr/engine.js"></script>
        <script type="text/javascript"
src="<%=request.getContextPath()%>/dwr/util.js"></script>
        <script type="text/javascript"
src="<%=request.getContextPath()%>/dwr/interface/AjaxUsersService.js"></
script>
        <script type="text/javascript">
         function check() {
           var username=document.all.username.value;
           if (username=="") {
             window.alert("用户名不能为空！");
             return;
           }
           AjaxUsersService.exist(username,show);
         }
         function show(data) {
           if (data) {
             window.alert("该用户名已被使用！");
             return;
           }
         }
        </script>
  </head>
  <body>
  <div align="center">
  <b>用户增加</b>
  <br />
  <br />
```

```
  <s:form action="usersAdd.action" method="post" theme="simple">
  <table style="border: thin dashed #008080;" width="500" align="center">
  <tr>
  <td style="width: 45%"> </td>
  <td style="width: 55%"> </td>
  </tr>
  <tr>
  <td align="right">用户名:</td>
  <td>
    <s:textfield name="users.username" label="用户名" size="10"
maxlength="10" onblur="check()" id="username"/>
  </td>
  </tr>
  <tr>
  <td align="right">用户类型:</td>
  <td>
    <s:select list="{'系统管理员','普通用户'}" value="'普通用户'"
name="users.usertype" label="用户类型"></s:select>
  </td>
  </tr>
  <tr>
  <td> </td>
  <td> </td>
  </tr>
  <tr>
  <td align="center" colspan="2">
  <s:submit value="确定"/> <s:reset value="重置" />
  </td>
  </tr>
  </table>
  </s:form>
  </div>
  </body>
</html>
```

(8) 在项目的 WebRoot 文件夹中添加一个新的 JSP 页面 usersSuccess.jsp，其代码如下：

```
<%@ page language="java" import="java.util.*" pageEncoding="UTF-8"%>
<%@ taglib prefix="s" uri="/struts-tags" %>
<html>
  <head>
    <title>操作信息</title>
    <link rel="stylesheet" href="./stylesheet.css"
type="text/css"></link>
  </head>
  <body>
    <img src="./images/LuRight.jpg">
    <FONT color="green">操作成功！</FONT>
    <s:action name="usersList" executeResult="true"
ignoreContextParams="false"></s:action>
```

```
  </body>
</html>
```

(9) 在项目的 WebRoot 文件夹中添加一个新的 JSP 页面 usersError.jsp，其代码如下：

```
<%@ page language="java" import="java.util.*" pageEncoding="UTF-8"%>
<%@ taglib prefix="s" uri="/struts-tags" %>
<html>
  <head>
    <title>操作信息</title>
    <link rel="stylesheet" href="./stylesheet.css"
type="text/css"></link>
  </head>
  <body>
     <img src="./images/LuWrong.jpg">
     <FONT color="red">操作失败！</FONT>
     <s:fielderror/>
     <s:action name="usersList" executeResult="true"
ignoreContextParams="false"></s:action>
  </body>
</html>
```

(10) 在项目的 WebRoot 文件夹中添加一个新的 JSP 页面 usersList.jsp，其代码如下：

```
<%@ page language="java" import="java.util.*" pageEncoding="UTF-8"%>
<%@ taglib prefix="s" uri="/struts-tags" %>
<html>
  <head>
    <title>系统用户</title>
    <link rel="stylesheet" href="./stylesheet.css"
type="text/css"></link>
  </head>
  <body>
  <div align="center">
  <b>用户列表</b><br />
  <table width="500" border="0">
  <tr>
  <td> <hr /> </td>
  </tr>
  <tr>
  <td>
    <table width="450" align="center" border="0">
        <tr>
        <td>
            <form action="usersList.action" method="post">
            用户名:<input type="text" name="users.username"
value="<s:property value="users.username" />" size="10" maxlength="10"/>
            <input type="submit" value="查询"/>
            </form>
        </td>
        <td align="right">
            <img alt="" src="./images/LuArrow.gif" align="absmiddle">
            <a href="usersAddView.action" target="rightFrame">增加</a>
```

```
        </td>
        </tr>
    </table>
    <table width="450" align="center" border="0">
        <tr>
            <td>
            查询结果：共 <s:property value="#request.page.totalSize"/> 项。
            </td>
        </tr>
    </table>
    <table width="450" align="center" border="1">
        <tr align="center">
            <td>用户名</td><td>用户类型</td><td>操作</td>
        </tr>
        <s:iterator value="#request.list" id="users">
        <tr align="center">
            <td><s:property value="#users.username"/></td>
            <td><s:property value="#users.usertype"/></td>
            <td align="center">
             <a href="usersDelete.action?users.username=<s:property
value="#users.username"/>" onClick="if(!confirm('确定删除吗？'))return
false;else return true;">删除</a>
             <a href="usersUpdateView.action?users.username=<s:property
value="#users.username"/>">修改</a>
             <a href="usersResetPwd.action?users.username=<s:property
value="#users.username"/>" onClick="if(!confirm('确定重置密码吗？'))return
false;else return true;">密码重置</a>
            </td>
        </tr>
        </s:iterator>
    </table>
    <table width="450" align="center" border="0">
        <tr align="center">
        <td>
            <s:set name="page" value="#request.page"></s:set>
                <s:if test="#page.totalSize!=0">
                    [第<s:property value="#page.pageNow"/>页/共<s:property
value="#page.totalPage"/>页]
                </s:if>
                <s:if test="#page.hasFirst">
                    <a
href="usersList.action?pageNow=1&users.username=<s:property
value="users.username"/>">首页</a>
                </s:if>
                <s:if test="#page.hasPre">
                    <a href="usersList.action?pageNow=<s:property value=
"#page.pageNow-1"/>&users.username=<s:property value="users.username"/>
">上一页</a>
                </s:if>
                <s:if test="#page.hasNext">
                    <a href="usersList.action?pageNow=<s:property
value="#page.pageNow+1"/>&users.username=<s:property
value="users.username"/>">下一页</a>
```

```
                </s:if>
                <s:if test="#page.hasLast">
                    <a href="usersList.action?pageNow=<s:property
value="#page.totalPage"/>&users.username=<s:property
value="users.username"/>">尾页</a>
                </s:if>
            </td>
            </tr>
        </table>
    </td>
    </tr>
    <tr>
    <td> <hr /> </td>
    </tr>
    </table>
    </div>
    </body>
</html>
```

2. 用户维护

在系统主界面中单击“用户维护”链接，若当前用户为普通用户，将打开相应的“无此权限”页面；反之，若当前用户为系统管理员，将打开图 10.16 所示的“用户列表”页面。该页面以分页的方式显示系统的有关用户记录(在此为每页显示 2 个用户记录)，并支持按用户名对系统用户进行模糊查询，同时提供了增加新用户以及对各个用户进行删除、修改或密码重置操作的链接。其中，“增加”链接的作用与系统主界面中的“用户增加”链接的作用是一样的。

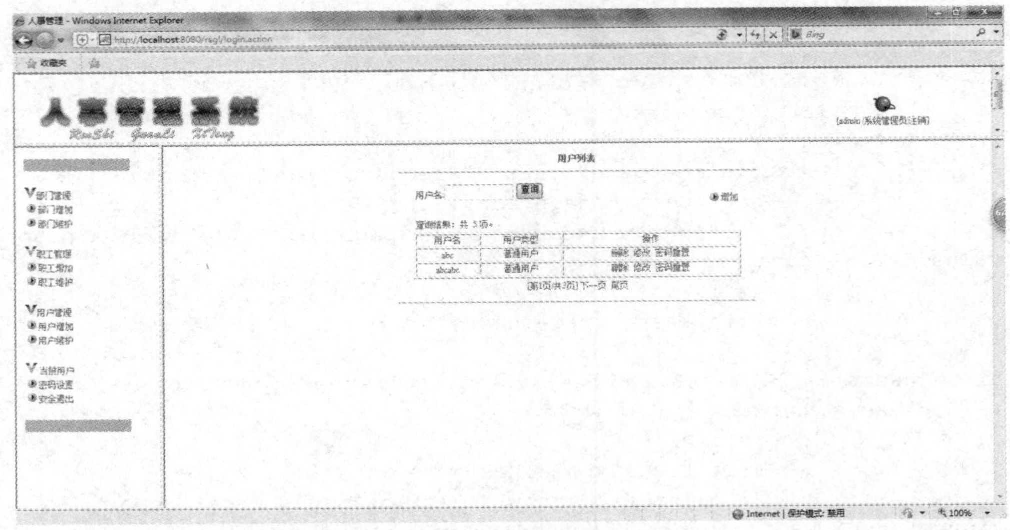

图 10.16 “用户列表”页面

要查询用户，只需在“用户列表”页面的“用户名”文本框中输入相应的查询条件，然后再单击“查询”按钮即可。

要删除用户，只需在用户列表中单击相应用户后的“删除”链接，并在弹出的图 10.17 所示的“确定删除吗？”提示框中单击“确定”按钮即可。若成功删除指定的用户，将显示图 10.18 所示的“操作成功！”页面；否则，将显示图 10.19 所示的“操作失败！”页面。

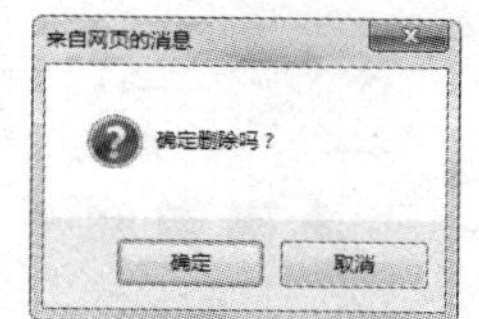

图 10.17　“确定删除吗？”提示框

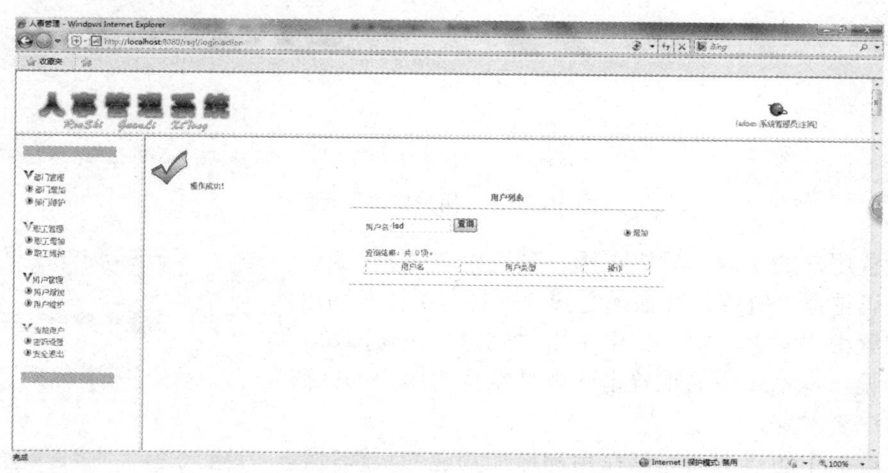

图 10.18　“操作成功！”页面

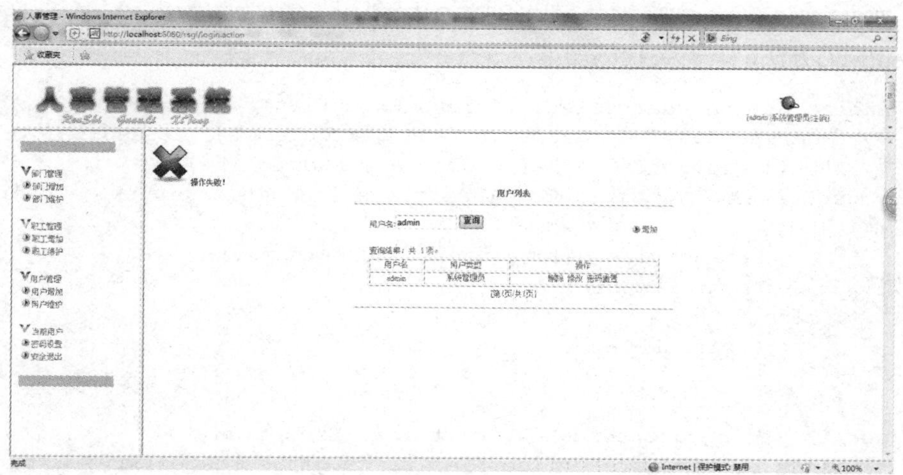

图 10.19　“操作失败！”页面

要修改用户，只需在用户列表中单击相应用户后的“修改”链接，打开图 10.20 所示的“用户修改”页面，并在其中进行相应的修改，最后单击“确定”按钮即可。若成功修改指定的用户，将显示相应的“操作成功”页面；否则，将显示相应的“操作失败”页面。

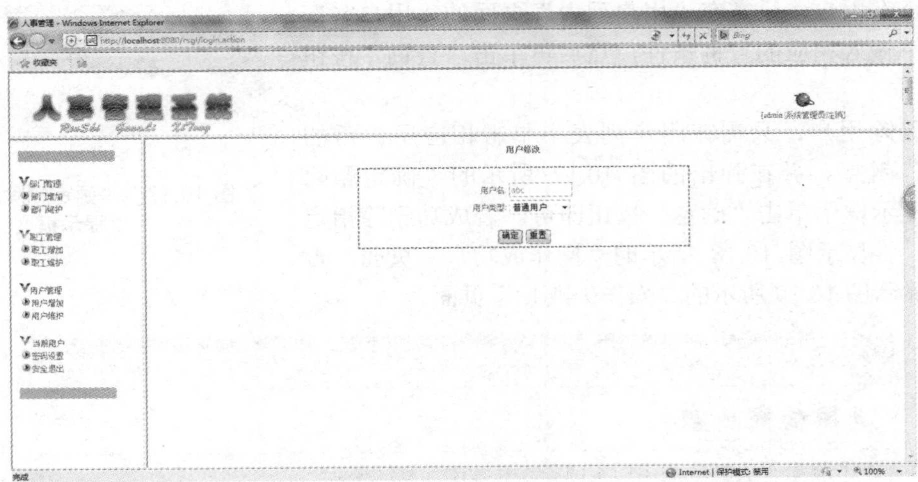

图 10.20 “用户修改”页面

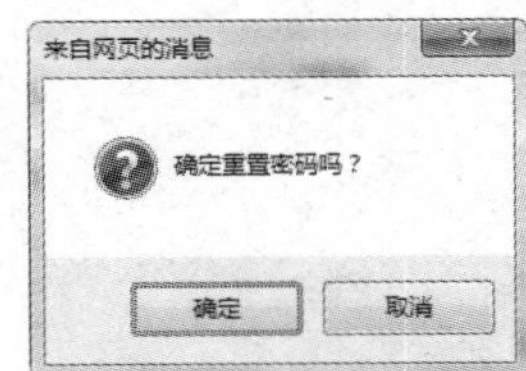

图 10.21 “确定重置密码吗”消息对话框

要重置用户的密码，只需在用户列表中单击相应用户后的“密码重置”链接，并在随之弹出的图 10.21 所示的“确定重置密码吗？”提示框中单击“确定”按钮即可。为简单起见，本系统中的重置密码就是将指定用户的密码修改为用户名本身。

用户维护中的查询功能已在用户增加功能的实现过程中一起实现了，其他功能的实现过程如下所述。

(1) 在 UsersAction 实现类中添加 usersDelete()、usersUpdateView()、usersUpdate()与 usersResetPwd()方法，其有关代码如下：

```
    public String usersDelete()throws Exception{
        Map session=(Map)ActionContext.getContext().getSession();
        if (users.getUsername().compareTo("admin")!=0 &&
users.getUsername().compareTo(((Users)session.get("user")).getUsername()
.trim())!=0){
            usersService.delete(users.getUsername());
            return SUCCESS;
        }else{
            return ERROR;
        }
    }
    public String usersUpdateView()throws Exception{
        this.users=usersService.find(users.getUsername());
        return SUCCESS;
    }
    public String usersUpdate()throws Exception{
        if (users.getUsername().trim().compareTo("admin")!=0){
            usersService.update(users);
            return SUCCESS;
        }else{
            return ERROR;
```

```
        }
    }
    public String usersResetPwd()throws Exception{
        users=usersService.find(users.getUsername());
        users.setPassword(users.getUsername());
        usersService.update(users);
        Map session=(Map)ActionContext.getContext().getSession();
        if
(users.getUsername().compareTo(((Users)session.get("user")).getUsername(
))==0){
            Users user=usersService.check(users.getUsername(),
users.getPassword());
            session.put("user", user);
        }
        return SUCCESS;
    }
```

(2) 在项目的 struts.xml 文件中完成名为 usersDelete、usersUpdateView、usersUpdate 与 usersResetPwd 的 Action 配置，其有关代码如下：

```
    <action name="usersDelete" class="usersAction" method="usersDelete">
        <result name="success">/usersSuccess.jsp</result>
        <result name="error">/usersError.jsp</result>
        <result name="none">/noAuthority.jsp</result>
        <interceptor-ref name="defaultStack"></interceptor-ref>
        <interceptor-ref name="loginFilter"/>
        <interceptor-ref name="authorityFilter"/>
    </action>
    <action name="usersUpdateView" class="usersAction"
     method="usersUpdateView">
        <result name="success">/usersUpdateView.jsp</result>
        <result name="none">/noAuthority.jsp</result>
        <interceptor-ref name="defaultStack"></interceptor-ref>
        <interceptor-ref name="loginFilter"/>
        <interceptor-ref name="authorityFilter"/>
    </action>
    <action name="usersUpdate" class="usersAction" method="usersUpdate">
        <result name="success">/usersSuccess.jsp</result>
        <result name="error">/usersError.jsp</result>
        <result name="none">/noAuthority.jsp</result>
        <interceptor-ref name="defaultStack"></interceptor-ref>
        <interceptor-ref name="loginFilter"/>
        <interceptor-ref name="authorityFilter"/>
    </action>
    <action name="usersResetPwd" class="usersAction"
     method="usersResetPwd">
        <result name="success">/usersSuccess.jsp</result>
        <result name="error">/usersError.jsp</result>
        <result name="none">/noAuthority.jsp</result>
        <interceptor-ref name="defaultStack"></interceptor-ref>
        <interceptor-ref name="loginFilter"/>
        <interceptor-ref name="authorityFilter"/>
    </action>
```

(3) 在项目的 WebRoot 文件夹中添加一个新的 JSP 页面 usersUpdateView.jsp，其代码如下：

```
<%@ page language="java" import="java.util.*" pageEncoding="UTF-8"%>
<%@ taglib prefix="s" uri="/struts-tags" %>
<% //request.setCharacterEncoding("UTF-8"); %>
<html>
  <head>
    <title>系统用户</title>
    <link rel="stylesheet" href="./stylesheet.css"
type="text/css"></link>
  </head>
  <body>
  <div align="center">
  <b>用户修改</b>
  <br />
  <br />
  <s:form action="usersUpdate.action" method="post" theme="simple">
  <table style="border: thin dashed #008080;" width="500" align="center">
  <tr>
  <td style="width: 45%"> </td>
  <td style="width: 55%"> </td>
  </tr>
  <tr>
  <td align="right">用户名:</td>
  <td>
    <s:textfield name="users.username" label="用户名" size="10"
maxlength="10" disabled="true"/>
  </td>
  </tr>
  <tr>
  <td align="right">用户类型:</td>
  <td>
    <s:select list="{'系统管理员','普通用户'}" value="users.usertype"
name="users.usertype" label="用户类型"></s:select>
  </td>
  </tr>
  <tr>
  <td> </td>
  <td>
    <s:hidden name="users.username"></s:hidden>
    <s:hidden name="users.password"></s:hidden>
  </td>
  </tr>
  <tr>
  <td align="center" colspan="2">
  <s:submit value="确定"/> <s:reset value="重置" />
  </td>
  </tr>
  </table>
```

```
  </s:form>
  </div>
  </body>
</html>
```

10.7.7　部门管理功能的实现

部门管理功能包括部门的增加与维护，而部门的维护又包括部门的查询、修改与删除。本系统规定，部门管理功能只能由系统管理员使用。

1. 部门增加

在系统主界面中单击“部门增加”链接，若当前用户为普通用户，将打开相应的“无此权限”页面；反之，若当前用户为系统管理员，将打开图 10.22 所示的“部门增加”页面。在其中输入部门的编号与名称后，再单击“确定”按钮，若能成功添加部门，将显示图 10.23 所示的“操作成功！”页面；否则，将显示图 10.24 所示的“操作失败！”页面。

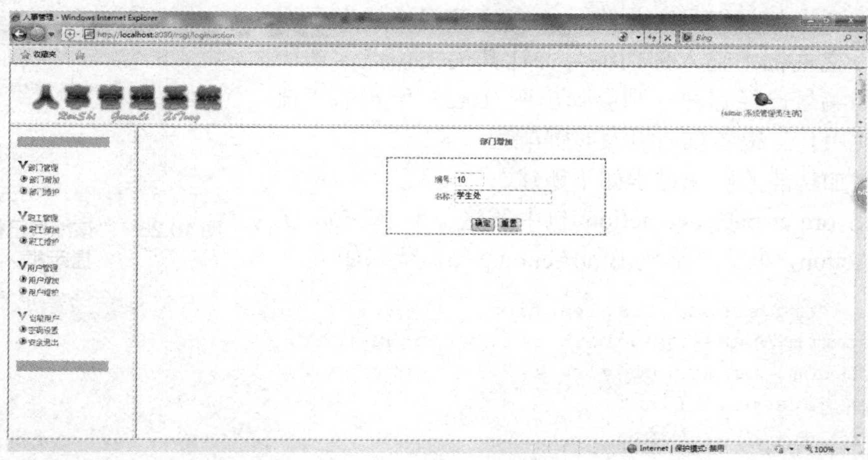

图 10.22　“部门增加”页面

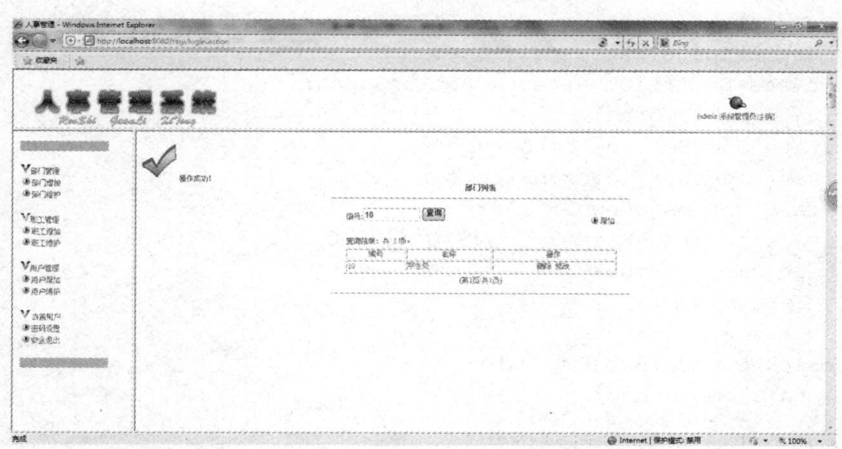

图 10.23　“操作成功！”页面

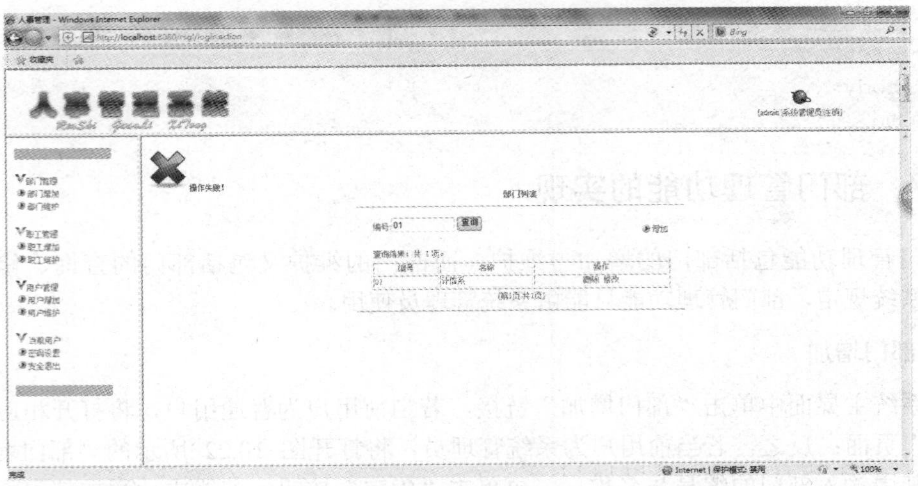

图 10.24 “操作失败！”页面

本系统要求部门编号必须唯一。在“部门增加”页面的“编号”文本框中输入部门编号并让其失去焦点，若所输入的部门编号已经存在，则会弹出图 10.25 所示的“该编号已被使用！”提示框，以及时提醒用户。

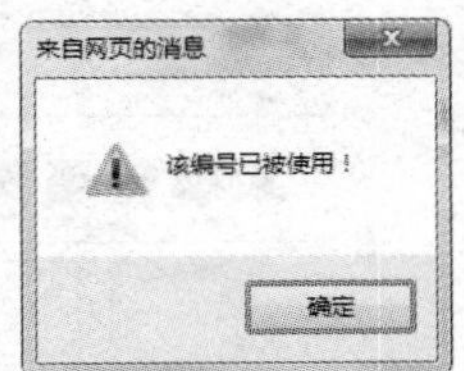

图 10.25 “该编号已被使用！”提示框

部门增加功能的实现过程如下所述。

(1) 在 org.etspace.rsgl.action 包中新建一个 Action 实现类 BmbAction，其文件名为 BmbAction.java，代码如下：

```
package org.etspace.rsgl.action;
import com.opensymphony.xwork2.ActionSupport;
import com.opensymphony.xwork2.ActionContext;
import java.util.List;
import java.util.Map;
import org.etspace.rsgl.vo.Bmb;
import org.etspace.rsgl.service.IBmbService;
import org.etspace.rsgl.service.IZgbService;
import org.etspace.rsgl.tool.Pager;
public class BmbAction extends ActionSupport {
    private Bmb bmb;
    private IBmbService bmbService;
    private IZgbService zgbService;
    private int pageNow=1;
    private int pageSize=2;  //通常应设为10
    public Bmb getBmb() {
        return bmb;
    }
    public void setBmb(Bmb bmb) {
        this.bmb = bmb;
    }
    public IBmbService getBmbService() {
        return bmbService;
```

```
    }
    public void setBmbService(IBmbService bmbService) {
        this.bmbService = bmbService;
    }
    public IZgbService getZgbService() {
        return zgbService;
    }
    public void setZgbService(IZgbService zgbService) {
        this.zgbService = zgbService;
    }
    public int getPageNow() {
        return pageNow;
    }
    public void setPageNow(int pageNow) {
        this.pageNow = pageNow;
    }
    public int getPageSize() {
        return pageSize;
    }
    public void setPageSize(int pageSize) {
        this.pageSize = pageSize;
    }
    public String execute()throws Exception{
        return SUCCESS;
    }
    public String bmbList()throws Exception{
        List list=bmbService.findAll(pageNow, pageSize, bmb);
        Map request=(Map)ActionContext.getContext().get("request");
        Pager page=new Pager(pageNow,pageSize,
        bmbService.findAllSize(bmb));
        request.put("list", list);
        request.put("page", page);
        return SUCCESS;
    }
    public String bmbAddView()throws Exception{
        return SUCCESS;
    }
    public String bmbAdd()throws Exception{
        if (bmb.getBmbh().length()!=0 && bmbService.find
        (bmb.getBmbh())==null){
            bmbService.save(bmb);
            return SUCCESS;
        }else{
            return ERROR;
        }
    }
}
```

(2) 在 Spring 的配置文件 applicationContext.xml 中添加一个 bean 元素，其实现类为 BmbAction，id 为 bmbAction。有关代码如下：

```
<bean id="bmbAction" class="org.etspace.rsgl.action.BmbAction"
scope="prototype">
    <property name="bmbService">
        <ref bean="bmbService"/>
    </property>
    <property name="zgbService">
        <ref bean="zgbService"/>
    </property>
</bean>
```

(3) 在项目的 struts.xml 文件中完成名为 bmbList、bmbAddView 与 bmbAdd 的 Action 配置，其有关代码如下：

```
<action name="bmbList" class="bmbAction" method="bmbList">
    <result name="success">/bmbList.jsp</result>
    <result name="none">/noAuthority.jsp</result>
    <interceptor-ref name="defaultStack"></interceptor-ref>
    <interceptor-ref name="loginFilter"/>
    <interceptor-ref name="authorityFilter"/>
</action>
<action name="bmbAddView" class="bmbAction" method="bmbAddView">
    <result name="success">/bmbAddView.jsp</result>
    <result name="none">/noAuthority.jsp</result>
    <interceptor-ref name="defaultStack"></interceptor-ref>
    <interceptor-ref name="loginFilter"/>
    <interceptor-ref name="authorityFilter"/>
</action>
<action name="bmbAdd" class="bmbAction" method="bmbAdd">
    <result name="success">/bmbSuccess.jsp</result>
    <result name="error">/bmbError.jsp</result>
    <result name="none">/noAuthority.jsp</result>
    <interceptor-ref name="defaultStack"></interceptor-ref>
    <interceptor-ref name="loginFilter"/>
    <interceptor-ref name="authorityFilter"/>
</action>
```

(4) 在项目的 dwr.xml 文件中添加一个 create 元素，其有关代码如下：

```
<create creator="spring" javascript="AjaxBmbService">
    <param name="beanName" value="bmbService"/>
    <include method="exist"/>
</create>
```

(5) 在项目的 WebRoot 文件夹中添加一个新的 JSP 页面 bmbAddView.jsp，其代码如下：

```
<%@ page language="java" import="java.util.*" pageEncoding="UTF-8"%>
<%@ taglib prefix="s" uri="/struts-tags" %>
<html>
  <head>
    <title>部门</title>
```

```
    <link rel="stylesheet" href="./stylesheet.css"
type="text/css"></link>
        <script type="text/javascript"
src="<%=request.getContextPath()%>/dwr/engine.js"></script>
        <script type="text/javascript"
src="<%=request.getContextPath()%>/dwr/util.js"></script>
        <script type="text/javascript"
src="<%=request.getContextPath()%>/dwr/interface/AjaxBmbService.js"></sc
ript>
        <script type="text/javascript">
         function check() {
           var bmbh=document.all.bmbh.value;
           if (bmbh=="") {
             window.alert("编号不能为空！");
             return;
           }
           AjaxBmbService.exist(bmbh,show);
         }
         function show(data) {
           if (data) {
             window.alert("该编号已被使用！");
             return;
           }
         }
        </script>
  </head>
  <body>
  <div align="center">
  <b>部门增加</b>
  <br />
  <br />
  <s:form action="bmbAdd.action" method="post" theme="simple">
  <table style="border: thin dashed #008080;" width="350" align="center">
  <tr>
  <td style="width: 30%"> </td>
  <td style="width: 70%"> </td>
  </tr>
  <tr>
  <td align="right">编号:</td>
  <td>
    <s:textfield name="bmb.bmbh" label="编号" size="2" maxlength="2"
onblur="check()" id="bmbh"/>
  </td>
  </tr>
  <tr>
  <td align="right">名称:</td>
  <td>
    <s:textfield name="bmb.bmmc" label="名称" size="20" maxlength="20"/>
  </td>
  </tr>
```

```
  <tr>
  <td> </td>
  <td> </td>
  </tr>
  <tr>
  <td align="center" colspan="2">
  <s:submit value="确定"/> <s:reset value="重置" />
  </td>
  </tr>
  </table>
  </s:form>
  </div>
  </body>
</html>
```

(6) 在项目的 WebRoot 文件夹中添加一个新的 JSP 页面 bmbSuccess.jsp，其代码如下：

```
<%@ page language="java" import="java.util.*" pageEncoding="UTF-8"%>
<%@ taglib prefix="s" uri="/struts-tags" %>
<html>
  <head>
    <title>操作信息</title>
    <link rel="stylesheet" href="./stylesheet.css"
type="text/css"></link>
  </head>
  <body>
    <img src="./images/LuRight.jpg">
    <FONT color="green">操作成功! </FONT>
    <s:action name="bmbList" executeResult="true"
ignoreContextParams="false"></s:action>
  </body>
</html>
```

(7) 在项目的 WebRoot 文件夹中添加一个新的 JSP 页面 bmbError.jsp，其代码如下：

```
<%@ page language="java" import="java.util.*" pageEncoding="UTF-8"%>
<%@ taglib prefix="s" uri="/struts-tags" %>
<html>
  <head>
    <title>操作信息</title>
    <link rel="stylesheet" href="./stylesheet.css"
type="text/css"></link>
  </head>
  <body>
    <img src="./images/LuWrong.jpg">
    <FONT color="red">操作失败! </FONT>
    <s:fielderror/>
```

```
    <s:action name="bmbList" executeResult="true"
ignoreContextParams="false"></s:action>
  </body>
</html>
```

(8) 在项目的 WebRoot 文件夹中添加一个新的 JSP 页面 bmbList.jsp，其代码如下：

```
<%@ page language="java" import="java.util.*" pageEncoding="UTF-8"%>
<%@ taglib prefix="s" uri="/struts-tags" %>
<html>
  <head>
    <title>部门</title>
    <link rel="stylesheet" href="./stylesheet.css"
type="text/css"></link>
  </head>
  <body>
  <div align="center">
  <b>部门列表</b><br />
  <table width="500" border="0">
  <tr>
  <td> <hr /> </td>
  </tr>
  <tr>
  <td>
    <table width="450" align="center" border="0">
        <tr>
        <td>
            <form action="bmbList.action" method="post">
            编号:<input type="text" name="bmb.bmbh" value="<s:property
value="bmb.bmbh" />" size="10" maxlength="10"/>
            <input type="submit" value="查询"/>
            </form>
        </td>
        <td align="right">
            <img alt="" src="./images/LuArrow.gif" align="absmiddle">
            <a href="bmbAddView.action" target="rightFrame">增加</a>
        </td>
        </tr>
    </table>
    <table width="450" align="center" border="0">
        <tr>
            <td>
            查询结果：共 <s:property value="#request.page.totalSize"/> 项。
            </td>
        </tr>
    </table>
    <table width="450" align="center" border="1">
        <tr align="center">
            <td>编号</td><td>名称</td><td>操作</td>
```

```
        </tr>
        <s:iterator value="#request.list" id="bmb">
        <tr>
            <td><s:property value="#bmb.bmbh"/></td>
            <td><s:property value="#bmb.bmmc"/></td>
            <td align="center">
             <a href="bmbDelete.action?bmb.bmbh=<s:property
value="#bmb.bmbh"/>" onClick="if(!confirm('确定删除吗? '))return
false;else return true;">删除</a>
             <a href="bmbUpdateView.action?bmb.bmbh=<s:property
value="#bmb.bmbh"/>">修改</a>
            </td>
        </tr>
        </s:iterator>
    </table>
    <table width="450" align="center" border="0">
        <tr align="center">
        <td>
            <s:set name="page" value="#request.page"></s:set>
                <s:if test="#page.totalSize!=0">
                    [第<s:property value="#page.pageNow"/>页/共<s:property
value="#page.totalPage"/>页]
                </s:if>
                <s:if test="#page.hasFirst">
                    <a href="bmbList.action?pageNow=1&bmb.bmbh=
<s:property value="bmb.bmbh"/>">首页</a>
                </s:if>
                <s:if test="#page.hasPre">
                    <a href="bmbList.action?pageNow=<s:property
value="#page.pageNow-1"/>&bmb.bmbh=<s:property value="bmb.bmbh"/>">上一页
</a>
                </s:if>
                <s:if test="#page.hasNext">
                    <a href="bmbList.action?pageNow=<s:property value=
"#page.pageNow+1"/>&bmb.bmbh=<s:property value="bmb.bmbh"/>">下一页</a>
                </s:if>
                <s:if test="#page.hasLast">
                    <a href="bmbList.action?pageNow=<s:property value=
"#page.totalPage"/>&bmb.bmbh=<s:property value="bmb.bmbh"/>">尾页</a>
                </s:if>
        </td>
        </tr>
    </table>
  </td>
  </tr>
  <tr>
  <td> <hr /></td>
  </tr>
```

```
  </table>
  </div>
  </body>
</html>
```

2. 部门维护

在系统主界面中单击“部门维护”链接，若当前用户为普通用户，将打开相应的“无此权限”页面；反之，若当前用户为系统管理员，将打开图 10.26 所示的“部门列表”页面。该页面以分页的方式显示系统的有关部门记录(在此为每页显示 2 个部门记录)，并支持按编号对部门进行模糊查询，同时提供了增加新部门以及对各个部门进行删除或修改操作的链接。其中，“增加”链接的作用与系统主界面中的“部门增加”链接的作用是一样的。

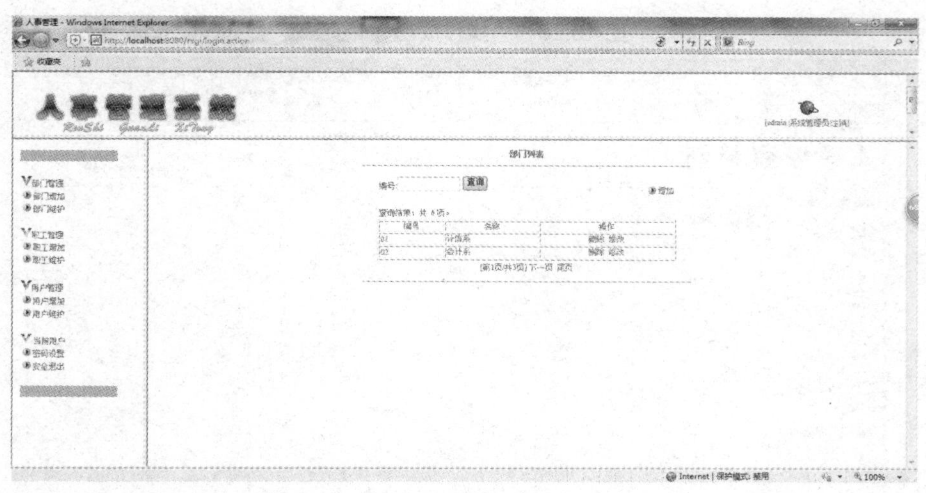

图 10.26　“部门列表”页面

要查询部门，只需在“部门列表”页面的“编号”文本框中输入相应的查询条件，然后单击“查询”按钮即可。

要删除部门，只需在部门列表中单击相应部门后的“删除”链接，并在弹出的图 10.27 所示的“确定删除吗？”提示框中单击“确定”按钮即可。若能成功删除指定的部门，将显示如图 10.28 所示的“操作成功！”页面；否则，将显示图 10.29 所示的“操作失败！”页面。

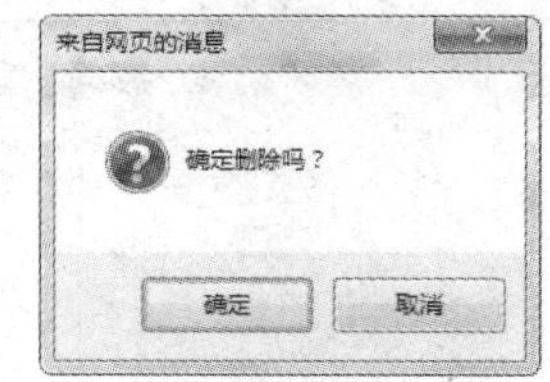

图 10.27　“确定删除吗？”提示框

要修改部门，只需在部门列表中单击相应部门后的“修改”链接，打开图 10.30 所示的“部门修改”页面，并在其中进行相应的修改，最后单击“确定”按钮即可。若能成功修改指定的部门，将显示相应的“操作成功！”页面；否则，将显示相应的“操作失败！”页面。

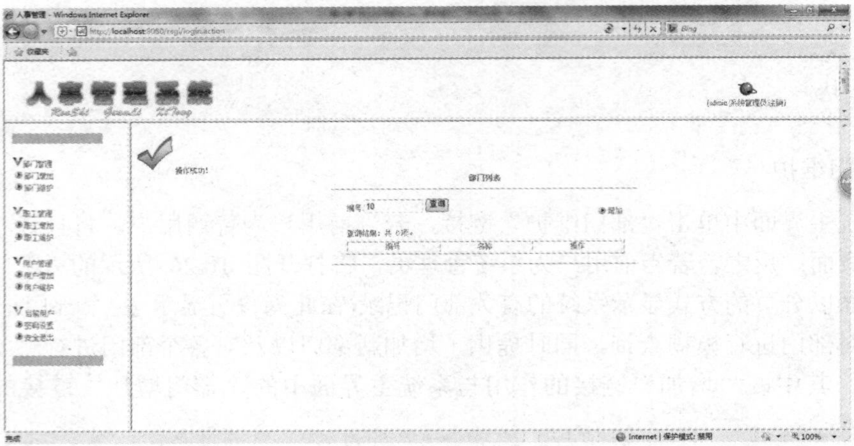

图 10.28 “操作成功！”页面

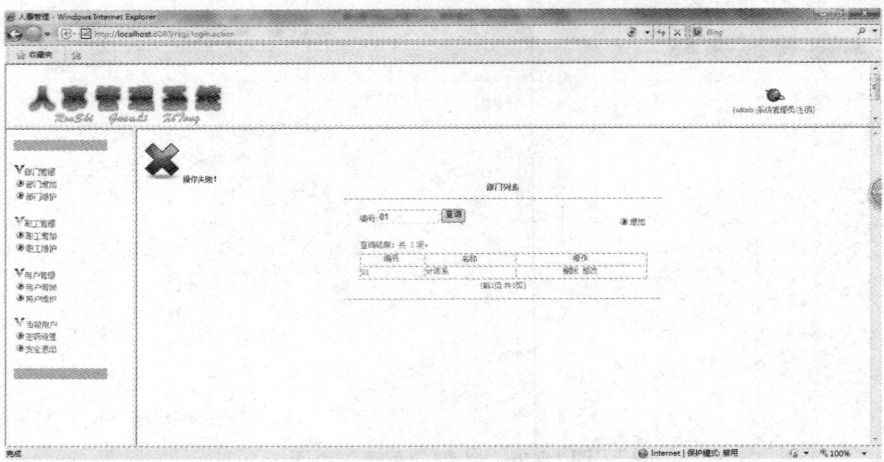

图 10.29 “操作失败！”页面

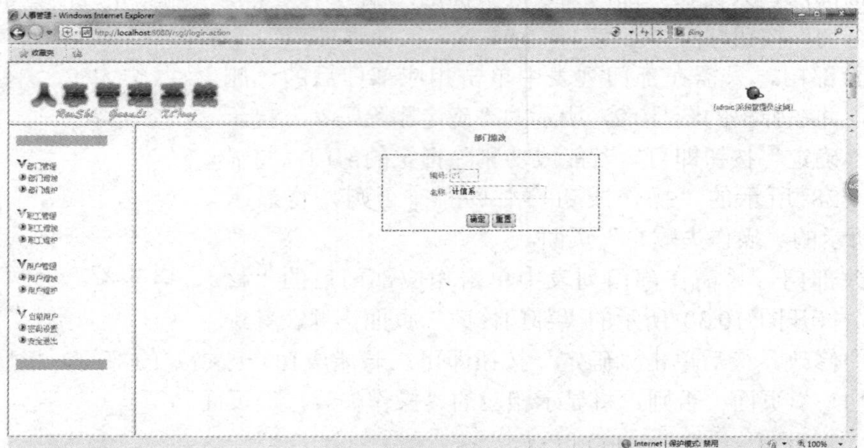

图 10.30 “部门修改”页面

部门维护中的查询功能已在部门增加功能的实现过程中一起实现了，其他功能的实现过程如下所述。

(1) 在 BmbAction 实现类中添加 bmbDelete()、bmbUpdateView()与 bmbUpdate()方法，其有关代码如下：

```
public String bmbDelete()throws Exception{
    List list=zgbService.getAllByBmbh(bmb.getBmbh());
    if(list.size()==0){
        bmbService.delete(bmb.getBmbh());
        return SUCCESS;
    }else{
        return ERROR;
    }
}
public String bmbUpdateView()throws Exception{
    this.bmb=bmbService.find(bmb.getBmbh());
    return SUCCESS;
}
public String bmbUpdate()throws Exception{
    bmbService.update(bmb);
    return SUCCESS;
}
```

(2) 在项目的 struts.xml 文件中完成名为 bmbDelete、bmbUpdateView 与 bmbUpdate 的 Action 配置，其有关代码如下：

```
<action name="bmbDelete" class="bmbAction" method="bmbDelete">
    <result name="success">/bmbSuccess.jsp</result>
    <result name="error">/bmbError.jsp</result>
    <result name="none">/noAuthority.jsp</result>
    <interceptor-ref name="defaultStack"></interceptor-ref>
    <interceptor-ref name="loginFilter"/>
    <interceptor-ref name="authorityFilter"/>
</action>
<action name="bmbUpdateView" class="bmbAction"
 method="bmbUpdateView">
    <result name="success">/bmbUpdateView.jsp</result>
    <result name="none">/noAuthority.jsp</result>
    <interceptor-ref name="defaultStack"></interceptor-ref>
    <interceptor-ref name="loginFilter"/>
    <interceptor-ref name="authorityFilter"/>
</action>
<action name="bmbUpdate" class="bmbAction" method="bmbUpdate">
    <result name="success">/bmbSuccess.jsp</result>
    <result name="error">/bmbError.jsp</result>
    <result name="none">/noAuthority.jsp</result>
    <interceptor-ref name="defaultStack"></interceptor-ref>
    <interceptor-ref name="loginFilter"/>
    <interceptor-ref name="authorityFilter"/>
</action>
```

(3) 在项目的 WebRoot 文件夹中添加一个新的 JSP 页面 bmbUpdateView.jsp，其代码如下：

```
<%@ page language="java" import="java.util.*" pageEncoding="UTF-8"%>
<%@ taglib prefix="s" uri="/struts-tags" %>
<html>
  <head>
    <title>部门</title>
    <link rel="stylesheet" href="./stylesheet.css"
type="text/css"></link>
  </head>
  <body>
  <div align="center">
  <b>部门修改</b>
  <br />
  <br />
  <s:form action="bmbUpdate.action" method="post" theme="simple">
  <table style="border: thin dashed #008080;" width="350" align="center">
  <tr>
  <td style="width: 30%"> </td>
  <td style="width: 70%"> </td>
  </tr>
  <tr>
  <td align="right">编号:</td>
  <td>
    <s:hidden name="bmb.bmbh"></s:hidden>
    <s:textfield name="bmb.bmbh" label="编号" size="2" maxlength="2"
disabled="true"/>
  </td>
  </tr>
  <tr>
  <td align="right">名称:</td>
  <td>
    <s:textfield name="bmb.bmmc" label="名称" size="20" maxlength="20"/>
  </td>
  </tr>
  <tr>
  <td> </td>
  <td> </td>
  </tr>
  <tr>
  <td align="center" colspan="2">
  <s:submit value="确定"/> <s:reset value="重置" />
  </td>
  </tr>
  </table>
  </s:form>
  </div>
  </body>
</html>
```

10.7.8　职工管理功能的实现

职工管理功能包括职工的增加与维护，而职工的维护又包括职工的查询、修改与删除。本系统规定，职工管理功能可由系统管理员或普通用户使用。

1. 职工增加

在系统主界面中单击“职工增加”链接，将打开图 10.31 所示的“职工增加”页面。在其中输入相应的职工信息后，再单击“确定”按钮，若能成功添加职工，将显示图 10.32 所示的“操作成功！”页面；否则，将显示图 10.33 所示的“操作失败！”页面。

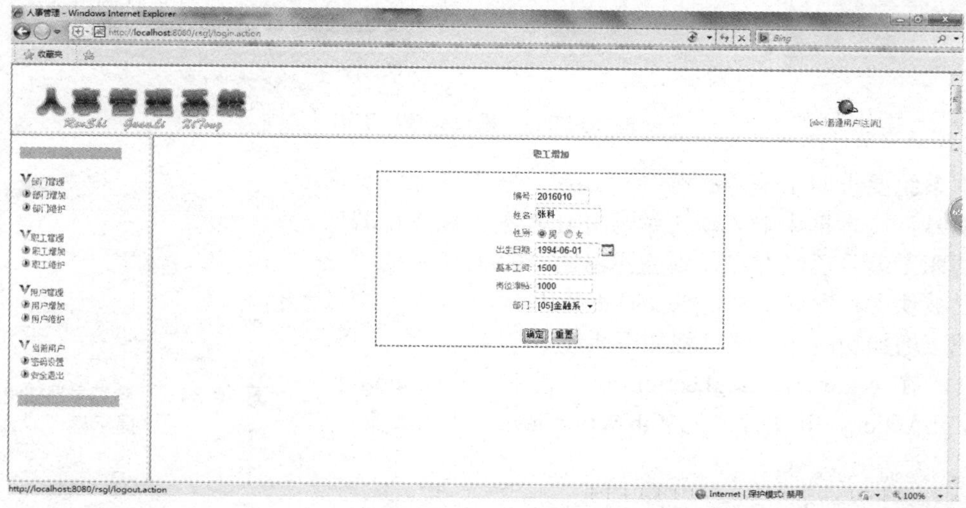

图 10.31　“职工增加”页面

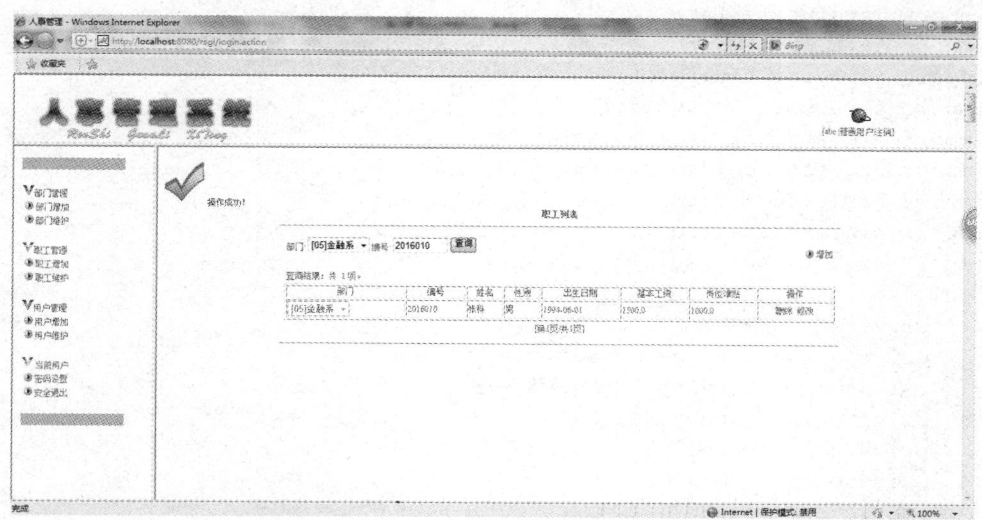

图 10.32　“操作成功”页面

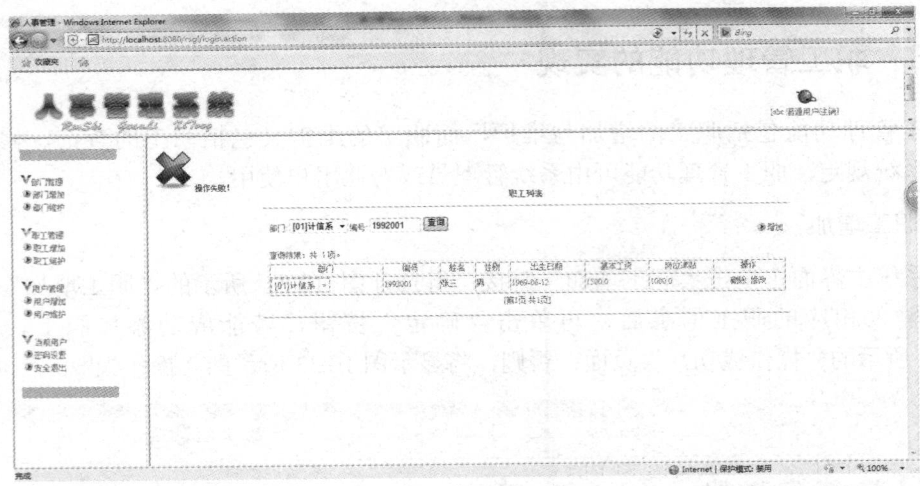

图 10.33 “操作失败”页面

本系统要求职工编号必须唯一。在“职工增加”页面的“编号”文本框中输入职工编号并让其失去焦点，若所输入的职工编号已经存在，则会弹出图 10.34 所示的“该编号已被使用！”提示框，以及时提醒用户。

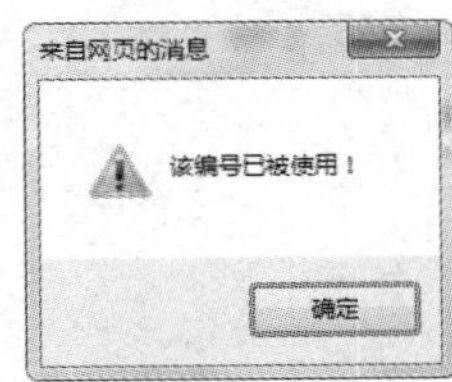

图 10.34 “该编号已被使用！”提示框

职工增加功能的实现过程如下所述。

(1) 在 org.etspace.rsgl.action 包中新建一个 Action 实现类 ZgbAction，其文件名为 ZgbAction.java，代码如下：

```
package org.etspace.rsgl.action;
import com.opensymphony.xwork2.ActionSupport;
import com.opensymphony.xwork2.ActionContext;
import java.util.List;
import java.util.Map;
import java.util.Iterator;
import java.util.ArrayList;
import org.etspace.rsgl.vo.Zgb;
import org.etspace.rsgl.vo.Bmb;
import org.etspace.rsgl.service.IZgbService;
import org.etspace.rsgl.service.IBmbService;
import org.etspace.rsgl.tool.Pager;
public class ZgbAction extends ActionSupport {
    private Zgb zgb;
    private IZgbService zgbService;
    private IBmbService bmbService;
    private int pageNow=1;
    private int pageSize=2;  //通常应设为10
    private List bmb;
    public Zgb getZgb() {
        return zgb;
    }
    public void setZgb(Zgb zgb) {
```

```
        this.zgb = zgb;
    }
    public IZgbService getZgbService() {
        return zgbService;
    }
    public void setZgbService(IZgbService zgbService) {
        this.zgbService = zgbService;
    }
    public IBmbService getBmbService() {
        return bmbService;
    }
    public void setBmbService(IBmbService bmbService) {
        this.bmbService = bmbService;
    }
    public int getPageNow() {
        return pageNow;
    }
    public void setPageNow(int pageNow) {
        this.pageNow = pageNow;
    }
    public int getPageSize() {
        return pageSize;
    }
    public void setPageSize(int pageSize) {
        this.pageSize = pageSize;
    }
    public void setBmb(List bmb){
        this.bmb=bmb;
    }
    public List getBmb(){
        return bmbService.getAll();
    }
    public String execute()throws Exception{
        return SUCCESS;
    }
    public String zgbList()throws Exception{
        List list=zgbService.findAll(pageNow, pageSize, zgb);
        Pager page=new Pager(pageNow,pageSize,
        zgbService.findAllSize(zgb));
        Map request=(Map)ActionContext.getContext().get("request");
        request.put("list", list);
        request.put("page", page);
        return SUCCESS;
    }
    public String zgbAddView()throws Exception{
        return SUCCESS;
    }
    public String zgbAdd()throws Exception{
        if (zgb.getBh().length()!=0 && zgbService.find(zgb.getBh())==null){
            zgbService.save(zgb);
```

```
            return SUCCESS;
        }else{
            return ERROR;
        }
    }
}
```

(2) 在 Spring 的配置文件 applicationContext.xml 中添加一个 bean 元素，其实现类为 ZgbAction，id 为 zgbAction，有关代码如下：

```
<bean id="zgbAction" class="org.etspace.rsgl.action.ZgbAction"
scope="prototype">
    <property name="zgbService">
        <ref bean="zgbService"/>
    </property>
    <property name="bmbService">
        <ref bean="bmbService"/>
    </property>
</bean>
```

(3) 在项目的 struts.xml 文件中完成名为 zgbList、zgbAddView 与 zgbAdd 的 Action 配置，其有关代码如下：

```
    <action name="zgbList" class="zgbAction" method="zgbList">
        <result name="success">/zgbList.jsp</result>
        <result name="none">/noAuthority.jsp</result>
        <interceptor-ref name="defaultStack"></interceptor-ref>
        <interceptor-ref name="loginFilter"/>
    </action>
    <action name="zgbAddView" class="zgbAction" method="zgbAddView">
        <result name="success">/zgbAddView.jsp</result>
        <result name="none">/noAuthority.jsp</result>
        <interceptor-ref name="defaultStack"></interceptor-ref>
        <interceptor-ref name="loginFilter"/>
    </action>
    <action name="zgbAdd" class="zgbAction" method="zgbAdd">
        <result name="success">/zgbSuccess.jsp</result>
        <result name="error">/zgbError.jsp</result>
        <result name="none">/noAuthority.jsp</result>
        <interceptor-ref name="defaultStack"></interceptor-ref>
        <interceptor-ref name="loginFilter"/>
    </action>
```

(4) 在项目的 dwr.xml 文件中添加一个 create 元素，其有关代码如下：

```
<create creator="spring" javascript="AjaxZgbService">
    <param name="beanName" value="zgbService"/>
    <include method="exist"/>
</create>
```

(5) 在项目的 WebRoot 文件夹中添加一个新的 JSP 页面 zgbAddView.jsp，其代码如下：

```
<%@ page language="java" import="java.util.*" pageEncoding="UTF-8"%>
<%@ taglib prefix="s" uri="/struts-tags" %>
<%@ taglib uri="/struts-dojo-tags" prefix="sd" %>
<html>
  <head>
    <title>职工</title>
    <link rel="stylesheet" href="./stylesheet.css" type="text/css"></link>
        <script type="text/javascript" src="<%=request.getContextPath()%>/
dwr/engine.js"></script>
        <script type="text/javascript" src="<%=request.getContextPath()%>/
dwr/util.js"></script>
        <script type="text/javascript" src="<%=request.getContextPath()%>/
dwr/interface/AjaxZgbService.js"></script>
        <script type="text/javascript">
         function check() {
           var bh=document.all.bh.value;
           if (bh=="") {
             window.alert("编号不能为空！");
             return;
           }
           AjaxZgbService.exist(bh,show);
         }
         function show(data) {
           if (data) {
             window.alert("该编号已被使用！");
             return;
           }
         }
        </script>
  </head>
  <sd:head/>
  <body>
  <div align="center">
  <b>职工增加</b>
  <br />
  <br />
  <s:form action="zgbAdd.action" method="post" theme="simple">
  <table style="border: thin dashed #008080;" width="500" align="center">
  <tr>
  <td style="width: 45%"> </td>
  <td style="width: 55%"> </td>
  <tr>
  <td align="right">编号:</td>
  <td>
    <s:textfield name="zgb.bh" label="编号" size="7" maxlength="7"
onblur="check()" id="bh"/>
  </td>
  </tr>
  <tr>
  <td align="right">姓名:</td>
```

```
  <td>
    <s:textfield name="zgb.xm" label="姓名" size="10" maxlength="10"/>
  </td>
  </tr>
  <tr>
  <td align="right">性别:</td>
  <td>
    <s:radio list="{'男','女'}" value="'男'" name="zgb.xb" label="性别">
</s:radio>
  </td>
  </tr>
  <tr>
  <td align="right">出生日期:</td>
  <td>
    <sd:datetimepicker name="zgb.csrq" id="csrq" displayFormat="yyyy-MM-dd">
</sd:datetimepicker>
  </td>
  </tr>
  <tr>
  <td align="right">基本工资:</td>
  <td>
    <s:textfield name="zgb.jbgz" label="基本工资" size="8" maxlength="8"/>
  </td>
  </tr>
  <tr>
  <td align="right">岗位津贴:</td>
  <td>
    <s:textfield name="zgb.gwjt" label="岗位津贴" size="8" maxlength="8"/>
  </td>
  </tr>
  <tr>
  <td align="right">部门:</td>
  <td>
    <select name="zgb.bm">
            <s:iterator value="#request.bmb" id="bmb">
                <option value="<s:property value="#bmb.bmbh"/>" <s:if
test="#bmb.bmbh==zgb.bm">selected="selected"</s:if>>
                [<s:property value="#bmb.bmbh"/>]<s:property
value="#bmb.bmmc"/></option>
            </s:iterator>
    </select>
  </td>
  </tr>
  <tr>
  <td> </td>
  <td> </td>
  </tr>
  <tr>
  <td align="center" colspan="2">
  <s:submit value="确定"/> <s:reset value="重置" />
```

```
  </td>
  </tr>
  </table>
  </s:form>
  </div>
  </body>
</html>
```

(6) 在项目的 WebRoot 文件夹中添加一个新的 JSP 页面 zgbSuccess.jsp，其代码如下：

```
<%@ page language="java" import="java.util.*" pageEncoding="UTF-8"%>
<%@ taglib prefix="s" uri="/struts-tags" %>
<html>
  <head>
    <title>操作信息</title>
    <link rel="stylesheet" href="./stylesheet.css"
type="text/css"></link>
  </head>
  <body>
    <img src="./images/LuRight.jpg">
    <FONT color="green">操作成功! </FONT>
    <s:action name="zgbList" executeResult="true"
ignoreContextParams="false"></s:action>
  </body>
</html>
```

(7) 在项目的 WebRoot 文件夹中添加一个新的 JSP 页面 zgbError.jsp，其代码如下：

```
<%@ page language="java" import="java.util.*" pageEncoding="UTF-8"%>
<%@ taglib prefix="s" uri="/struts-tags" %>
<html>
  <head>
    <title>操作信息</title>
    <link rel="stylesheet" href="./stylesheet.css"
type="text/css"></link>
  </head>
  <body>
    <img src="./images/LuWrong.jpg">
    <FONT color="red">操作失败! </FONT>
    <s:fielderror/>
    <s:action name="zgbList" executeResult="true"
ignoreContextParams="false"></s:action>
  </body>
</html>
```

(8) 在项目的 WebRoot 文件夹中添加一个新的 JSP 页面 zgbList.jsp，其代码如下：

```
<%@ page language="java" import="java.util.*" pageEncoding="UTF-8"%>
<%@ taglib prefix="s" uri="/struts-tags" %>
<html>
  <head>
```

```
    <title>职工</title>
    <link rel="stylesheet" href="./stylesheet.css"
type="text/css"></link>
  </head>
  <body>
  <div align="center">
  <b>职工列表</b><br />
  <table width="800" border="0">
  <tr>
  <td><hr /></td>
  </tr>
  <tr>
  <td>
    <table width="780" align="center" border="0">
         <tr>
         <td>
              <form action="zgbList.action" method="post">
              部门:
              <select name="zgb.bm">
                   <option value=""></option>
                   <s:iterator value="#request.bmb" id="bmb">
                        <option value="<s:property value="#bmb.bmbh"/>"
<s:if test="#bmb.bmbh==zgb.bm">selected="selected"</s:if>>
                        [<s:property value="#bmb.bmbh"/>]<s:property
value="#bmb.bmmc"/></option>
                   </s:iterator>
              </select>
              编号:
              <input type="text" name="zgb.bh" value="<s:property
value="zgb.bh" />" size="7" maxlength="7"/>
              <input type="submit" value="查询"/>
              </form>
         </td>
         <td align="right">
              <img alt="" src="./images/LuArrow.gif" align="absmiddle">
              <a href="zgbAddView.action" target="rightFrame">增加</a>
         </td>
         </tr>
     </table>
     <table width="780" align="center" border="0">
         <tr>
              <td>
              查询结果: 共 <s:property value="#request.page.totalSize"/> 项。
              </td>
         </tr>
     </table>
     <table width="780" align="center" border="1">
         <tr align="center">
              <td>部门</td><td>编号</td><td>姓名</td><td>性别</td><td>出生日期
</td><td>基本工资</td><td>岗位津贴</td><td>操作</td>
```

```
        </tr>
        <s:iterator value="#request.list" id="zgb">
        <tr>
            <td>
            <select disabled>
                <s:iterator value="#request.bmb" id="bmb">
                    <option value="<s:property value="#bmb.bmbh"/>" <s:if
test="#bmb.bmbh==#zgb.bm">selected="selected"</s:if>>
                    [<s:property value="#bmb.bmbh"/>]<s:property
value="#bmb.bmmc"/></option>
                </s:iterator>
            </select>
            </td>
            <td><s:property value="#zgb.bh"/></td>
            <td><s:property value="#zgb.xm"/></td>
            <td><s:property value="#zgb.xb"/></td>
            <td><s:property value="%{getText('{0,date,yyyy-MM-
dd}',{#zgb.csrq})}"/></td>
            <td><s:property value="#zgb.jbgz"/></td>
            <td><s:property value="#zgb.gwjt"/></td>
            <td align="center">
             <a href="zgbDelete.action?zgb.bh=<s:property
value="#zgb.bh"/>" onClick="if(!confirm('确定删除吗？'))return false;else
return true;">删除</a>
             <a href="zgbUpdateView.action?zgb.bh=<s:property
value="#zgb.bh"/>">修改</a>
            </td>
        </tr>
        </s:iterator>
    </table>
    <table width="780" align="center" border="0">
        <tr align="center">
        <td>
            <s:set name="page" value="#request.page"></s:set>
                <s:if test="#page.totalSize!=0">
                    [第<s:property value="#page.pageNow"/>页/共<s:property
value="#page.totalPage"/>页]
                </s:if>
                <s:if test="#page.hasFirst">
                    <a href="zgbList.action?pageNow=1&zgb.bm=<s:property
value="zgb.bm"/>&zgb.bh=<s:property value="zgb.bh"/>">首页</a>
                </s:if>
                <s:if test="#page.hasPre">
                    <a href="zgbList.action?pageNow=<s:property
value="#page.pageNow-1"/>&zgb.bm=<s:property
value="zgb.bm"/>&zgb.bh=<s:property value="zgb.bh"/>">上一页</a>
                </s:if>
                <s:if test="#page.hasNext">
```

```
                    <a href="zgbList.action?pageNow=<s:property
value="#page.pageNow+1"/>&zgb.bm=<s:property
value="zgb.bm"/>&zgb.bh=<s:property value="zgb.bh"/>">下一页</a>
                </s:if>
                <s:if test="#page.hasLast">
                    <a href="zgbList.action?pageNow=<s:property
value="#page.totalPage"/>&zgb.bm=<s:property
value="zgb.bm"/>&zgb.bh=<s:property value="zgb.bh"/>">尾页</a>
                </s:if>
        </td>
        </tr>
    </table>
  </td>
  </tr>
  <tr>
  <td><hr /></td>
  </tr>
  </table>
  </div>
  </body>
</html>
```

2. 职工维护

在系统主界面中单击“职工维护”链接，将打开图 10.35 所示的“职工列表”页面。该页面以分页的方式显示系统的有关职工记录(在此为每页显示 2 个职工记录)，并支持按编号与所在部门对职工进行模糊查询，同时提供了增加新职工以及对各个职工进行删除或修改操作的链接。其中，“增加”链接的作用与系统主界面中的“职工增加”链接的作用是一样的。

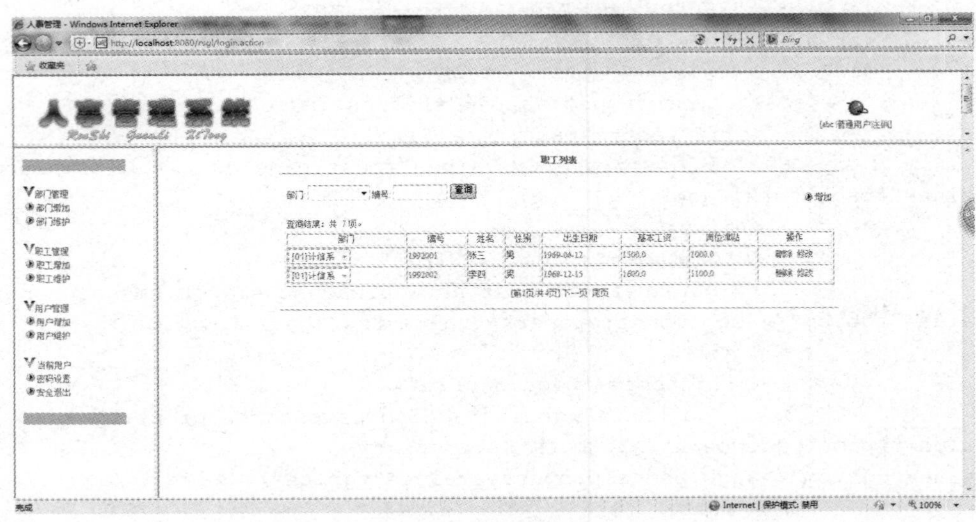

图 10.35　“职工列表”页面

要查询职工，只需在“职工列表”页面的“编号”文本框中输入相应的查询条件，或在“部门”下拉列表框中选中相应的部门，然后单击“查询”按钮即可。

要删除职工，只需在职工列表中单击相应职工后的“删除”链接，并在随之弹出的图 10.36 所示的“确定删除吗？”提示框中单击“确定”按钮即可。若能成功删除指定的职工，将显示相应的“操作成功”页面；否则，将显示相应的“操作失败”页面。

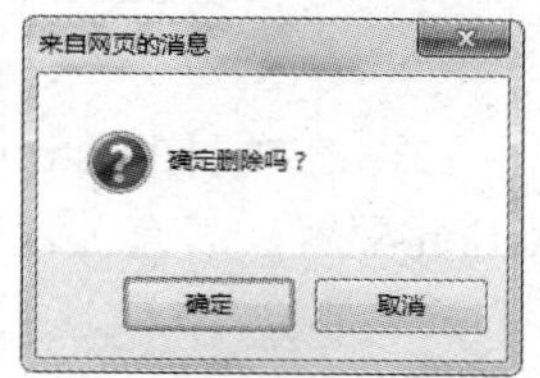

图 10.36　“确定删除吗”消息对话框

要修改职工，只需在职工列表中单击相应职工后的“修改”链接，打开图 10.37 所示的“职工修改”页面，并在其中进行相应的修改，最后单击“确定”按钮即可。若能成功修改指定的职工，将显示相应的“操作成功”页面；否则，将显示相应的“操作失败”页面。

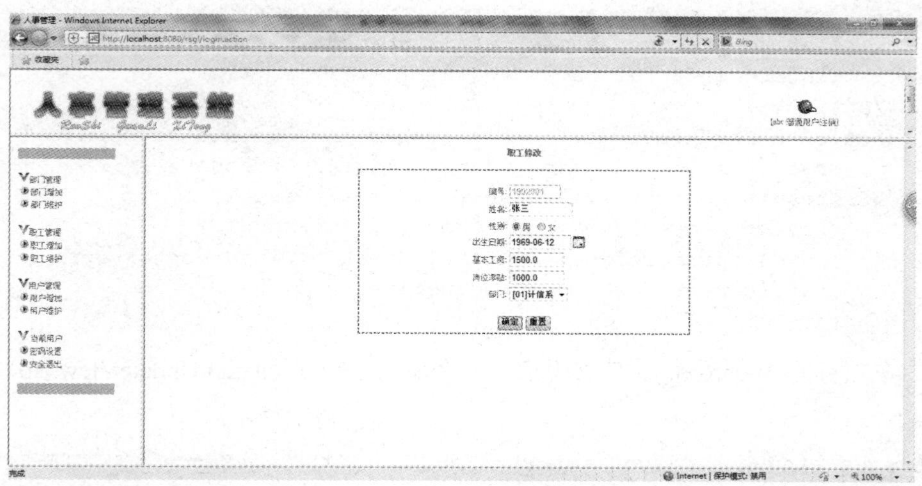

图 10.37　“职工修改”页面

职工维护中的查询功能已在职工增加功能的实现过程中一起实现了，其他功能的实现过程如下所述。

(1) 在 ZgbAction 实现类中添加 zgbDelete()、zgbUpdateView()与 zgbUpdate()方法，其有关代码如下：

```
public String zgbDelete()throws Exception{
    zgbService.delete(zgb.getBh());
    return SUCCESS;
}
public String zgbUpdateView()throws Exception{
    this.zgb=zgbService.find(zgb.getBh());
    return SUCCESS;
}
public String zgbUpdate()throws Exception{
    zgbService.update(zgb);
```

```
        return SUCCESS;
    }
```

(2) 在项目的 struts.xml 文件中完成名为 zgbDelete、zgbUpdateView 与 zgbUpdate 的 Action 配置，其有关代码如下：

```
    <action name="zgbDelete" class="zgbAction" method="zgbDelete">
        <result name="success">/zgbSuccess.jsp</result>
        <result name="error">/zgbError.jsp</result>
        <result name="none">/noAuthority.jsp</result>
        <interceptor-ref name="defaultStack"></interceptor-ref>
        <interceptor-ref name="loginFilter"/>
    </action>
    <action name="zgbUpdateView" class="zgbAction"
method="zgbUpdateView">
        <result name="success">/zgbUpdateView.jsp</result>
        <result name="none">/noAuthority.jsp</result>
        <interceptor-ref name="defaultStack"></interceptor-ref>
        <interceptor-ref name="loginFilter"/>
    </action>
    <action name="zgbUpdate" class="zgbAction" method="zgbUpdate">
        <result name="success">/zgbSuccess.jsp</result>
        <result name="error">/zgbError.jsp</result>
        <result name="none">/noAuthority.jsp</result>
        <interceptor-ref name="defaultStack"></interceptor-ref>
        <interceptor-ref name="loginFilter"/>
    </action>
```

(3) 在项目的 WebRoot 文件夹中添加一个新的 JSP 页面 zgbUpdateView.jsp，其代码如下：

```
<%@ page language="java" import="java.util.*" pageEncoding="UTF-8"%>
<%@ taglib prefix="s" uri="/struts-tags" %>
<%@ taglib uri="/struts-dojo-tags" prefix="sd" %>
<html>
  <head>
    <title>职工</title>
    <link rel="stylesheet" href="./stylesheet.css"
type="text/css"></link>
  </head>
  <sd:head/>
  <body>
  <div align="center">
  <b>职工修改</b>
  <br />
  <br />
  <s:form action="zgbUpdate.action" method="post" theme="simple">
  <table style="border: thin dashed #008080;" width="500" align="center">
  <tr>
  <td style="width: 45%"> </td>
  <td style="width: 55%"> </td>
```

```
  </tr>
  <tr>
  <td align="right">编号:</td>
  <td>
    <s:hidden name="zgb.bh"></s:hidden>
    <s:textfield name="zgb.bh" label="编号" size="7" maxlength="7"
disabled="true"/>
  </td>
  </tr>
  <tr>
  <td align="right">姓名:</td>
  <td>
    <s:textfield name="zgb.xm" label="姓名" size="10" maxlength="10"/>
  </td>
  </tr>
  <tr>
  <td align="right">性别:</td>
  <td>
    <s:radio list="{'男','女'}" value="zgb.xb" name="zgb.xb" label="性别">
</s:radio>
  </td>
  </tr>
  <tr>
  <td align="right">出生日期:</td>
  <td>
    <sd:datetimepicker name="zgb.csrq" id="csrq" displayFormat="yyyy-MM-
dd"></sd:datetimepicker>
  </td>
  </tr>
  <tr>
  <td align="right">基本工资:</td>
  <td>
    <s:textfield name="zgb.jbgz" label="基本工资" size="8" maxlength="8"/>
  </td>
  </tr>
  <tr>
  <td align="right">岗位津贴:</td>
  <td>
    <s:textfield name="zgb.gwjt" label="岗位津贴" size="8" maxlength="8"/>
  </td>
  </tr>
  <tr>
  <td align="right">部门:</td>
  <td>
    <select name="zgb.bm">
        <s:iterator value="#request.bmb" id="bmb">
            <option value="<s:property value="#bmb.bmbh"/>" <s:if
test="#bmb.bmbh==zgb.bm">selected="selected"</s:if>>
            [<s:property value="#bmb.bmbh"/>]<s:property
value="#bmb.bmmc"/></option>
```

```
        </s:iterator>
    </select>
  </td>
  </tr>
  <tr>
  <td> </td>
  <td> </td>
  </tr>
  <tr>
  <td align="center" colspan="2">
  <s:submit value="确定"/> <s:reset value="重置" />
  </td>
  </tr>
  </table>
  </s:form>
  </div>
  </body>
</html>
```

本 章 小 结

本章以一个简单的人事管理系统为例，首先介绍系统的功能与用户，然后介绍轻量级 Java EE 系统的分层模型、开发模式与开发顺序，接着介绍系统数据库结构的设计与项目总体架构的具体搭建方法，最后依次介绍系统持久层、业务层、表示层的设计与实现方法。通过本章的学习，读者应熟练掌握系统的分析与设计方法，并能使用 JSP+Struts 2+Spring+Hibernate+DWR+DAO+Service 模式开发相应的 Web 应用系统或轻量级 Java EE 系统。

思 考 题

1. 轻量级 Java EE 系统的分层模型包括哪三层？请简述之。
2. 请简述使用 Struts 2、Spring、Hibernate 与 DWR 框架开发 Web 应用系统的基本方案。
3. 开发一个大型 Java EE 项目的基本顺序是什么？
4. 如何完成 Java EE 项目的前期搭建工作？
5. 持久层开发的主要工作是什么？
6. 业务层开发的主要工作是什么？
7. 表示层开发的主要工作是什么？

附录

实验指导

实验 1　Web 项目的创建与部署

一、实验目的与要求

(1) 掌握 Java EE 开发环境的搭建方法。
(2) 熟悉 Java EE 开发环境，掌握其常用功能与基本操作。
(3) 掌握 Java Web 项目的创建与部署方法。
(4) 掌握 Java Web 项目的常用管理操作。

二、实验内容

(1) 搭建 Java EE 开发环境。
(2) 创建一个 Java Web 项目。
(3) 设计一个“Hi,Java EE!”页面，如图 P1.1 所示。
(4) 设计一个可动态显示问候语及当前时间的 JSP 页面 Hello.jsp 如图 P1.2 所示。要求：时间在 6 点以前显示“早上好！”，在 6 点至 12 点之间显示“上午好！”，在 12 点至 14 点之间显示“中午好！”，在 14 点至 18 点之间显示“下午好！”，在 18 点之后显示“晚上好！”。

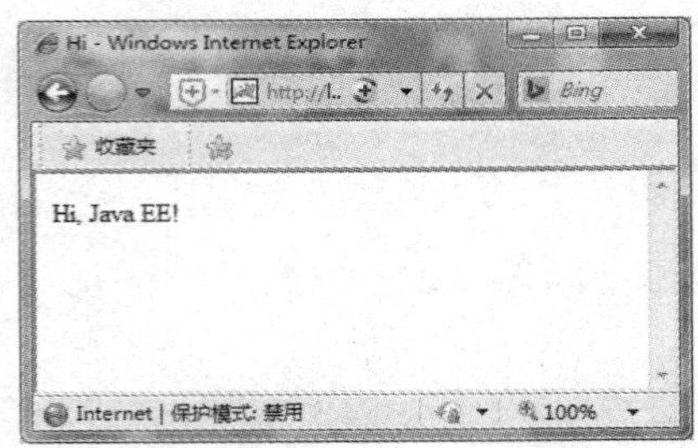

图 P1.1　Hi 页面

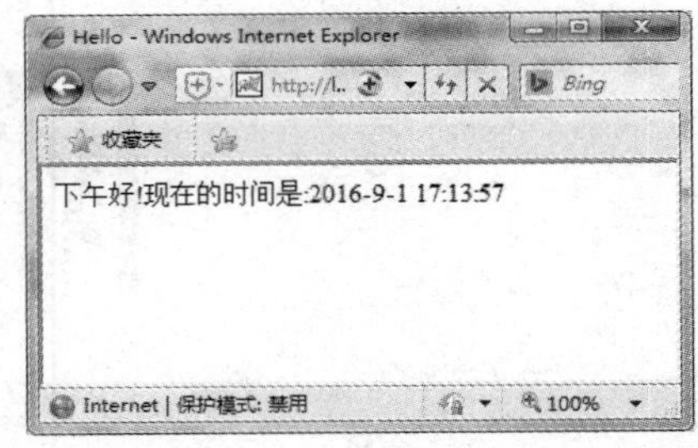

图 P1.2　Hello 页面

(5) 部署所创建的 Web 项目，并通过浏览器直接访问 Hello.jsp 页面。
(6) 导出所创建的 Web 项目。
(7) 删除所创建的 Web 项目。
(8) 导入所创建的 Web 项目。

实验 2　JSP 的应用

一、实验目的与要求

参考教材的有关内容与示例，按要求编写并调试相应的程序，理解并掌握 JSP 的基本应用技术。

二、实验内容

(1) 完成教材第 2 章 2.4.1 小节的系统登录案例。

(2) 完成教材第 2 章 2.4.2 小节的简易聊天室案例。

(3) 设计一个学生成绩增加页面(见图 P2.1)，单击“确定”按钮后可在其处理页面中显示所输入的有关信息(见图 P2.2)。在此，假定班级只有两个，即 02 计应一(其编号为 200201011)与 02 计应二(其编号为 200201012)。

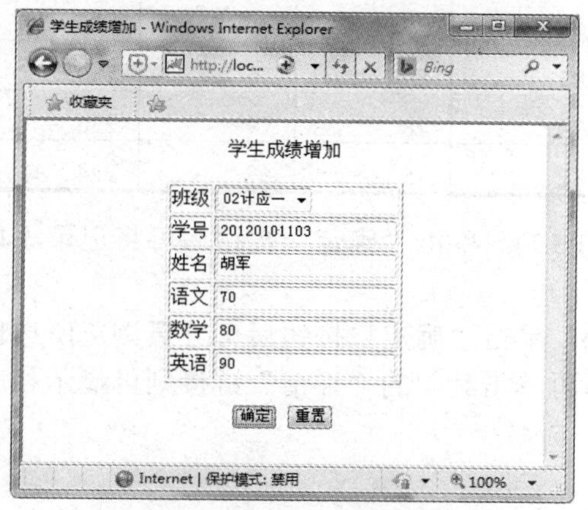

图 P2.1 学生成绩增加页面

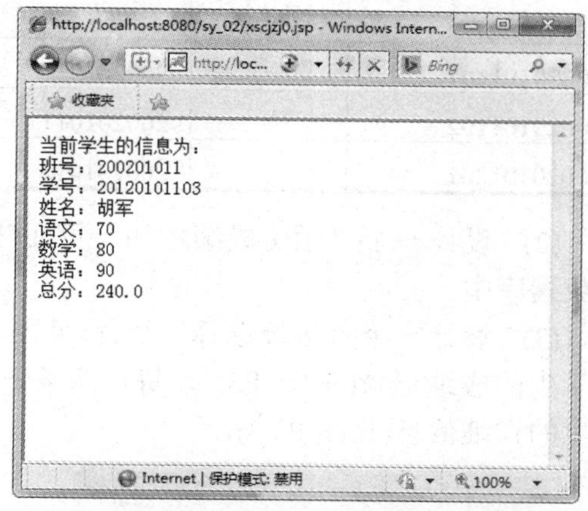

图 P2.2 学生信息页面

实验 3 JDBC 的应用

一、实验目的与要求

参考教材的有关内容与示例，按要求编写并调试相应的程序，理解并掌握 JSP 的基本应用技术。

二、实验内容

(1) 在 SQL Server 中创建一个学生数据库 student，并在其中创建班级表 class 与成绩表 score。

① class 表的字段名依次为 bjbh(班级编号)、bjmc(班级名称)，其类型分别为 char(9)、varchar(20)，主键为 bjbh，有关记录见表 P3.1。

② score 表的字段名依次为 xh(学号)、xm(姓名)、yw(语文)、sx(数学)、yy(英语)、zf(总分)、bh(班号)，除 xh、xm、bh 的类型分别为 char(11)、char(10)、char(9)外，其余各字段的类型均为 decimal(6,2)，主键为 xh，有关记录见表 P3.2。

表 P3.1　班级表 class 的记录

班级编号	班级名称
200201011	02 计应一
200201012	02 计应二

表 P3.2　成绩表 score 的记录

学　号	姓　名	班　号	语　文	数　学	英　语	总　分
20020101101	张三	200201011	85	89	76	250
20020101102	李四	200201011	79	78	79	236
20020101201	王五	200201012	82	75	67	224

(2) 设计一个“学生成绩增加”页面(见图 P3.1)，单击“确定”按钮后可将记录添加到数据库中。

(3) 设计一个“班级选择”页面(见图 P3.2)，单击“确定”按钮后可分页浏览所选班级学生的成绩记录(见图 P3.3，每页两条记录)，再单击其中的“详情”链接则可显示相应学生的详细信息(见图 P3.4)。

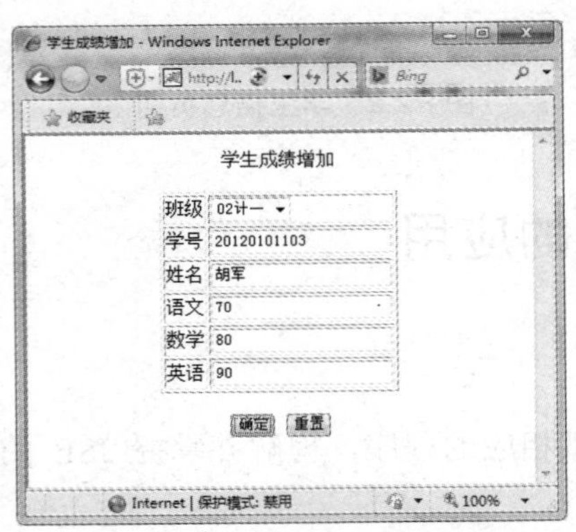

图 P3.1　“学生成绩增加”页面

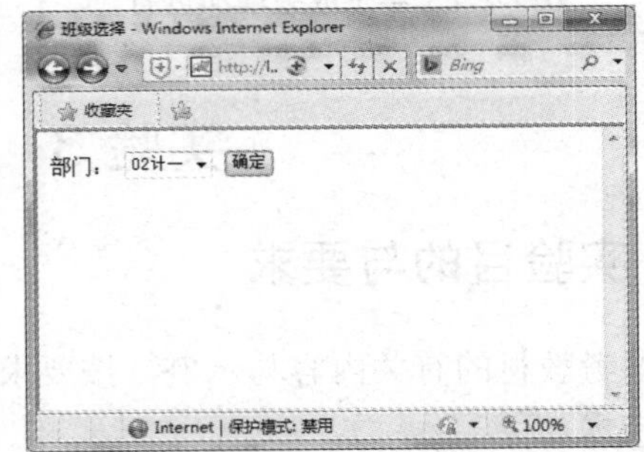

图 P3.2　“班级选择”页面

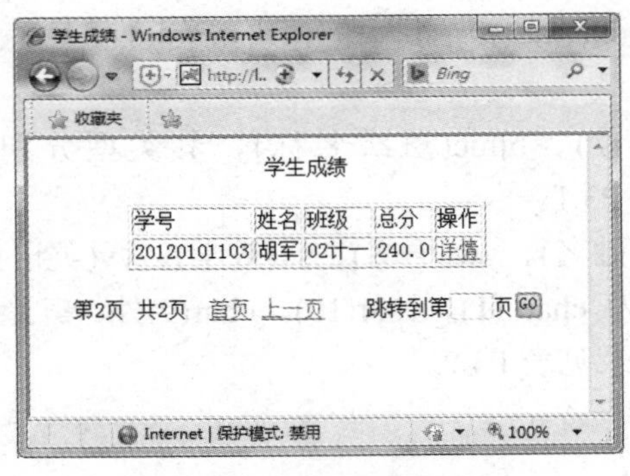

图 P3.3　成绩列表页面

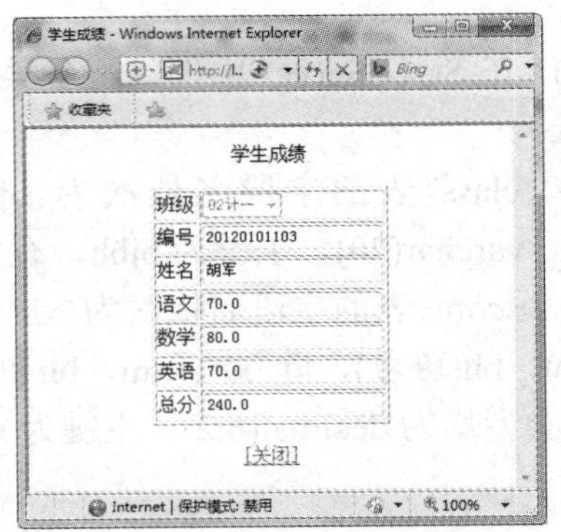

图 P3.4　“学生成绩”页面

(4) (选做)在第 3 题的基础上实现学生成绩记录的修改功能。

(5) (选做)在第 3 题的基础上实现学生成绩记录的删除功能。

实验 4　JavaBean 的应用

一、实验目的与要求

参考教材的有关内容与示例，按要求编写并调试相应的程序，理解并掌握 JavaBean 的基本应用技术。

二、实验内容

(1) 完成教材第 4 章的【实例 4-1】与【实例 4-2】。

(2) 完成教材第 4 章的【实例 4-3】。

(3) 设计一个封装数据库访问操作 JavaBean。

(4) 设计一个“学生成绩增加”页面(见图 P4.1)，单击“确定”按钮后可将记录添加到数据库中。要求使用(3)所创建的 JavaBean。

图 P4.1　“学生成绩增加”页面

(5) 设计一个“班级选择”页面(见图 P4.2)，单击“确定”按钮后可分页浏览所选班级学生的成绩记录(见图 P4.3，每页两条记录)，再单击其中的“详情”链接则可显示相应学生的详细信息(见图 P4.4)。要求使用(3)所创建的 JavaBean。

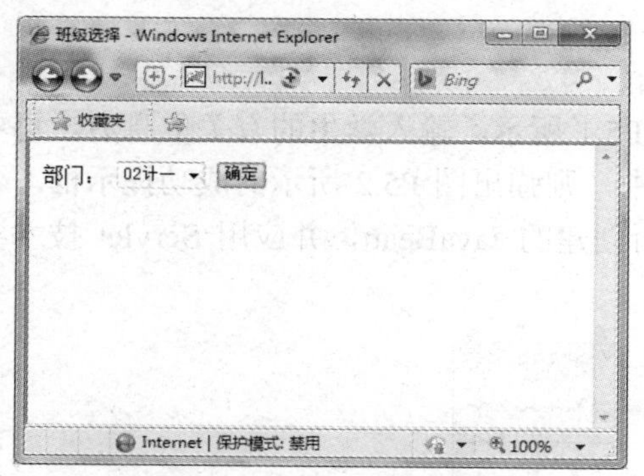

图 P4.2　“班级选择”页面

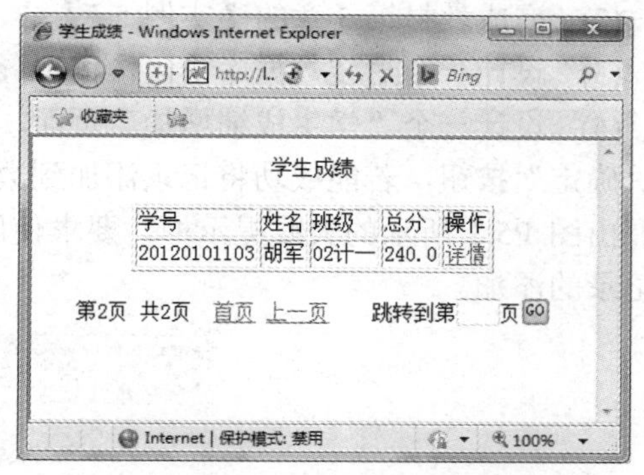

图 P4.3　成绩列表页面

(6) (选做)在(5)的基础上实现学生成绩记录的修改功能。要求使用(3)所创建的 JavaBean。

(7) (选做)在(5)的基础上实现学生成绩记录的删除功能。要求使用(3)所创建的 JavaBean。

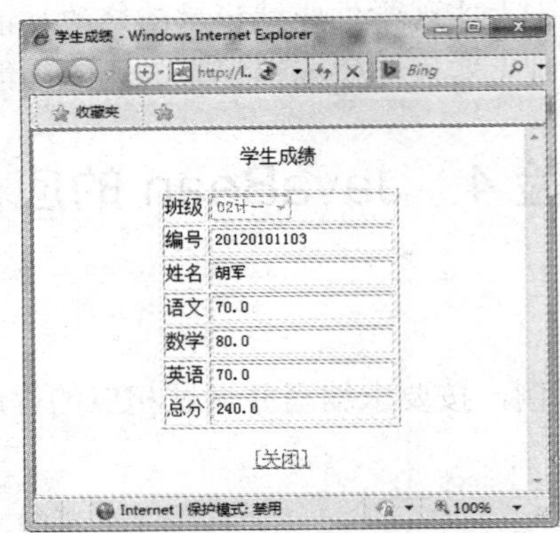

图 P4.4　“学生成绩”页面

实验 5　Servlet 的应用

一、实验目的与要求

参考教材的有关内容与示例，按要求编写并调试相应的程序，理解并掌握 Servlet 的基本应用技术。

二、实验内容

(1) 完成教材第 5 章的【实例 5-1】~【实例 5-4】。

(2) 完成教材第 5 章的【实例 5-5】。

(3) 设计一个封装数据库访问操作 JavaBean。

(4) 设计一个“学生成绩增加”页面，如图 P5.1 所示，输入学生的有关信息后，再单击“确定”按钮，若能成功将记录添加到数据库中，则弹出图 P5.2 所示的成功提示框，否则弹出图 P5.3 所示的失败提示框。要求使用(3)所创建的 JavaBean，并应用 Servlet 技术实现记录的添加。

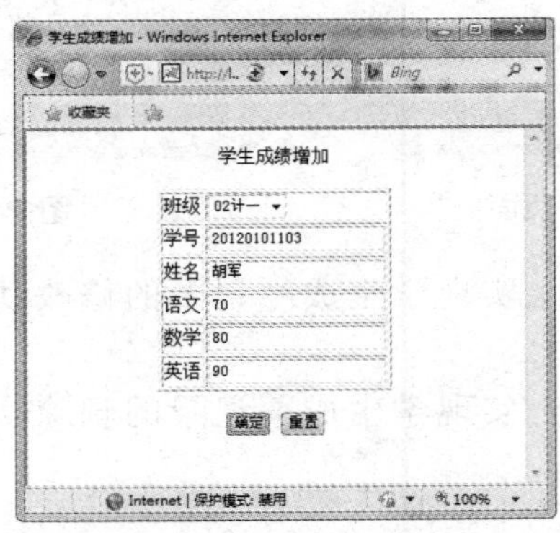

图 P5.1　“学生成绩增加”页面

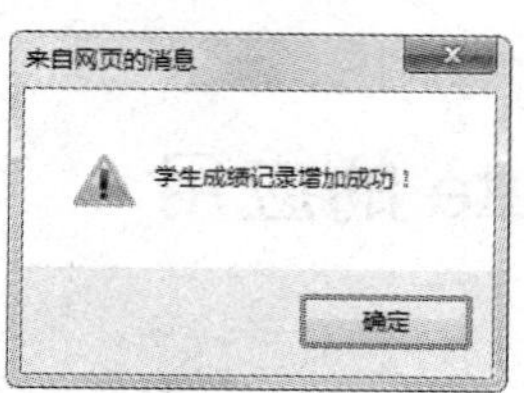

图 P5.2　成功提示框

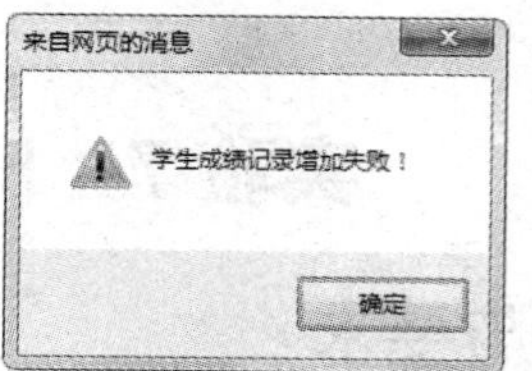

图 P5.3　失败提示框

实验 6　Struts 2 的应用

一、实验目的与要求

参考教材的有关内容与示例，按要求编写并调试相应的程序，理解并掌握 Struts 2 框架的基本应用技术。

二、实验内容

(1) 完成教材第 6 章的【实例 6-1】。

(2) 完成教材第 6 章的【实例 6-2】。

(3) 设计一个封装数据库访问操作 JavaBean。

(4) 设计一个“学生成绩增加”页面，如图 P6.1 所示，输入学生的有关信息后，再单击“确定”按钮，若能成功将记录添加到数据库中，则跳转到相应成功页面(见图 P6.2)，否则跳转到相应失败页面(见图 P6.3)。要求使用(3)所创建的 JavaBean，并应用 Struts 2 框架实现记录的添加。

图 P6.1　“学生成绩增加”页面

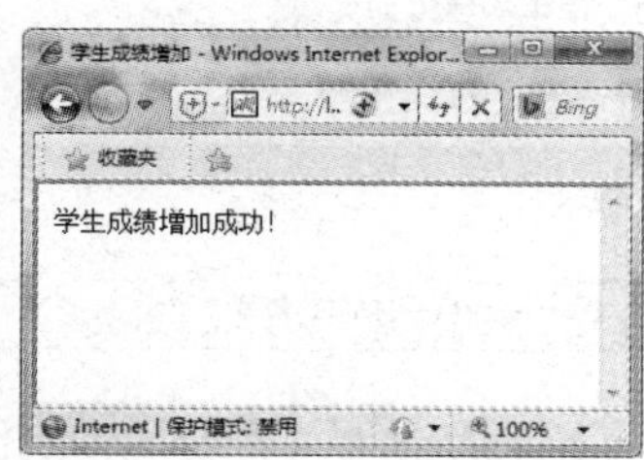

图 P6.2　成功页面

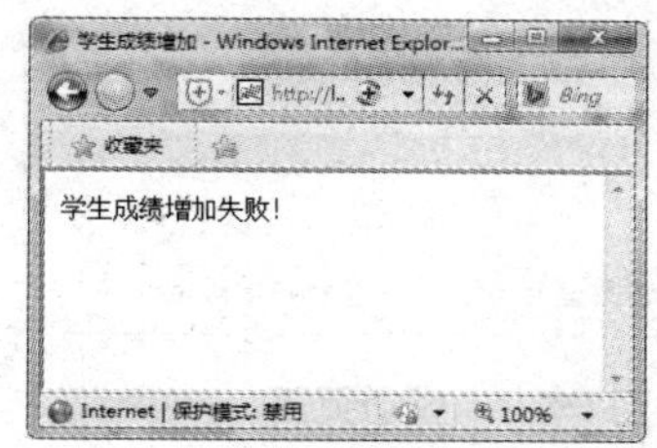

图 P6.3　失败页面

实验 7　Hibernate 的应用

一、实验目的与要求

参考教材的有关内容与示例，按要求编写并调试相应的程序，理解并掌握 Hibernate 框架的基本应用技术以及 JSP+Struts 2+Hibernate+DAO 模式的应用开发技术。

二、实验内容

设计一个“学生成绩增加”页面(见图 P7.1)，输入学生的有关信息后，再单击“确定”按钮，若能成功将记录添加到数据库中，则跳转到相应成功页面(见图 P7.2)，否则跳转到相应失败页面(见图 P7.3)。要求使用 Struts 2+Hibernate+DAO 模式(为方便起见，可先将数据库 student 中的表 class 更名为 classes)。

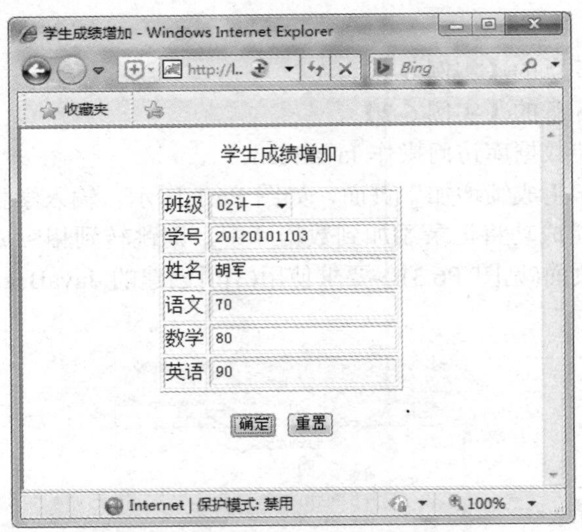

图 P7.1　“学生成绩增加”页面

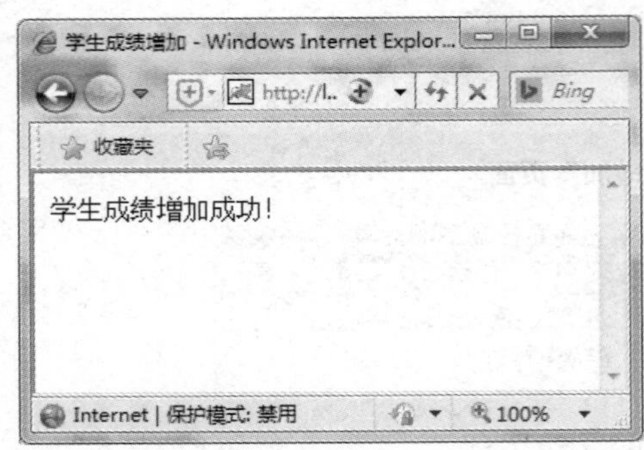

图 P7.2　成功页面

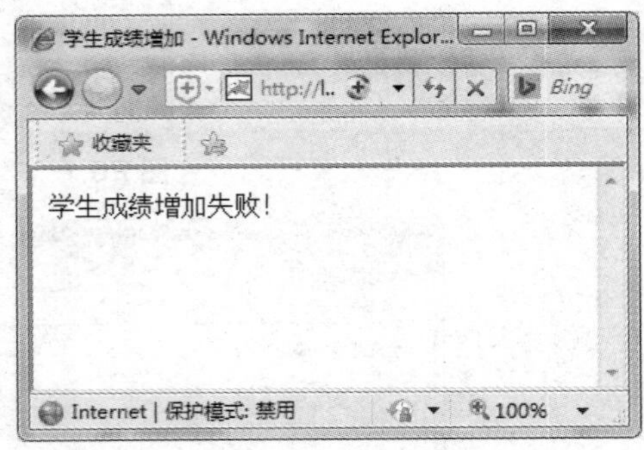

图 P7.3　失败页面

实验 8　Spring 的应用

一、实验目的与要求

参考教材的有关内容与示例，按要求编写并调试相应的程序，理解并掌握 Spring 框架的基本应用技术以及 JSP+Struts2+ Spring+Hibernate+DAO 模式的应用开发技术。

二、实验内容

设计一个“学生成绩增加”页面(见图 P8.1)，输入学生的有关信息后，再单击“确定”按钮，若能成功将记录添加到数据库中，则跳转到相应成功页面(见图 P8.2)，否则跳转到相应失败页面(见图 P8.3)。要求使用 JSP+Struts 2+Spring+Hibernate+DAO 模式。

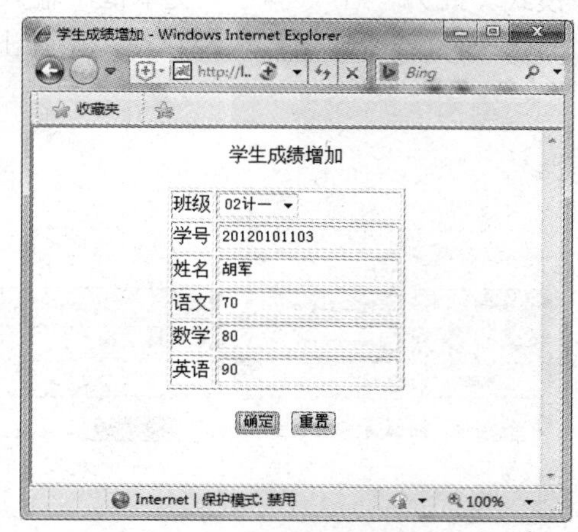

图 P8.1　“学生成绩增加”页面

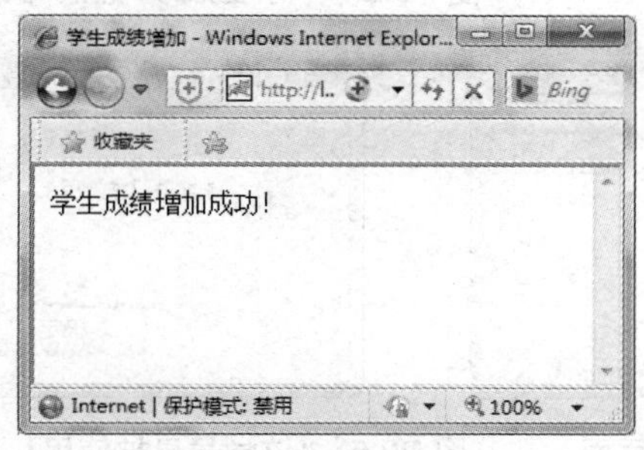

图 P8.2　成功页面

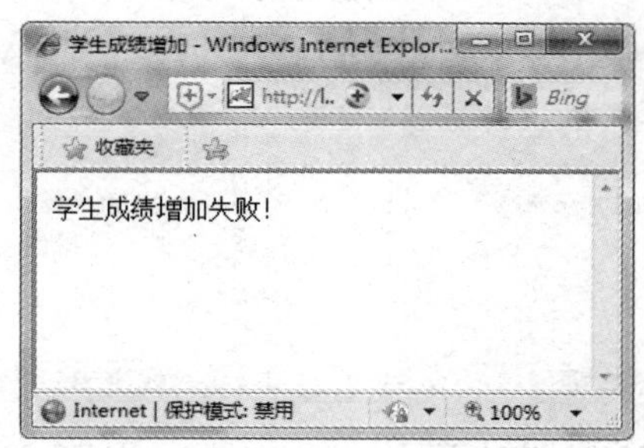

图 P8.3　失败页面

实验 9　Ajax 的应用

一、实验目的与要求

参考教材的有关内容与示例，按要求编写并调试相应的程序，理解并掌握 Ajax 与 DWR 框架的基本应用技术以及 JSP+Struts 2+Spring+Hibernate+DAO+DWR 模式的应用开发技术。

二、实验内容

(1) 完成教材第 9 章的【实例 9-1】。

(2) 设计一个“当前时间”页面(见图 P9.1)，单击“更新时间”按钮时可局部刷新页面中所显示的“当前时间”。

(3) 设计一个“学生成绩增加”页面(见图 P9.2)，输入学生的有关信息后，再单击“确定”按钮，若能成功将记录添加到数据库中，则跳转到相应成功页面(见图 P9.3)，否则跳转到相应失败页面(见图 P9.4)。要求使用 JSP+Struts 2+Spring+Hibernate+DAO+DWR 模式。此外，在“学号”文本框中输入学号并让其失去焦点时，应即时验证所输入的学号的唯一性(若所输学号已经存在，则弹出图 P9.5 所示的“该学号已被使用！”提示框)。

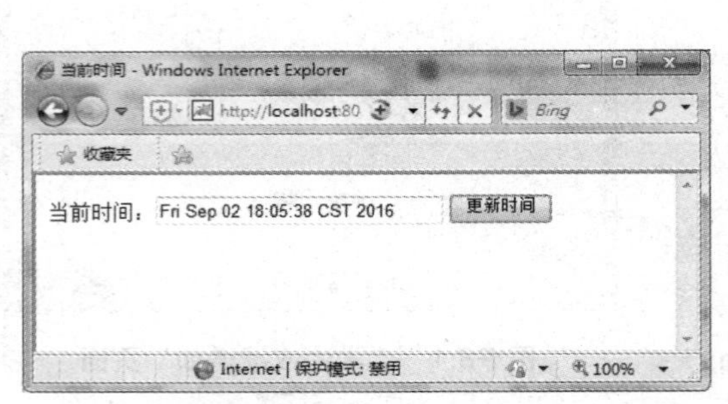

图 P9.1　“当前时间”页面

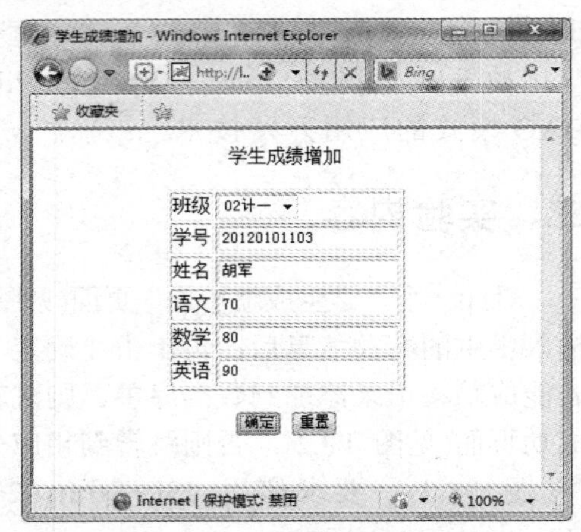

图 P9.2　“学生成绩增加”页面

图 P9.3　成功页面

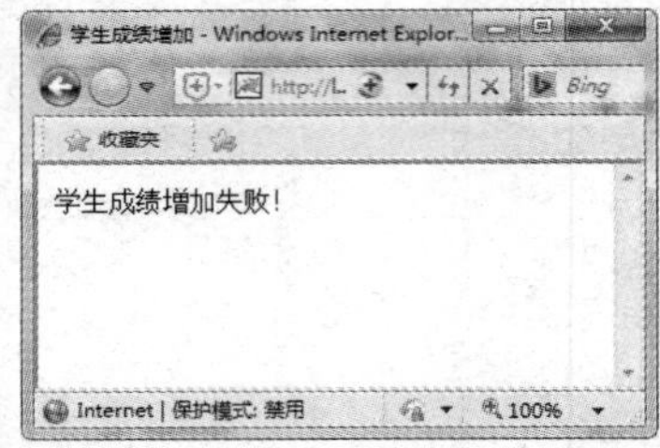

图 P9.4　失败页面

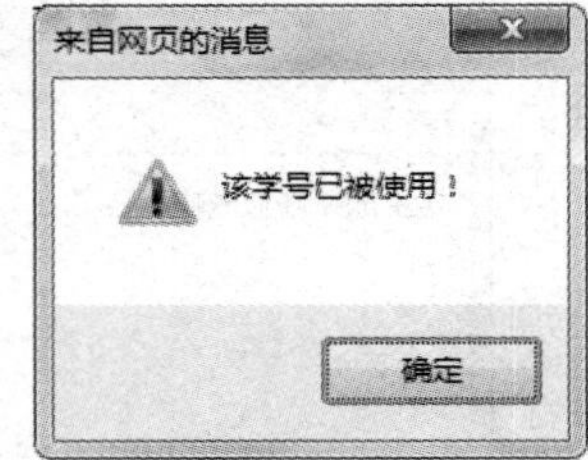

图 P9.5　“该学号已被使用！”提示框

实验 10　Web 应用系统的设计与实现

一、实验目的与要求

通过 Web 应用系统(学生成绩管理系统)的实际开发，切实掌握基于 Java EE 的 Web 应用系统的开发技术，进一步熟悉 Struts 2、Spring、Hibernate 与 DWR 等框架的整合应用技术与轻量级 Java EE 系统的开发模式。

二、实验内容

参照教学案例，使用 JSP+Struts 2+Spring+Hibernate+DWR+DAO+Service 模式，自行设计并实现一个基于 Web 的学生成绩管理系统。系统的基本功能如下。

(1) 班级管理。包括班级的查询、增加、修改、删除等。

(2) 学生管理。包括学生的查询、增加、修改、删除等。

(3) 课程管理。包括课程的查询、增加、修改、删除等。

(4) 成绩管理。包括成绩的查询、增加、修改、删除等。

(5) 用户管理。包括用户的查询、增加、修改、删除等。

注意： 系统用户的类型不同，其操作权限也应有所不同。此外，要确保班级、学生、课程及用户的唯一性。

参 考 文 献

[1] 郑阿奇. Java EE 基础实用教程[M]. 北京：电子工业出版社，2009.

[2] 刘素芳. JSP 动态网站开发案例教程[M]. 北京：机械工业出版社，2012.

[3] 郑阿奇. Java EE 项目开发教程[M]. 2 版. 北京：电子工业出版社，2013.

[4] 俞东进，任祖杰. Java EE Web 应用开发基础[M]. 北京：电子工业出版社，2012.

[5] 郑阿奇. Java EE 实用教程[M]. 2 版. 北京：电子工业出版社，2015.

[6] 聂艳明，等. Java EE 开发技术与实践教程[M]. 北京：电子工业出版社，2014.

[7] 卢守东，等. 《Java Web 应用开发技术》教学策略[J]. 福建电脑， 2016，32(4)：76，96.